Cardiac Gene Expression

METHODS IN MOLECULAR BIOLOGY™

John M. Walker, SERIES EDITOR

409. **Immunoinformatics:** *Predicting Immunogenicity* In Silico, edited by *Darren R. Flower, 2007*
408. **Gene Function Analysis,** edited by *Michael Ochs, 2007*
407. **Stem Cell Assays,** edited by *Vemuri C. Mohan, 2007*
406. **Plant Bioinformatics:** *Methods and Protocols,* edited by *David Edwards, 2007*
405. **Telomerase Inhibition:** *Strategies and Protocols,* edited by *Lucy Andrews and Trygve O. Tollefsbol, 2007*
404. **Topics in Biostatistics,** edited by *Walter T. Ambrosius, 2007*
403. **Patch-Clamp Methods and Protocols,** edited by *Peter Molnar and James J. Hickman, 2007*
402. **PCR Primer Design,** edited by *Anton Yuryev, 2007*
401. **Neuroinformatics,** edited by *Chiquito J. Crasto, 2007*
400. **Methods in Lipid Membranes,** edited by *Alex Dopico, 2007*
399. **Neuroprotection Methods and Protocols,** edited by *Tiziana Borsello, 2007*
398. **Lipid Rafts,** edited by *Thomas J. McIntosh, 2007*
397. **Hedgehog Signaling Protocols,** edited by *Jamila I. Horabin, 2007*
396. **Comparative Genomics,** *Volume 2,* edited by *Nicholas H. Bergman, 2007*
395. **Comparative Genomics,** *Volume 1,* edited by *Nicholas H. Bergman, 2007*
394. **Salmonella:** *Methods and Protocols,* edited by *Heide Schatten and Abe Eisenstark, 2007*
393. **Plant Secondary Metabolites,** edited by *Harinder P. S. Makkar, P. Siddhuraju, and Klaus Becker, 2007*
392. **Molecular Motors:** *Methods and Protocols,* edited by *Ann O. Sperry, 2007*
391. **MRSA Protocols,** edited by *Yinduo Ji, 2007*
390. **Protein Targeting Protocols, Second Edition,** edited by *Mark van der Giezen, 2007*
389. *Pichia* **Protocols, Second Edition,** edited by *James M. Cregg, 2007*
388. **Baculovirus and Insect Cell Expression Protocols, Second Edition,** edited by *David W. Murhammer, 2007*
387. **Serial Analysis of Gene Expression (SAGE):** *Digital Gene Expression Profiling,* edited by *Kare Lehmann Nielsen, 2007*
386. **Peptide Characterization and Application Protocols,** edited by *Gregg B. Fields, 2007*
385. **Microchip-Based Assay Systems:** *Methods and Applications,* edited by *Pierre N. Floriano, 2007*
384. **Capillary Electrophoresis:** *Methods and Protocols,* edited by *Philippe Schmitt-Kopplin, 2007*
383. **Cancer Genomics and Proteomics:** *Methods and Protocols,* edited by *Paul B. Fisher, 2007*

382. **Microarrays, Second Edition:** *Volume 2, Applications and Data Analysis,* edited by *Jang B. Rampal, 2007*
381. **Microarrays, Second Edition:** *Volume 1, Synthesis Methods,* edited by *Jang B. Rampal, 2007*
380. **Immunological Tolerance:** *Methods and Protocols,* edited by *Paul J. Fairchild, 2007*
379. **Glycovirology Protocols**, edited by *Richard J. Sugrue, 2007*
378. **Monoclonal Antibodies:** *Methods and Protocols,* edited by *Maher Albitar, 2007*
377. **Microarray Data Analysis:** *Methods and Applications,* edited by *Michael J. Korenberg, 2007*
376. **Linkage Disequilibrium and Association Mapping:** *Analysis and Application,* edited by *Andrew R. Collins, 2007*
375. **In Vitro Transcription and Translation Protocols:** *Second Edition,* edited by *Guido Grandi, 2007*
374. **Quantum Dots:** *Applications in Biology,* edited by *Marcel Bruchez and Charles Z. Hotz, 2007*
373. **Pyrosequencing® Protocols**, edited by *Sharon Marsh, 2007*
372. **Mitochondria: Practical Protocols**, edited by *Dario Leister and Johannes Herrmann, 2007*
371. **Biological Aging:** *Methods and Protocols,* edited by *Trygve O. Tollefsbol, 2007*
370. **Adhesion Protein Protocols,** *Second Edition,* edited by *Amanda S. Coutts, 2007*
369. **Electron Microscopy:** *Methods and Protocols, Second Edition,* edited by *John Kuo, 2007*
368. **Cryopreservation and Freeze-Drying Protocols,** *Second Edition,* edited by *John G. Day and Glyn Stacey, 2007*
367. **Mass Spectrometry Data Analysis in Proteomics,** edited by *Rune Matthiesen, 2007*
366. **Cardiac Gene Expression:** *Methods and Protocols,* edited by *Jun Zhang and Gregg Rokosh, 2007*
365. **Protein Phosphatase Protocols:** edited by *Greg Moorhead, 2007*
364. **Macromolecular Crystallography Protocols:** *Volume 2, Structure Determination,* edited by *Sylvie Doublié, 2007*
363. **Macromolecular Crystallography Protocols:** *Volume 1, Preparation and Crystallization of Macromolecules,* edited by *Sylvie Doublié, 2007*
362. **Circadian Rhythms:** *Methods and Protocols,* edited by *Ezio Rosato, 2007*
361. **Target Discovery and Validation Reviews and Protocols:** *Emerging Molecular Targets and Treatment Options, Volume 2,* edited by *Mouldy Sioud, 2007*
360. **Target Discovery and Validation Reviews and Protocols:** *Emerging Strategies for Targets and Biomarker Discovery, Volume 1,* edited by *Mouldy Sioud, 2007*
359. **Quantitative Proteomics by Mass Spectrometry**, edited by *Salvatore Sechi, 2007*

METHODS IN MOLECULAR BIOLOGY™

Cardiac Gene Expression

Methods and Protocols

Edited by

Jun Zhang

Division of Cardiology, CVRL/Geffen School of Medicine at UCLA, Los Angeles, CA

and

Gregg Rokosh

Division of Cardiology, University of Louisville, Louisville, KY

HUMANA PRESS ✴ TOTOWA, NEW JERSEY

© 2007 Humana Press Inc.
999 Riverview Drive, Suite 208
Totowa, New Jersey 07512

www.humanapress.com

All rights reserved. No part of this book may be reproduced, stored in a retrieval system, or transmitted in any form or by any means, electronic, mechanical, photocopying, microfilming, recording, or otherwise without written permission from the Publisher. Methods in Molecular Biology™ is a trademark of The Humana Press Inc.

All papers, comments, opinions, conclusions, or recommendations are those of the author(s), and do not necessarily reflect the views of the publisher.

This publication is printed on acid-free paper. ∞
ANSI Z39.48-1984 (American Standards Institute)
Permanence of Paper for Printed Library Materials.

Cover illustration: From Fig. 5D (Background) and Fig. 6D (Inset) of Chapter 9, "*In Situ* Hybridization: *A Technique to Study Localization of Cardiac Gene Expression*," by Thierry P. Calmels and David Mazurais.

Cover Design by Nancy K. Fallatt

Production Editor: Jim Geronimo

For additional copies, pricing for bulk purchases, and/or information about other Humana titles, contact Humana at the above address or at any of the following numbers: Tel.: 973-256-1699; Fax: 973-256-8341; E-mail: orders@humanapr.com; or visit our Website: www.humanapress.com

Photocopy Authorization Policy:
Authorization to photocopy items for internal or personal use, or the internal or personal use of specific clients, is granted by Humana Press Inc., provided that the base fee of US $30.00 per copy is paid directly to the Copyright Clearance Center at 222 Rosewood Drive, Danvers, MA 01923. For those organizations that have been granted a photocopy license from the CCC, a separate system of payment has been arranged and is acceptable to Humana Press Inc. The fee code for users of the Transactional Reporting Service is: [1-58829-352-1/07 $30.00].

Printed in the United States of America. 10 9 8 7 6 5 4 3 2 1

eISBN 1-59745-030-8

Library of Congress Cataloging in Publication Data

Cardiac gene expression : methods and protocols / edited by Jun Zhang and Gregg Rokosh.
       p. ; cm. — (Methods in molecular biology ; v. 366)
   Includes bibliographical references and index.
   ISBN 1-58829-352-1 (alk. paper)
   1. Myocardium—Laboratory manuals. 2. Gene expression—Laboratory manuals. 3. Genetic regulation—Laboratory manuals. I. Zhang, Jun, 1968- II. Rokosh, Gregg. III. Series: Methods in molecular biology (Clifton, N.J.) ; v. 366.
   [DNLM: 1. Gene Expression Regulation—Laboratory Manuals. 2. Gene Expression Profiling—Laboratory Manuals. 3. Myocardium—Laboratory Manuals. QU 25 C267 2007]
   QP113.2C364 2007
   612.1'7—dc22
                                                                                       2006008866

Preface

The past decade has ushered in enormous changes in how we perceive and study changes in gene expression in the heart. Early in the 1990s, the human genome project was just getting underway and establishing methods with the sensitivity to measure changes in the expression of genes with low copy number was an accomplishment. We all experienced some trepidation when the first news of microarrays arrived espousing the ability to measure changes in expression of hundreds to tens of thousands of genes (the whole genome) at once. This high throughput method was an astonishing jump in our approach to biological science. At the same time Steve Fodor and Pat Brown published papers describing two completely different approaches to measuring the expression changes of large numbers of genes at the same time. Thus began the microarray era and as a consequence the beginning of an era with a host of new approaches in pursuit of understanding the role and regulation of gene expression in cell biology and pathology including driving forward the field of bioinformatics.

The array, no pun intended, of contributions contained in *Cardiac Gene Expression: Methods and Protocols* an edition of Humana's Methods in Molecular series, address both new and established methods that researchers in the cardiac field will certainly find useful as a reference for the development of projects and training. Our aim in this compilation was to provide insight and details for a comprehensive range of methods that will serve both startup and sophisticated users alike. Sections cover expression profiling by microarray Section I, targeted analysis of gene expression (Section II), transcription factor DNA binding and regulation of promoter activity (Section III), *in silico* approaches to identifying functional *cis* regulatory elements and regulation of cardiac gene expression (Section IV), *in silico* and mass spectrometry methods to identify sequence nucleotide polymorphisms (SNPs) (Section V), and to bring findings from the above studies to the next level overexpression of genes in vivo and isolated myocytes and cardiac-specific targeted gene deletion (Section VI).

Section I, Cardiac Gene Expression Profiling: the Global Perspective. Five chapters describe several different approaches to examining and identifying changes in gene expression in the transcriptome as well as analytic approaches. Methods and analysis have improved significantly as many investigators have strived to increase array reliability and reproducibility. Section II, Cardiac Gene Regulation: Gene-Specific mRNA Measurement in the Myocardium, follows accordingly with chapters outlining more sensitive and gene targeted expression methods that are more conducive for follow up studies to verify and further characterize those important findings from array experiments or those of your favorite gene. Underlying mechanisms of gene regulation can be studied using methods that focus on the interaction of transcription factors with their cognate *cis* binding elements and how these *cis* elements impact overall promoter activity in Section III, Cardiac Gene Regulation: Promoter Characterization in the Myocardium. Changes in gene expression reflect the combined effects of transcriptional enhancers and repressors that serve to precisely control the level of expression of thousands of genes from conception to death. Studies that focus on how the interaction between transcription factors and their cognate *cis* DNA elements regulate gene expression were provided some assistance recently with the completion of the Human Genome Project in 2003 in addition to several other genomes with more coming available at a rapid pace. New analytical approaches to decipher the functional elements buried in the 3 billion nucleotides of the human genome and other model genomes are described in Sections IV, *In Silico* Assessment of Regulatory *cis*-Elements and Gene Regulation, and V, Cardiac Single Nucleotide Polymorphisms. One important aspect of understanding the importance of available sequence is being able to sift through sequence and reliably identify and distinguish functional regulatory elements from nonfunctional elements. Pennachio and colleagues at Lawrence Livermore Labs have simplified this task by using a comparative approach. By using available genome sequence for several different species across evolution this approach was able to reliably predict the functionality of elements according to their evolutionary conservation. Resources for the analysis of gene regulation data and SNPs will provide essential functionality for the understanding of changes in gene expression and effects of SNPs on gene function and expression. With the identification of exciting new targets one begins to think of the functional aspects and begins to plan experiments to validate hypotheses. Section VI, Gene Overexpression and Targeting in the Myocardium, highlights methods that facilitate overexpression or cardiac specific targeted deletion of your favorite gene in the heart

Thus, this array of contributions provides an array of methods that will take the investigator through screening, analysis, characterization, and functional confirmation of novel genes or old genes with a new function serving as a template for a solid research program.

Jun Zhang
Gregg Rokosh

Contents

Preface v
Contributors ix

PART I. CARDIAC GENE EXPRESSION: THE GLOBAL PERSPECTIVE

1 Microarray Analysis of Gene Expression in Murine
   Cardiac Graft Infiltrating Cells
   Yurong Liang, Xin Lu, and David L. Perkins 3

2 Expression Profiling Using Affymetrix GeneChip®
   Probe Arrays
   Martina Schinke-Braun and Jennifer A. Couget 13

3 Serial Analysis of Gene Expression (SAGE):
   A Useful Tool to Analyze the Cardiac Transcriptome
   **Kirill V. Tarasov, Sheryl A. Brugh, Yelena S. Tarasova,
   and Kenneth R. Boheler** 41

4 Functional Genomics by cDNA Subtractive Hybridization
   Christophe Depre 61

5 Statistical Methods in Cardiac Gene Expression Profiling:
   From Image to Function
   Sek Won Kong 75

PART II. CARDIAC GENE REGULATION: GENE-SPECIFIC mRNA MEASUREMENT IN THE MYOCARDIUM

6 Measurement of Cardiac Gene Expression by Reverse
   Transcription Polymerase Chain Reaction (RT-PCR)
   Nicola King 109

7 Quantitative (Real-Time) RT-PCR in Cardiovascular Research
   Kevin John Ashton and John Patrick Headrick 121

8 RNase Protection Assay for Quantifying Gene
   Expression Levels
   Yongxia Qu and Mohamed Boutjdir 145

9 In Situ Hybridization: A Technique to Study Localization
   of Cardiac Gene Expression
   Thierry P. Calmels and David Mazurais 159

Part III. Cardiac Gene Regulation: Promoter Characterization in the Myocardium

10 Characterization of cis-Regulatory Elements and Transcription Factor Binding: *Gel Mobility Shift Assay*
   Jim Jung-Ching Lin, Shaun E. Grosskurth, Shannon M. Harlan, Elisabeth A. Gustafson-Wagner, and Qin Wang 183

11 Mapping Transcriptional Start Sites and *In Silico* DNA Footprinting
   Martin E. Cullen and Paul J. R. Barton 203

12 Characterization of Cardiac Gene Promoter Activity: *Reporter Constructs and Heterologous Promoter Studies*
   Hsiao-Huei Chen and Alexandre F. R. Stewart 217

Part IV. *In Silico* Assessment of Regulatory cis-Elements and Gene Regulation

13 Comparative Genomics: *A Tool to Functionally Annotate Human DNA*
   Jan-Fang Cheng, James R. Priest, and Len A. Pennacchio 229

14 Developing Computational Resources in Cardiac Gene Expression
   Michael B. Bober and Raimond Winslow 253

Part V. Cardiac Single Nucleotide Polymorphisms

15 *In Silico* Analysis of SNPs and Other High-Throughput Data
   Neema Jamshidi, Thuy D. Vo, and Bernhard O. Palsson 267

16 Discovery and Identification of Sequence Polymorphisms and Mutations with MALDI-TOF MS
   Dirk van den Boom and Mathias Ehrich 287

Part VI. Gene Overexpression and Targeting in the Myocardium

17 Conditional Targeting: *Inducible Deletion by Cre Recombinase*
   Kelly R. O'Neal and Ramtin Agah 309

18 Cardiomyocyte Preparation, Culture, and Gene Transfer
   Alexander H. Maass and Massimo Buvoli 321

19 Adeno-Associated Viral Vector–Delivered Hypoxia-Inducible Gene Expression in Ischemic Hearts
   Hua Su and Yuet Wai Kan 331

20 Lentivirus-Mediated Gene Expression
   Jing Zhao and Andrew M. L. Lever 343

Index 357

Contributors

RAMTIN AGAH • *The Program in Human Molecular Biology and Genetics, University of Utah, Salt Lake City, UT, USA (Present address: Altos Cardiovascular Institute, Mountain View, CA, USA)*
KEVIN JOHN ASHTON • *Heart Foundation Research Centre, Griffith University, Southport, Australia*
PAUL J. R. BARTON • *Heart Science Centre, National Heart and Lung Institute, Imperial College London, Harefield, Middlesex, UK*
MICHAEL B. BOBER • *Division of Medical Genetics, A.I. DuPont Hospital for Children, Wilmington, DE, USA*
KENNETH R. BOHELER • *The Laboratory of Cardiovascular Science, The National Institute on Aging, NIH, Baltimore, MD, USA*
MOHAMED BOUTJDIR • *VA New York Harbor Healthcare System, SUNY Downstate Medical Center, Brooklyn, NY, and NYU School of Medicine, New York, NY, USA*
SHERYL A. BRUGH • *The Laboratory of Cardiovascular Science, The National Institute on Aging, NIH, Baltimore, MD, USA*
MASSIMO BUVOLI • *Department of Molecular Cellular and Developmental Biology, University of Colorado at Boulder, Boulder, CO, USA*
THIERRY P. CALMELS • *Bioprojet Biotech, Pharmacology Department, Saint Grégoire Cedex, France*
HSIAO-HUEI CHEN • *University of Ottawa Health Research Institute, Ottawa, Canada*
JAN-FANG CHENG • *Genomics Division, Lawrence Berkeley National Laboratory, Berkeley, CA, USA, and Joint Genome Institute, U.S. Department of Energy, Walnut Creek, CA, USA*
JENNIFER A. COUGET • *Harvard University Bauer Center for Genomics Research, Cambridge, MA, USA*
MARTIN E. CULLEN • *Heart Science Centre, National Heart and Lung Institute, Imperial College London, Harefield, Middlesex, UK*
CHRISTOPHE DEPRE • *Cardiovascular Research Institute, University of Medicine and Dentistry New Jersey, New Jersey Medical School, Newark, NJ, USA*
MATHIAS EHRICH • *SEQUENOM Inc., San Diego, CA, USA*
SHAUN E. GROSSKURTH • *Department of Biological Sciences, University of Iowa, Iowa City, IA, USA*
ELISABETH A. GUSTAFSON-WAGNER • *Department of Biological Sciences, University of Iowa, Iowa City, IA, USA*

SHANNON M. HARLAN • *Department of Biological Sciences, University of Iowa, Iowa City, IA, USA*

JOHN PATRICK HEADRICK • *Heart Foundation Research Centre, Griffith University, Southport, Australia*

NEEMA JAMSHIDI • *Department of Bioengineering, University of California, San Diego, San Diego, CA, USA*

YUET WAI KAN • *Cardiovascular Research Institute/Department of Medicine/ Department of Laboratory Medicine, University of California, San Francisco, San Francisco, CA, USA*

NICOLA KING • *Bristol Heart Institute, Department Clinical Science Medicine at South Bristol, Faculty of Medicine and Dentistry, University of Bristol, Bristol, UK*

SEK WON KONG • *Department of Cardiology, Children's Hospital Boston, Harvard Medical School, Boston, MA, USA*

ANDREW M. L. LEVER • *University of Cambridge, Department of Medicine, Addenbrooke's Hospital, Hill Road, Cambridge, UK*

YURONG LIANG • *Laboratory of Molecular Immunology, Brigham & Women's Hospital and Harvard School of Public Health, Harvard Medical School, MA, USA*

JIM JUNG-CHING LIN • *Department of Biological Sciences, University of Iowa, Iowa City, IA, USA*

XIN LU • *Laboratory of Molecular Immunology, Brigham & Women's Hospital and Harvard School of Public Health, Harvard Medical School, MA, USA*

ALEXANDER H. MAASS • *Department of Medicine, University of Wuerzburg, Wuerzburg, Germany*

DAVID MAZURAIS • *Université de Rennes I, Unité INRA-SCRIBE, Campus Beaulieu, Rennes Cedex, France*

KELLY R. O'NEAL • *The Program in Human Molecular Biology and Genetics, University of Utah, Salt Lake City, UT, USA*

BERNHARD O. PALSSON • *Department of Bioengineering, University of California, San Diego, San Diego, CA, USA*

LEN A. PENNACCHIO • *Genomics Division, Lawrence Berkeley National Laboratory, Berkeley, CA, USA and Joint Genome Institute, U.S. Department of Energy, Walnut Creek, CA, USA*

DAVID L. PERKINS • *Laboratory of Molecular Immunology, Brigham & Women's Hospital and Harvard School of Public Health, Harvard Medical School, MA, USA*

JAMES R. PRIEST • *Genomics Division, Lawrence Berkeley National Laboratory, Berkeley, CA, USA and Joint Genome Institute, U.S. Department of Energy, Walnut Creek, CA, USA*

Contributors

YONGXIA QU • *VA New York Harbor healthcare System, SUNY Downstate Medical Center, Brooklyn, NY, USA*

MARTINA SCHINKE-BRAUN • *Novartis Institutes for BioMedical Research, Cambridge, MA, USA*

ALEXANDRE F. R. STEWART • *University of Ottawa Heart Institute, Ottawa, Canada*

HUA SU • *Cardiovascular Research Institute/Department of Medicine, University of California, San Francisco, San Francisco, CA, USA*

KIRILL V. TARASOV • *The Laboratory of Cardiovascular Science, The National Institute on Aging, NIH, Baltimore, MD, USA*

YELENA S. TARASOVA • *The Laboratory of Cardiovascular Science, The National Institute on Aging, NIH, Baltimore, MD, USA*

DIRK VAN DEN BOOM • *SEQUENOM Inc., San Diego, CA, USA*

THUY D. VO • *Department of Bioengineering, University of California, San Diego, San Diego, CA, USA*

QIN WANG • *Department of Pharmacology, Vanderbilt University Medical Center, Nashville, TN, USA*

RAIMOND WINSLOW • *Center for Cardiovascular Bioinformatics and Modeling, Johns Hopkins University, and The Whitaker Biomedical Engineering Institute, Baltimore, MD, USA*

JING ZHAO • *Department of Medicine, Addenbrooke's Hospital, University of Cambridge, Hill Road, Cambridge, UK*

I

CARDIAC GENE EXPRESSION
THE GLOBAL PERSPECTIVE

1

Microarray Analysis of Gene Expression in Murine Cardiac Graft Infiltrating Cells

Yurong Liang, Xin Lu, and David L. Perkins

Summary

Microarray technology can rapidly generate large databases of gene expression profiles. Our laboratory has applied these techniques to analyze differential gene expression in cardiac tissue and cells based on mouse heart transplantation. We have analyzed the different gene expression profiles such as stress or injury including ischemia following transplantation. We also have investigated the role of infiltrating inflammatory cells during graft rejection by purifying subsets of infiltrating cells using GFP transgenic mice and detailed all technical experiences and issues. The purpose of this study is to assist researchers to simplify the process of analyzing large database using microarray technology.

Key Words: Gene expression; microarray; bioinformatics; heart transplantation; mouse.

1. Introduction

The analysis of gene expression profiles using microarray technology is a powerful approach to investigate the functions of specific tissues or cells. Our laboratory has applied these techniques to analyze differential gene expression in cardiac tissue and cells in a model of murine heart transplantation *(1–4)*. Specifically, we have analyzed the response by cardiac cells to various forms of stress or injury including ischemia following transplantation *(5)*. In addition, we have investigated the role of infiltrating inflammatory cells during graft rejection by purifying subsets of infiltrating cells. Using current microarray technology, it is possible to analyze approx 45,000 probe sets representing known mouse genes or expressed sequence tags (ESTs). The ability to perform global analyses of gene expression creates the potential to analyze complex biological systems. These methods could be applied to other questions of cardiac development or disease.

2. Materials

1. Collagenase II (Gibco) and pancreatin (Sigma).
2. D-phosphate-buffered saline (PBS; Gibco).
3. Tri Reagent (Gibco-BRL Life Technologies, Rockville, MD).
4. Dnase I (Invitrogen).
5. SuperScript II (Invitrogen).
6. ALTRA flow cytometer (Beckman Coulter).
7. SYBR Green PCR Master Mix (Applied Biosystems, Foster City, CA).
8. GeneAmp 5700 Sequence Detection System (Applied Biosystems).
9. SuperScript Choice system (Gibco-BRL Life Technologies) and T7-(dT) polymerase (Gensetoligos, La Jolla, CA).
10. BioArray High Yield RNA Transcript Labeling Kit (Enzo Diagnostics, Farmingdale, NY).
11. RNeasy mini kit (Qiagen, Valencia, CA).
12. Affymetrix GeneChip Software.

3. Methods

3.1. Vascularized Heterotopic Cardiac Transplantation

1. Murine hearts are transplanted as previously described *(6,7)*.
2. Briefly, hearts are harvested from freshly sacrificed donors and immediately transplanted into 8- to 12-wk-old recipients that are anesthesized via intraperitoneal injection with 100 mg/kg of ketamine and 20 mg/kg of xylazine.
3. The donor aorta is attached to the recipient abdominal aorta by end-to-side anastamosis, and the donor pulmonary artery is attached to the recipient vena cava by end-to-side anastomosis.
4. All surgical procedures should be completed in less than 45 min from the time that the donor heart is harvested to ensure similar ischemia times. Donor hearts that do not beat immediately after reperfusion or that stop within 1 d following transplantation should be excluded (>98% of all grafts function at 1 d following transplantation).

3.2. Single Cell Suspension

1. Donor grafts are harvested at the indicated time following transplantation and processed to prepare a single-cell suspension using collagenase and pancreatin digestion.
2. The graft heart is harvested following cold saline perfusion.
3. Hearts are minced to fine fragments with a scalpel or razor blade.
4. The heart tissue is digested four times with 0.5% collagenase II (Gibco) and 2.5% pancreatin (Sigma) in 37°C water for 7 min (*see* **Note 1**).

5. The cell suspension should be filtered and washed twice with 2% FCS D-PBS solution.
6. Resuspend cells, add 2 mL 2% FAS solution, and perform flow cytometry analysis.

3.3. Cell Sorting

Graft infiltrating cells have been shown to play important roles in triggering immune responses during graft rejection after transplantation and other inflammatory diseases such as myocarditis. To determine whether gene expression differences were expressed in infiltrating inflammatory or stromal cells, we used microarray technology to analyze the gene expression profile. To purify cell populations of infiltrating or stromal cells, we purified cell subsets by fluorescence-activated cell sorting (FACS) based on expression of green fluorescent protein (GFP) or fluorescent labeled monoclonal antibodies (*see* **Note 2**). Gene expression can be analyzed by DNA microarrays or real-time polymerase chain reaction (PCR) in the purified cell populations.

3.3.1. Analysis of Graft Infiltrating Cells

Because of technical difficulties, methods of purifying infiltrating cells often isolate a small percentage of the total population of infiltrating cells. To improve specificity and yield, we have developed a protocol using donor or recipient mice containing a transgene that constitutively expresses the GFP in all cells. These cells have greater than three logs of green fluorescence, making purification by FACS efficient and quantitative. As previously reported, we can purify sufficient infiltrating cells to perform microarray analysis from small numbers of mice. For example, our typical yield is from 10^6 (at early time points) to 10^7 (at late time points) infiltrating cells per graft (*see* **Note 3**). Thus, we can harvest sufficient cells from a single mouse at d 7 following transplantation to obtain sufficient RNA for microarray analysis. An advantage of this approach is that infiltrating cells can be analyzed without requiring amplification of RNA.

3.4. RNA Extraction

Total RNA is isolated from tissues or purified cell populations using TRIZOL reagent (Gibco-BRL Life Technologies). RNA purity is determined initially by 260/280 = 1.85 to 2.01 and by scanning with an Agilent 2100 Bioanalyzer using the RNA 6000 Nano LabChip®. RNA samples not meeting these basic parameters of quality should be excluded from the study.

3.5. DNA Microarrays

1. The initial step of cDNA synthesis is performed using Affymetrix protocols with the T7 dT Primer (100 pM) 5'-GGCCAGTGAATTGTAATACGACTCACTATA GGGAGGCGG-(dT) 24-3'.

2. In vitro transcription and preparation of labeled RNA is performed using the Enzo BioArray High Yield RNA Transcription Labeling Kit.
3. The in vitro transcription sample is cleaned with standard Affymetrix protocols and quantified on a Bio-Tek UV Plate Reader.
4. Twenty micrograms of in vitro transcription material is the nominal amount hybridized to the GeneChip® arrays, an amount easily obtainable from graft tissue.
5. The arrays are incubated in a model 320 (Affymetrix) hybridization oven at a constant temperature of 45°C overnight.
6. Preparation of the microarray for scanning is performed with Affymetrix wash protocols on a model 450 Fluidics station.
7. Scanning is performed on an Affymetrix model 3000 scanner with an autoloader.
8. Chip library files specific to each array and necessary for scan interpretation are stored on the computer workstation controlling the scanner.

3.6. Real-Time Quantitative PCR

Quantification of differentially expressed genes detected by DNA microarrays can be confirmed by real-time PCR. RNA is prepared from each sample of tissue or purified cells and analyzed by real-time PCR.

1. Briefly, isolated RNA is reverse transcribed using SuperScript II RNase Reverse Transcriptase (Gibco, Carlsbad, CA).
2. All primer pairs are designed with the Primer Express software (Applied Biosystems).
3. During primer testing, nontemplate controls and dissociation curves are used to detect primer-dimer conformation and nonspecific amplification.
4. Direct detection of the PCR product is monitored by fluorescence of SYBR Green induced by binding to double-stranded DNA.
5. Reactions are performed in a MicroAmp Optical 96-well reaction plate (Applied Biosystems) using each primer pair in 5 µL of cDNA mix, 5 µL of primer, and 10 µL of SYBR Green Master Mix (Applied Biosystems) per well.
6. The gene-specific PCR products are continuously measured by means of the GeneAmp 5700 Sequence Detection System (Applied Biosystems) during 40 cycles (*see* **Note 4**).
7. The threshold cycle (which equals the PCR cycle at which an increase in reporter fluorescence first exceeds a baseline signal) of each target product is determined and normalized to the amplification plot of GAPDH.
8. All experiments are run in triplicate, and the thermal cycling parameters are maintained at constant values.
9. Fold change is calculated relative to control cycle threshold (C_T). The C_T value is defined as the number of PCR cycles required for the fluorescence signal to exceed the detection threshold value. With a PCR efficiency of 100%, the C_T values of two separate genes can be compared (ΔC_T); the fold difference = $2^{-(C_T - C_{T\,control})}$.

3.7. Bioinformatics

3.7.1. Data Management

1. For each microarray, we store the expression level of approx 45,000 probe sets linked with the experimental group as the class label.
2. Experimental data are catalogued in a manner consistent with the Minimum Information About a Microarray Experiment (MIAME) checklist published by the Microarray Gene Expression Data Society (MGED) *(8)*. Documentation should include the experimental design, samples used, sample preparation and labeling, hybridization procedures and parameters, measurement data and specifications, and array design as described in the checklist on the MGED website (www.mged.org/Workgroups/MIAME/miame_checklist.html).
3. All data are available upon request. All samples from experiments are assigned a sequential number with each individual aliquot/sample given an extension of this number. Therefore, each aliquot/sample can be individually tracked and linked with our microarray data. Experimental data including date, mouse strains, date of birth, weight, and sex for both the donor and recipient, as well as time of harvest, are stored for each experiment.
4. All microarray data including .dat files should be backed up and accessible by password-protected Internet access.

3.7.2. Low-Level Data Processing

1. Raw microarray data are normalized and processed by the Affymetrix Microarray Data Analysis Suite (MAS5.0), GCRMA, or dChip software.
2. The quantitative RNA level is computed from the signal strength of the 11 pairs of perfect match (PM) and mismatch (MM) probe pairs representing each gene, where MM probes act as specificity controls that allow the direct subtraction of background and cross-hybridization signals.
3. In the analysis with MAS5, each array is normalized to a standard of 2500 units per probe set. To determine the quantitative RNA level, the averages of the differences (avg diff) representing PM – MM for each gene-specific probe set are calculated. The expression of each probe set is categorized as present (P), marginal (M), or absent (A).
4. We also tried to use the rank invariant set normalization and model-based expression algorithm by dChip software as well as the GCRMA algorithm implemented in BioConductor (http://www.bioconductor.org/) to perform the normalization and calculate the expression levels. As previously reported, our comparison of dChip and GCRMA showed that GCRMA identified a greater number of significantly modulated genes.
5. To eliminate noise and facilitate future gene selection procedures, a filtering process based on a coefficient of variation (CV) is applied to the whole data set of 45,000 probe sets. Nondifferentially expressed genes with a low CV and nonexpressing genes with low expression levels across all the microarrays are considered noninformative and are excluded in the subsequent analyses. The class

labels are masked during the gene filtering step to prevent bias in the gene selection process.

3.8. Algorithms to Cluster Gene Expression Profiles

There are multiple algorithms and software that perform clustering analysis of gene expression data. Two examples that we have used in our laboratory are hierarchical clustering and self-organizing maps (SOMs).

3.8.1. Hierarchical Clustering Dendrograms

Hierarchical clustering analysis can be performed using either commercial or free access software, such as Cluster and TreeView *(9)* (courtesy of M. Eisen, Lawrence Livermore Radiation Laboratory, Berkeley, CA) and GeneCluster (courtesy of Whitehead Institute for Biomedical Institute, Cambridge, MA) software. Briefly, dissimilarity between groups is determined by calculation of a difference metric, such as Euclidean distance or the Pearson correlation coefficient, between each series of values.

3.8.2. Self-Organizing Maps

1. SOMs can be generated by GeneCluster among the experimental groups.
2. The number of maps is selected empirically to eliminate clusters with few genes or large standard deviations. The number of epochs (iterations) of the algorithms is selected to minimize the standard deviations (SDs) of the groupings and is limited only by computer time. For example, we commonly used between 100 and 5000 epochs.
3. Using multiple heuristic observations, the goal is to generate maps in which increased number of nodes produced clusters with low number of genes, whereas decreased number of nodes produced larger SDs. Also, increasing the number of epochs (= 500) should not produce substantial changes in the number of clusters or SDs.

3.9. Statistics

For analysis of gene expression data between two experimental groups, the correlation or regression analysis between the response variable and the expression level is calculated.

1. The expression levels of each gene, under different experimental conditions or different groups of samples, are compared by the statistical methods just given.
2. The p-values are calculated and adjusted by false discovery rate (FDR) control. Calculation of the p-value and FDR adjustment is conducted via R, using functions provided by the packages from BioConductor (http://www.bioconductor.org/). Genes whose FDR adjusted p-values below a specified level are selected as differentially regulated and are used in follow-up studies.

Microarray Analysis of Gene Expression

3. Alternatively, for the two-class problem, genes can be selected by building a classification model, e.g., the support vector machine (SVM) model, estimating the prediction error by cross-validation, and also selecting important genes by evaluating their relative contribution to the classification model. The RSVM algorithm was developed by Zhang and Lu *(10)* and has been used successfully in protein marker identification problems (http://www.stanford.edu/group/wonglab/RSVMpage/R-SVM.html).

3.9.1. Selection of Significantly Differentially Expressed Genes

Our previous studies compared the power to detect significantly modulated genes in duplicate versus quadruplicate microarrays, analyzing independent samples. These results established that quadruplicates allowed detection of greater numbers of modulated genes. Based on the number of significantly regulated genes detected, a cost analysis indicated that the most efficient approach would be to analyze quadruplicate samples. Therefore, quadruplicate microarrays are recommended to estimate the individual and replicate variations, to ensure the maintenance of quality control standards, and to determine whether the experimental variation exceeds the technical variation for our data.

1. The data can be classified according to the categories of the experimental groups, and an initial two-way comparison can be performed between groups. In these studies the expression of each gene in the experimental group is compared with the corresponding expression samples in the control group.
2. We can identify differentially expressed genes by applying FDR adjustment to the raw *p*-values of all genes, and we can select genes that are differentially expressed, with the FDR controlled below tuned criterion, e.g., 5%.
3. Using this approach, we can select subsets of genes that are significantly differentially expressed in the experimental and control groups.

3.9.2. Randomization Test of the Gene Selection Procedure

1. The gene selection procedure should be validated by a randomization test.
2. The class labels of the microarray are randomly permuted, and the same gene selection procedure and Gene Ontology (GO) annotation analysis are implemented.
3. This randomization is performed multiple times and the number of genes, as well as highly concentrated GO terms, if any, is compared with the discoveries based on true class label.

3.9.3. Statistical Validation

1. The predictive power of a certain subset of genes, with a certain prediction model, can be estimated by cross-validation.
2. In cross-validation, we will leave a small number of samples out, e.g., leave one out when the sample size is very small, or leave 10 to 20% out when we have a moderate number of samples; then we build a predictive model based on the

selected genes of the training set and predict the class labels of the left out samples.
3. This is equivalent to testing the model using another independent test set, but the cross-validation procedure is performed multiple times rather than only one splitting of the whole data set. In this way, we can get a better estimation of the predicting error rates.
4. When doing leave-one-out cross-validation, the number of iterations typically is the same as the sample size, i.e., leave out one sample each time. When doing cross-validation by leaving out a portion of samples each time, the number of iterations is no less than the sample size; the only limitation is the computational power available.

3.10. Biological Interpretation

Microarray technology can rapidly generate large databases of gene expression profiles. The challenge in array studies is to link the descriptive expression profiles with relevant biological processes or clinical diagnoses. A common analytical approach has been to select a few genes with the greatest change in expression and correlate them with a specific disease or biological phenotype. However, this approach eliminates >99% of the data from subsequent analysis and may ignore important biological observations. For example, is a gene upregulated early, but to a low level, less important (or a less effective therapeutic target) than a gene upregulated late, but to a high level? Thus, many studies have used arbitrary criteria, such as the ratio of expression, to focus on only a few of the observed changes.

3.10.1. Biological Validation

In addition to the statistical validation of our results, we also assess the biological interpretation of our selected genes based on determination of the biological processes, molecular functions, and cellular components, as defined by the GO database. In association with each gene-based classification, we validate the biological significance of our candidate lists of differentially expressed genes by GO annotation.

1. The list of genes identified as differentially regulated are analyzed by GO annotations to find highly concentrated biological processes, cellular components, and molecular functions.
2. The GO analysis is performed by GeneMerge (http://www.oeb.harvard.edu/hartl/lab/publications/GeneMerge/GeneMerge.html) *(11)* and GeneNotes (http://combio.cs.brandeis.edu/GeneNotes) software.
3. The pathways that evolved in the regulated genes are found by matching the gene list with a pathway database, e.g., the Kyoto Encyclopedia of Genes and Genomes (KEGG) pathway database.

4. Information from other public databases will also be integrated whenever necessary by GeneNotes software. The databases include, but are not limited to, chromosome mapping, gene annotation, homologous genes, unigenes, RefSeq, protein sequences, protein-protein interaction, and PubMed literature.
5. For regulatory motif finding, we cluster genes by their expression profile using GeneCluster, GenePattern (http://www.broad.mit.edu/cancer/software/software.html), and TightCluster (http://www.pitt.edu/~ctseng/research/tightClust_download.html) *(12)*.
6. Motifs are found from the clustered genes by *de novo* motif-finding algorithms, including BioProspector *(13)*, MDscan *(14)*, and CompareProspector *(15)*; they are validated by the TransFac database.

3.11. Quality Control

1. Various sources of experimental noise will inevitably be introduced into the data set; thus we must assess the noise before interpreting the data with classification information.
2. The level of experimental noise should be estimated by replicates of microarrays on samples that are processed independently (including RNA processing and microarray hybridization).
3. With this approach, the within-replicate experimental variation level can be estimated and compared with individual variations and between-group variations. Only when the replicate variation is significantly smaller than the individual variation and in turn the individual variation is significantly smaller than the between-group variations, can any differences observed by comparing different groups be considered significant.
4. Another source of variation is the batch effect. Because of the ongoing experimental design, it is not feasible to collect all samples and perform the microarray studies simultaneously. Therefore, it is essential to monitor possible batch effects to avoid/correct any bias between batches.

4. Notes

1. Enzyme digestion methods vary. This modified method contributes to a higher yield of viable cardiac cells during our previous experiments.
2. Each gate can clearly show the two populations of GFP+ and GFP– cells. The purity of each isolated cell fraction was >99% by FACS.
3. Every purified cell population should be 10,000 or more to get enough RNA. During RNA extraction, you may not be able to see the pellet at the final step.
4. Because of the lower concentration of cDNA (≤ 1.0 μg/μL), we used 50 cycles for the primer pair amplification to obtain good production, instead of the 40 cycles we usually use for real-time PCR.

Acknowledgments

We thank Walter Zybko PhD, John Daley, and Suzan Lazo-Kallanian for technical support.

References

1. Christopher, K., Mueller, T. F., Ma, C., Liang, Y., and Perkins, D. L. (2002) Analysis of the innate and adaptive phases of allograft rejection by cluster analysis of transcriptional profiles. *J. Immunol.* **169**, 522–530.
2. Christopher, K., Mueller, T. F., DeFina, R., et al. (2003) The graft response to transplantation: a gene expression profile analysis. *Physiol. Genomics* **15**, 52–64.
3. DeFina, R. A., Liang, Y., He, H., et al. (2004) Analysis of immunoglobulin and T-cell receptor gene deficiency in graft rejection by gene expression profiles. *Transplantation* **77**, 580–586.
4. Liang, Y., Christopher, K., DeFina, R., et al. (2003) Analysis of cytokine functions in graft rejection by gene expression profiles. *Transplantation* **76**, 1749–1758.
5. He, H., Stone, J. R., and Perkins, D. L. (2002) Analysis of robust innate immune response after transplantation in the absence of adaptive immunity. *Transplantation* **73**, 853–861.
6. Corry, R. J., Winn, H. J., and Russell, P. S. (1973) Heart transplantation in congenic strains of mice. *Transplant. Proc.* **5**, 733–735.
7. McKee, C. M., Defina, R., He, H., Haley, K. J., Stone, J. R., and Perkins, D. L. (2002) Prolonged allograft survival in TNF receptor 1-deficient recipients is due to immunoregulatory effects, not to inhibition of direct antigraft cytotoxicity. *J. Immunol.* **168**, 483–489.
8. Brazma, A., Hingamp, P., Quackenbush, J., et al. (2001) Minimum information about a microarray experiment (MIAME)—toward standards for microarray data. *Nat. Genet.* **29**, 365–371.
9. Eisen, J. S. (1998) Genetic and molecular analyses of motoneuron development. *Curr. Opin. Neurobiol.* **8**, 697–704.
10. Zhang, X., Lu, X., Shi, Q., et al. (2006) Recursive SVM feature selection and sample classification for mass-spectrometry and micrarray data. *BMC Bioinformatics* **7**, 197.
11. Castillo-Davis, C. I. and Hartl, D. L. (2003) Gene Merge—post-genomic analysis, data mining, and hypothesis testing. *Bioinformatics* **19**, 891–892.
12. Tseng, G. C. and Wong, W. H. (2005) Tight clustering: a resampling-based approach for identifying stable and tight patterns in data. *Biometrics* **61**, 10–16.
13. Liu, X., Brutlag, D. L., and Liu, J. S. (2001) BioProspector: discovering conserved DNA motifs in upstream regulatory regions of co-expressed genes. *Pac. Symp. Biocomput.* 127–138.
14. Liu, X. S., Brutlag, and Liu, J. S. (2002) An algorithm for finding protein-DNA binding sites with applications to chromatin-immunoprecipitation microarray experiments. *Nat. Biotechnol.* **20**, 835–839.
15. Liu, Y., Wei, L., Batzoglou, S., Brutlag, D. L., Liu, J. S., and Liu, X. S. (2004) A suite of web-based programs to search for transcriptional regulatory motifs. *Nucleic Acids Res.* **32**, W204–207.

2

Expression Profiling
Using Affymetrix GeneChip® Probe Arrays

Martina Schinke-Braun and Jennifer A. Couget

Summary

Large-scale microarray expression profiling studies have helped us to understand basic biological processes and to classify and predict the prognosis of cancers; they have also accelerated the identification of new drug targets. Affymetrix GeneChip® probe arrays are high-density oligonucleotide microarrays that are available for many prokaryotic and eukaryotic species. Affymetrix human and mouse whole-genome microarrays analyze the expression level of up to 47,000 transcripts and variants. Each transcript is measured by 11 probe pairs, which consist of a perfect match 25mer oligonucleotide (PM) and a 25mer mismatch oligonucleotide (MM) that contains a single base pair mismatch in the central position. The PM/MM design is used for identification and subtraction of nonspecific hybridization and background signals.

Advantages of Affymetrix GeneChip arrays include highly standardized array fabrication, as well as standardized target preparation, hybridization, and processing protocols, resulting in low technical variability and good reproducibility between experiments. This chapter describes the standard assay procedures for isolating total RNA from heart tissue, generating a biotin-labeled target for expression analysis, processing of Affymetrix GeneChip probe arrays using the Affymetrix instrument system, and quality control measures.

Key Words: Affymetrix; GeneChip; expression profiling; RNA amplification; RNA quality; Agilent Bioanalyzer; in vitro transcription; heart.

1. Introduction

Although many microarray technology platforms are available, Affymetrix GeneChip microarrays are still the most commonly used; Affymetrix has been a pioneer in the microarray field, and its technological strength has been consistently ranked among the top three in the pharmaceutical and biotech sector by MIT's *Technology Review*. Affymetrix GeneChip microarrays are also

From: *Methods in Molecular Biology, vol. 366: Cardiac Gene Expression: Methods and Protocols*
Edited by: J. Zhang and G. Rokosh © Humana Press Inc., Totowa, NJ

superior to other technology platforms because their technology is well established and standardized protocols and an automated instrument setup for microarray processing are used.

The generation of gene expression microarray data using Affymetrix GeneChip probe arrays follows a simple procedure consisting of six major steps *(1)*. The single most important step in ensuring a successful GeneChip experiment is the preparation of a clean, intact RNA sample. After the RNA sample passes a quality assessment, the labeling process is initiated *(2)*. Briefly, the first step of the labeling procedure is the synthesis of double-stranded cDNA from the RNA sample using reverse transcriptase and an oligo-dT primer *(3)*. Next, the cDNA serves as a template in an in vitro transcription (IVT) reaction that produces amplified amounts of biotin-labeled antisense mRNA, which is also referred to as labeled cRNA or the microarray target. Prior to hybridization, the cRNA is fragmented to 50- to 200-base fragments using heat and Mg^{2+}, which facilitates efficient hybridization *(4)*. The cRNA is added to the hybridization cocktail, which contains salts, blocking agents, and bacterial RNA spike-in controls. This cocktail is injected into the GeneChip array hybridization chamber and hybridized at 45°C for 16 h *(5)*. After hybridization, the GeneChip array is subjected to a series of washing and staining steps involving a fluorescent molecule that binds to biotin and a signal amplification step *(6)*. The GeneChip array is then scanned with a confocal laser scanner, and the image of the distribution pattern of fluorescent signals on the microarray is recorded. Control measures are implemented at each step to ensure good-quality data.

Although processing of Affymetrix GeneChip probe arrays is highly standardized and automated, substantial variation has been observed owing to factors such as RNA isolation batches, hybridization day, wash and reagent batches, operator, and lot-to-lot array variation *(1)*. The effects of the processing may be even larger than the biological effects being studied. To keep variability to a minimum, it is advantageous to label and process the samples for a given experiment at the same time. If the number of samples in an experiment is too large, care must be taken not to confound the treatments with the processing. However, if sample preparation and processing batches are being tracked and incorporated in the design and analysis of microarray experiments, their impact on the data can be removed using statistical techniques such as analysis of variance.

The protocols in this chapter that pertain to sample labeling and Affymetrix GeneChip microarray processing are condensed versions excerpted from the GeneChip Expression Analysis Technical Manual (with permission from Affymetrix). The purpose is to give an overview of the process and point out some of the important tricks and details. For the detailed protocols, it is strongly recommended to refer to the appropriate Affymetrix manuals.

2. Materials
2.1. Heart Tissue Isolation
1. Phosphate-buffered saline (PBS), pH 7.4, at 4°C.
2. 70% Ethanol.
3. Sterile gauze.
4. Liquid nitrogen.

2.2. RNA Extraction from Heart Tissue
1. Power homogenizer (Omni PCR Tissue Homogenizing Kits, Omni International, or similar) with 7-mm saw-toothed generator probe.
2. Small stainless-steel spatula.
3. Filter plugged, sterile, RNase-free pipet tips of various volumes.
4. Disposable polystyrene sterile 75-mm transparent loose-cap tubes (Fisher Scientific, cat. no. 4-956-30).
5. Sterile, RNase-free Eppendorf tubes.
6. TRIZOL reagent (Invitrogen, cat. no. 15596-026).
7. Chloroform.
8. Isopropanol.
9. Ethanol (100%).
10. Diethyl pyrocarbonate (DEPC)-treated, autoclaved water (Ambion, cat. no. 9920).
11. Phenol/chloroform/isoamyl alcohol (25:24:1), pH 7.9 (Ambion, cat. no. 9730).
12. Isoamyl alcohol/chloroform (1:24).
13. PBS, pH 7.4, at 4°C.
14. RNeasy MiniKit (Qiagen, cat. no. 74104); contains 50 RNeasy mini spin columns, collection tubes, and RNase-free reagents and buffers.

2.3. RNA Quality Control and Quantitation Using the Agilent Bioanalyzer
1. Bioanalyzer RNA Nano Kit (Agilent, cat. no. 5064-8229). Includes 25 chips and reagents (gel, dye, filter columns, RNA Nano marker).
2. RNase Zap (Ambion).
3. DEPC water (Ambion).
4. Agilent 2100 Bioanalyzer.
5. Chip priming station (Agilent, cat. no. 5065-4401).
6. Vortexer (e.g., IKA model MS1) with vortex mixer adapter (Agilent, cat. no. 5022-2190).

2.4. RNA Amplification and Labeling Using Affymetrix One-Cycle Labeling Reagents
For steps 1–5, of **Subsection 3.5.**

1. One-Cycle Target Labeling and Control Reagents (Affymetrix, cat. no. 900493). This kit contains all required labeling and control reagents to perform 30 one-cycle labeling reactions. All components (IVT Labeling Kit, One-Cycle cDNA Synthesis Kit, GeneChip Sample Cleanup Module, Poly-A RNA Control Kit,

and Hybridization Control Kit) can also be purchased separately. The Oligo-dT primer (50 μM, high-performance liquid chromatography [HPLC] purified, Affymetrix, cat. no. 900375) is also included and has the sequence:

5' GGCCAGTGAATTGTAATACGACTCACTATAGGGAGGCGG-(dT)$_{24}$ 3'.

2. Absolute ethanol (stored at –20°C for RNA precipitation; stored at room temperature for use with the GeneChip Sample Cleanup Module).
3. 80% ethanol (stored at –20°C for RNA precipitation; store ethanol at room temperature for use with the GeneChip Sample Cleanup Module).
4. 3 M Sodium acetate (NaOAc; Sigma-Aldrich, cat. no. S7899 or similar).
5. 1 N NaOH.
6. 1 N HCl.

2.4.1. Fragmentation of Labeled cRNA

1. 5X Fragmentation Buffer: 200 mM Tris-acetate, pH 8.1, 500 mM KOAc, 150 mM MgOAc in DEPC-treated water; filter through a 0.2-μm filter, aliquot, and store at room temperature. This buffer is included in the Affymetrix GeneChip Clean-up Module (Affymetrix, cat. no. 900371).
2. DEPC-water (Ambion, cat. no. 9920 or similar).
3. Labeled cRNA from **Subheading 2.3.**, **item 5**.
4. 0.5 mL RNase-free tubes (Ambion, cat. no. 12250 or similar).

2.5. Controls

2.5.1. Poly-A RNA Spike-In Controls

1. 1 Poly-A RNA Control Kit (Affymetrix, cat. no. 900433).

2.5.2. Hybridization Controls

1. GeneChip Eukaryotic Hybridization Control Kit, Affymetrix, cat. no. 900454 (30 reactions) or cat. no. 900457 (150 reactions); contains control cRNA and control Oligo B2.
2. 3 nM Control Oligo B2 (Affymetrix, cat. no. 900301).

2.6. Preparation of the Hybridization Cocktail and Hybridization

Note that Affymetrix now offers a GeneChip Hybridization, Wash, and Stain Kit that provides all necessary reagents required to complete the hybridization, wash, and staining processes for GeneChip brand arrays in cartridge format. The kit includes pre-formulated solutions to process 30 arrays on the GeneChip Fluidics Station 400 or 450.

1. Nuclease-free water (Ambion, cat. no. 9930).
2. 50 mg/mL Bovine serum albumin (BSA) solution (Invitrogen Life Technologies, cat. no. 15561-020).
3. Herring sperm DNA (Promega, cat. no. D1811).
4. 5 M NaCl, RNase-free, DNase-free (Ambion, cat. no. 9760G).

5. MES hydrate SigmaUltra (Sigma-Aldrich, cat. no. M5287).
6. MES sodium salt (Sigma-Aldrich, cat. no. M5057).
7. 0.5 M EDTA, pH 8.0 (Ambion, cat. no. 9260G).
8. Dimethyl sulfoxide (DMSO; Sigma-Aldrich, cat. no. D5879).
9. 10% Surfact-Amps 20 (Tween-20; Pierce Chemical, cat. no. 28320).
10. 12X MES stock buffer: 1.22 M MES, 0.89 M [Na$^+$]. 35.2 g MES hydrate, 96.65 g MES sodium salt, nuclease-free water to 500 mL; pH should be 6.5 to 6.7. Sterile filter through 0.2-µm filter, and store protected from light at 4°C.
11. 2X MES hybridization buffer: 100 mM MES, 1 M [Na$^+$], 20 mM EDTA, 0.01% Tween-20. 8.3 mL 12X MES buffer, 17.7 mL 5 M NaCl, 4 mL 0.5 M EDTA pH 8.0, 0.1 mL 10% Tween-20, 19.9 mL nuclease-free water. Store at 4°C, and discard if yellow.
12. 1.7 mL SafeSeal tubes (RNase-free; Sorenson, cat. no. 11510).
13. 0.1 to 10 µL sterile, filter-free pipet tips with sharp tips, not beveled.
14. Lab tape or Tough Spots (Diversified Biotech, T-Spots).

2.7. Washing, Staining, and Scanning of Affymetrix GeneChip Probe Arrays

1. Water, molecular biology grade (BioWhittaker Molecular Applications/Cambrex, cat. no. 51200).
2. Distilled water (Invitrogen, cat. no. 15230-147).
3. 50 mg/mL BSA solution (Invitrogen, cat. no. 15561-020).
4. R-Phycoerythrin Streptavidin (Molecular Probes, cat. no. S-866).
5. 5 M NaCl, RNase-free, DNase-free (Ambion, cat. no. 9760G).
6. PBS, pH 7.2 (Invitrogen Life Technologies, cat. no. 20012-027).
7. 20X SSPE: 3 M NaCl, 0.2 M NaH$_2$PO$_4$, 0.02 M EDTA (BioWhittaker Molecular Applications/Cambrex, cat. no. 51214).
8. Goat IgG, reagent grade (Sigma-Aldrich, cat. no. I 5256); prepare 10 mg/mL stock by resuspending 50 mg in 5 mL of 150 mM NaCl. Aliquot and store at –20°C.
9. Antistreptavidin antibody (goat), biotinylated (Vector, cat. no. BA-0500). Prepare 0.5 mg/mL stock in sterile, nuclease-free water.
10. 10% Surfact-Amps 20 (Tween-20) (Pierce Chemical, cat. no. 28320).
11. 1.5-mL Sterile, RNase-free, microcentrifuge vials (USA Scientific, cat. no. 1415-2600).
12. Lab tape or Tough Spots (Diversified Biotech, T-Spots).
13. Wash Buffer A (nonstringent Wash Buffer): 300 mL of 20X SSPE, 1 mL of 10% Tween-20, 699 mL water. Filter through a 0.2-µm filter, and store protected from light at 4°C.
14. Wash Buffer B (stringent Wash Buffer): 83.3 mL of 12X MES stock buffer, 5.2 mL of 5 M NaCl, 1 mL of 10% Tween-20, 910.5 mL of water. Filter through a 0.2-µm filter, and store protected from light at 4°C.
15. 2X Stain Buffer: 41.7 mL of 12X MES stock buffer, 92.5 mL of 5 M NaCl, 2.5 mL of 10% Tween-20, 113.3 mL of water. Filter through a 0.2-µm filter, and store protected from light at 4°C.

3. Methods
3.1. Experimental Design

Microarray experimental design is a critical and mostly overlooked key element in a successful microarray experiment. The requirements for experimental design are no different for cardiovascular experiments than for any other field of application. The best designed microarray experiments begin with determining the objectives of the experiment and some well-defined goals. An experimental design that addresses a key hypothesis rather than being overly complex minimizes the arrays required and simplifies the data analysis. If several questions are to be addressed at once, the approach has to have enough statistical power, e.g., sufficient replicates, to answer all questions. This means that for each new variable added to a design, the required number of arrays is multiplied. Pilot microarray studies that focus on a single variable versus a control state are useful for first-time microarray users. They can help to identify problems related to the biological sample, the procedures, and the data analysis. Refining methods after a small-scale study is far cheaper and more effective than complex mathematical fixes after the fact. Pilot studies also provide a good estimate of the variance of gene expression, which is useful in determining how many replicates the experiment's key questions will require.

3.2. Heart Tissue Isolation from Mice or Rats

1. Anesthetize or sacrifice the animal according to the animal care and use guidelines of your institution and National Institute of Health standards.
2. Generously wet the chest of the animal with 70% alcohol, and wipe with sterile gauze to prevent the transfer of animal hair to the instruments and inside the body cavity.
3. Beginning 1 in. from the base of the sternum, open the body cavity up to the salivary glands.
4. Remove the sternum and move the lung to the side. Carefully lift up the heart with forceps and clip the dorsal aorta. Quickly remove the heart.
5. Adult heart tissue is rinsed in ice-cold PBS to remove blood, briefly blotted on a piece of gauze and immediately snap frozen in liquid nitrogen (*see* **Note 1**). Embryonic tissue is dissected in ice-cold PBS, and heart tissue is frozen in liquid nitrogen or on dry ice immediately after dissection.

3.3. RNA Extraction from Cardiac Tissues

Total RNA is isolated using the TRIZOL reagent, a monophasic solution of phenol and guanidine isothiocyanate, following the instructions of the manufacturer (*see* **Note 2**). Depending on the amount of starting material, the isolated RNA is then further purified by a phenol-chloroform extraction followed by ethanol precipitation, or by binding to a silica gel-based membrane using RNeasy columns.

1. Prechill 100 mL PBS and a 75-mm tube with TRIZOL reagent on ice.
2. Weigh the tissue while taking care to keep it frozen (*see* **Note 3**). An adult mouse heart weighs about 100 to 170 mg. Homogenize tissue samples in 1 mL of TRIZOL reagent per 50 to 100 mg of tissue using a power homogenizer with a sawtooth generator probe (*see* **Note 4**). The sample volume should not exceed 10% of the volume of TRIZOL reagent used for homogenization.
3. Transfer samples into 75-mm tubes filled with the appropriate amount of room temperature (RT) TRIZOL reagent using a prechilled spatula (*see* **Note 5**). The minimum volume needed for these tubes is 1 mL.
4. Homogenize at full speed for 3 × 20 s while the sample is still frozen. Allow the sample to cool between homogenization steps.
5. Let the sample sit at room temperature for 5 min after homogenization to allow for complete dissociation of protein complexes.
6. Clean the generator probe in between samples by running the generator at full speed in the following solvents:

    30 s in 100% ethanol.
    30 s in chilled PBS 1.
    30 s in chilled PBS 2.
    10 s in chilled TRIZOL reagent.

    Dry well with Kimwipes

7. Add 0.2 mL of chloroform per 1 mL of TRIZOL reagent, and shake vigorously for 15 s.
8. Immediately transfer the contents into an RNase-free Eppendorf tube.
9. Incubate at RT for 3 min and centrifuge at 12,000g for 15 min at 4°C. Centrifugation separates the lower, red phenol-chloroform phase from the interphase and the clear aqueous upper phase, which contains the RNA.
10. Transfer the aqueous phase to a new Eppendorf tube.

    Important: For small starting amounts of tissue (e.g., cardiac needle biopsies) further purify the RNA by performing an additional phenol-chloroform extraction following **steps 11** to **16** in **Subheading 3.3.1.** below. For larger tissue starting amounts (e.g., using a whole adult mouse heart), use RNeasy columns and follow **steps 17** to **35** in **Subheading 3.3.2.**

3.3.1. Phenol-Chloroform Extraction

11. Add an equal amount (600 µL, if 1 mL of TRIZOL was used for the initial homogenization) of phenol-chloroform and vortex for 1 min. Centrifuge at top speed for 5 min at RT.
12. To remove phenol, transfer the aqueous phase to a new Eppendorf tube. Add an equal amount (600 µL) of isoamyl alcohol/chloroform and vortex for 1 min. Centrifuge at top speed for 5 min at RT.

13. To precipitate the RNA, transfer the aqueous phase to a new microcentrifuge tube and add 0.6 to 0.7 vol of isopropanol (approx 700 µL for 1 mL of TRIZOL in the initial homogenization). Mix by inverting 6 to 8 times and incubate at RT for 10 min.
14. Centrifuge at 12,000g for 10 min at 4°C, discard the supernatant, and wash the pellet with 1 mL of 75% ethanol in DEPC-treated water.
15. Centrifuge at 7500g for 5 min at 4°C, discard the supernatant, and allow the pellet to air-dry for 10 min at RT.
16. Dissolve the pellet in DEPC-treated water to a final concentration of 1 µg/µL. The expected yield of RNA is about 1 µg of RNA per mg of tissue.

3.3.2. Purification of RNA Using RNeasy Columns

17. The RNeasy spin columns are used to purify RNA after TRIZOL extraction (*see* **Subheading 3.3., steps 1 to 10**) when the amount of starting material was at least 50 mg of tissue. Guanidine isothiocyanate (GITC)-containing lysis buffer and ethanol are added to the sample to create conditions that promote selective binding of RNA to the RNeasy membrane. Contaminants are washed away, and the purified RNA is then eluted from the membrane in DEPC-treated water.
18. Precipitate the RNA by adding 0.6 to 0.7 vol of isopropanol (approx 700 µL for 1 mL of TRIZOL in the initial homogenization) to the supernatant of **step 11** in **Subheading 3.3.1.** Mix by inverting 6 to 8 times and incubate at RT for 10 min.
19. Centrifuge at 12,000g for 10 min at 4°C, discard the supernatant, and wash the pellet with 1 mL of 75% ethanol in DEPC-treated water.
20. Centrifuge at 7500g for 5 min at 4°C, discard the supernatant, and allow the pellet to air-dry for 10 min at RT.
21. Estimate the amount of RNA in the pellet based on the assumption that 1 µg RNA can be extracted per mg tissue. The binding capacity of an RNeasy mini spin column is 100 µg of RNA. Therefore, dissolve up to 100 µg RNA in 100 µL RNase-free water. If the expected RNA yield is larger than 100 µg, use multiple RNeasy columns per sample.

The following steps are carried out at RT. Work quickly through the whole cleanup procedure to minimize the risk of RNA degradation.

22. Add 10 µL β-mercaptoethanol (β-ME) per 1 mL buffer RLT before use.

    Caution: β-ME is toxic; dispense in a fume hood and wear appropriate protective clothing.

    Buffer RLT is stable for 1 mo after addition of β-ME.
23. Add 350 µL Buffer RLT to the 100 µL RNA-containing solution, and mix thoroughly.
24. Add 250 µL ethanol (96–100%) to the diluted RNA, and mix thoroughly by pipeting.
25. Immediately apply the sample (700 µL) to an RNeasy mini column placed in a 2-mL collection tube (supplied with the kit).

26. Close the tube and centrifuge for 15 s at ≥8000g (≥10,000 rpm). Discard the flowthrough and collection tube.
27. Buffer RPE is supplied as a concentrate. Before using for the first time, add 4 vol of ethanol (96–100%), as indicated on the bottle, to obtain a working solution.
28. Transfer the RNeasy column into a new 2-mL collection tube. Pipet 500 µL Buffer RPE onto the RNeasy column. Close the tube and centrifuge for 15 s at ≥8000g (≥10,000 rpm) to wash the column. Discard the flowthrough, and reuse the collection tube.
29. Add another 500 µL Buffer RPE to the RNeasy column. Close the tube and centrifuge for 2 min at ≥8000g (≥10,000 rpm) to dry the RNeasy silica-gel membrane.
30. Discard the flowthrough, and reuse the collection tube. Centrifuge the column in the collection tube in a microcentrifuge at full speed for 1 min. It is important to dry the RNeasy silica-gel membrane since residual ethanol will interfere with downstream reactions.
31. After the centrifugation, place the RNeasy mini column in a new 1.5-mL collection tube (supplied with the kit).
32. Depending on the expected RNA yield, add 30 to 50 µL RNase-free water directly onto the RNeasy silica-gel membrane.
33. Close the tube and centrifuge for 1 min at ≥8000g (≥10,000 rpm) to elute the RNA.
34. Repeat the elution step (**step 33**) with another 30 to 50 µL RNase-free water, and elute into the same collection tube.
35. Determine the concentration, yield, and purity (*see* **Subheading 3.4.**) and freeze the samples at –80°C until the RNA is to be used for the labeling process (*see* **Note 6**).

3.4. RNA Quality Control and Quantitation

Impurities in RNA samples have an adverse effect on both the labeling efficiency and the stability of the fluorescent labels that are used. Therefore, the success of a microarray experiment largely depends on the quality of the prepared RNA. It is important to check the quality of the RNA after each thawing and before starting the labeling reaction, since impure RNA samples might degrade during the thawing process.

One measure of RNA purity is the ratio of absorbance readings at 260 and 280 nm. The A260:A280 ratio for RNA samples of acceptable purity should be between 1.8 and 2. RNA preparations should be free of contaminating proteins and organic solvents such as phenol or ethanol, and salts; they should also show no signs of degradation. A high-quality preparation of mammalian total RNA is characterized by two bright bands at approx 4.5 and 1.9 kb, representing 28S and 18S ribosomal RNA, and the absence of genomic DNA. A simple test for genomic DNA contamination is to use your RNA directly as a template in a polymerase chain reaction (PCR) reaction with primers for any well-characterized gene, such as β-actin or GAPDH. The primer positions should not span a large intron and should be chosen to amplify a short fragment (<1 kb in length). RNA subjected to a reverse transcription reaction can be used as a positive control. If the PCR produces a visible band on an ethidium bromide-stained agarose gel, the RNA preparation contains genomic DNA. For a successful microarray experiment, the genomic DNA content in the RNA preparation should be less than 0.001% and should not produce a visible band after 35 PCR cycles.

The integrity of the RNA can be determined by running an aliquot of the RNA preparation on a 1% denaturing agarose gel, or, more quantitatively, using the Agilent 2100 Bioanalyzer and RNA 6000 Nano Labchip or RNA 6000 Pico Labchip. This method requires only nanogram amounts of RNA and is therefore especially suitable if the amount of RNA is limited. The protocol below provides instructions for determining the integrity and concentration of RNA samples on the Agilent 2100 Bioanalyzer using the RNA Nano Protocols. The manufacturer's detailed protocol is freely available through their website at www.chem.agilent.com.

3.4.1. Preparing the Reagents

1. Prepare the gel by placing 400 μL of RT RNA 6000 Nano gel matrix onto a spin filter column.
2. Place the spin filter in an Eppendorf tube and spin for 10 min at 1500g. The filtered gel should be stored at 4°C and used within a month after preparation.
3. Prepare the gel-dye mix by adding 2 μL of dye to 130 μL of filtered gel. All reagents should be at RT. Vortex and spin down at 13,000g for 10 min. To protect the gel-dye mix from light, cover the tube with foil. The gel-dye mix should be used within 1 wk from the date of preparation.
4. Prepare the RNA 6000 ladder by denaturing an aliquot of the ladder for 2 min at 70°C. Snap-cool on ice. For each chip, 1 μL of the ladder is needed.

3.4.2. Preparing the Bioanalyzer

The electrodes should be decontaminated before each run.

1. Place a washing chip (electrode cleaner) filled with 350 μL RNaseZap in the Bioanalyzer and close the lid for 1 min. Remove the washing chip from the Bioanalyzer.
2. Place another washing chip filled with 350 μL DEPC-water in the Bioanalyzer, and close the lid for 30 s.
3. Remove the washing chip and leave the lid open for 10 s to allow the electrodes to dry.

3.4.3. Loading the Chip

Remove the chip from packaging and inspect the under-side wafer of the chip for any defects.

1. Place the chip into the priming station. Pull out the syringe to the 1-mL mark.
2. Pipet 9 µL of gel-dye mix into the well marked with a back-circled "G." Always place the pipet tip into the center and bottom of each well when dispensing.
3. Snap the priming station lid closed.
4. Push the plunger of the syringe down to 0.2 mL, so that it fits snugly under the silver stopper. Press down slowly and steadily on the plunger when priming.
5. Wait exactly 30 s.
6. Squeeze the silver stopper to release the plunger. It should come up to at least 0.7 mL within 1 to 2 s. If it does not prime well, check that the gasket is clean. Change the gasket if the problem persists.
7. Pull the plunger up the rest of the way to 1 mL. Be careful to avoid a negative pressure vacuum, which can lead to bubbles.
8. Lift the priming station lid. Hold the chip in place with one hand and lift the lid with the other to prevent dislodging of the chip from the priming station.
9. Pipette 9 µL of gel-dye mix into each one of the other two wells labeled "G."
10. Pipette 5 µL of the RNA 6000 Nano Marker into the well marked with the ladder symbol and into each of the 12 sample wells.
11. Do not leave any wells empty. Add 6 µL of the RNA Nano Marker to each unused well (*see* **Note 7**).

3.4.4. Starting the Chip Run

1. The quantitative measurement range for total RNA concentrations is 25 to 500 ng/µL, and that for mRNA concentrations is 25 to 250 ng/µL. For qualitative measurements to check for RNA integrity only, the range is 5 to 500 ng/µL.
2. To minimize secondary structure, heat denature (70°C, 2 min) the samples before loading on the chip.
3. Add 1 µL of sample into each well. Empty wells should be filled with 6 µL water or the Nano marker. Accurate pipeting is very important. Use properly calibrated pipets. You may pipet up and down gently to mix samples in the wells.
4. Pipet 1 µL of the ladder into the ladder well.
5. Vortex the chip for 1 min at about 512*g*. Use a piece of tape or an elastic band to secure the chip in the adapter. This will prevent the chip from falling out even if it has been inserted into the vortexer adapter in the wrong orientation.
6. Use the chip within 5 min of preparation to prevent evaporation. Cover the chip if it will be left standing for any length of time
7. Launch the 2100 Expert software.
8. Place the chip into the Bioanalyzer and close the lid. A chip symbol will appear on the screen.
9. Click the *Electrophoresis* button, click *Assay*, click *RNA*, and select *Eukaryote Total RNA Nano* from the menu.

10. If necessary, edit the number of samples using the menu on the lower left.
11. Press the *Start* button in the upper right to start the chip run.
12. Edit the sample names by clicking the blue *Data File* link or click the *Data and Assay* context.
13. Select the *Chip Summary* tab, and enter the sample names into the table.
14. Press *Apply* in the lower right hand corner when you are finished.
15. Click the *Instrument* context to review the live trace of your samples.
16. The approximate run time per sample is 90 s. A full chip run takes about 20 min.

3.4.5. Cleaning and Maintenance

1. Remove the chip from the Bioanalyzer immediately after the run is completed. Do not let the sample dry on the electrodes.
2. After each run, follow the cleaning **steps 1** to **3** in **Subheading 3.4.2.**, "Preparing the Bioanalyzer."
3. The focus lens should be cleaned monthly with isopropanol.
4. The electrodes should be cleaned at least quarterly by sonication.

3.4.6. Interpreting RNA Nano Results

1. To check the results of the run, select the *Gel* or *Electropherogram* tab in the *Data and Assay* context.
2. Select the ladder. The electropherogram of the ladder should show six RNA peaks (depending on the peak finding settings, the last peak might not be detected) and one marker peak. All seven peaks should be well resolved.
3. To review the data for a specific sample, select the sample name and highlight the *Results* subtab. Major features of an electropherogram for a successful eukaryotic total RNA run are sharp and well-resolved 18S and 28S ribosomal RNA peaks. There should be no or only a small peak for the 5S ribosomal RNA. The baseline should be flat. Smaller additional peaks and a hump in the baseline are indicative of degradation (*see* **Note 8**).
4. Microarray gene expression experiments should not be performed with poor-quality total RNA samples.

3.5. RNA Amplification and Labeling

Total RNA or mRNA (*see* **Note 9**) is first reverse-transcribed using a T7 promoter-containing oligo(dT) primer in the first-strand cDNA synthesis reaction. Following RNase H-mediated second-strand cDNA synthesis, the double-stranded cDNA is purified and serves as a template in the subsequent in vitro transcription (IVT) reaction. The IVT reaction is carried out in the presence of T7 RNA polymerase and a biotinylated nucleotide analog/ribonucleotide mix for complementary RNA (cRNA) amplification and biotin labeling.

3.5.1. First-Strand cDNA Synthesis

1. Perform the steps outlined in **Subheading 3.6.1.** if poly-A RNA spike-in controls are being used.
2. Briefly spin down all tubes in the kit before using the reagents.
3. For the first-strand cDNA synthesis, program a thermal cycler with heated lid as follows: 70°C for 10 min, 4°C hold, 42°C for 2 min, 42°C for 1 h, 4°C hold. The 4°C holds are for reagent addition steps.
4. Pipet the total RNA sample (5 µg) into an RNase-free thin-wall 0.2-mL PCR tube. If the reaction includes the poly-A RNA controls, the volume of the total RNA should comprise no more than 8 µL (*see* **Note 10**).
5. Add 2 µL of the appropriately diluted poly-A RNA controls (*see* **Subheading 3.6.1.**), 2 µL of 50 µM T7-Oligo(dT) Primer (included), and RNase-free water to a final volume of 12 µL.
6. Gently flick the tube a few times to mix, and then centrifuge briefly (approx 5 s) to collect the reaction at the bottom of the tube.
7. Perform the first incubation step in the thermal cycler, e.g., incubate for 10 min at 70°C. Cool the sample after the incubation at 4°C at least 2 min.
8. Prepare sufficient First-Strand Mastermix for all RNA samples: 4 µL 5X First Strand Reaction Mix, 2 µL 0.1 M dithiothreitol (DTT), 1 µL 10 mM dNTP.
9. Add 7 µL First-Strand Mastermix to each RNA/T7-Oligo(dT) Primer mix for a final volume of 19 µL while the samples are on the 4°C hold step. Mix well by flicking the tubes a few times, and centrifuge briefly to collect the reaction at the bottom of the tube.
10. Incubate for 2 min at 42°C.
11. Add 1 µL SuperScript II reverse transcriptase to each RNA sample for a final volume of 20 µL.
12. Incubate at 42°C for 1 h.
13. Place the samples on ice for at least 2 min, and immediately proceed with the second-strand synthesis.

3.5.2. Second-Strand cDNA Synthesis

1. For the second-strand cDNA synthesis, program a thermal cycler as follows: 16°C for 2 h, 4°C hold, 16°C for 5 min, 4°C hold.
2. Prepare a second-strand mastermix for all the samples. The following recipe is for a single reaction: 91 µL RNase-free water, 30 µL 5X second-strand reaction mix, 30 µL 10 mM dNTP, 1 µL *E. coli* DNA ligase, 4 µL *E. coli* DNA polymerase I, 1 µL RNaseH. The total volume is 130 µL.

    Mix well and centrifuge briefly to collect the mix at the bottom of the tube.
3. Add 130 µL of the second-strand mastermix to each first-strand synthesis reaction from **Subheading 3.5.1.** Mix well and briefly centrifuge.
4. Return the tubes to the thermal cycler and incubate for 2 h at 16°C.
5. Add 2 µL of T4 DNA polymerase to each sample and continue incubation at 16°C for another 5 min.

6. Add 10 µL of 0.5 M EDTA and place the reactions on ice. The total volume is now 162 µL.
7. Do not leave the samples on ice for a prolonged period. Samples may be stored at −20°C for later use. However, it is recommended to proceed directly to **Subheading 3.5.3.** just below to avoid unnecessary freeze/thaw cycles that may impact the integrity of the cDNA.

3.5.3. Cleanup of Double-Stranded cDNA

The components needed for cleaning up the cDNA are supplied with the GeneChip Sample Cleanup Module. Add 24 mL of 96 to 100% ethanol to the cDNA Wash Buffer prior to the first use.

1. Add 600 µL of cDNA Binding Buffer to the second strand synthesis reaction and mix by vortexing. Check that the color of the mixture is yellow, similar to the cDNA Binding Buffer. Add 10 µL of 3 M NaOAc, pH 5.0, if the color is orange or violet.
2. Apply 500 µL of the sample to the cDNA Cleanup Spin Column sitting in a 2-mL Collection Tube (supplied), and centrifuge for 1 min at ≥8000g (≥10,000 rpm).
3. Discard the flowthrough, reload the spin column with the remaining mixture (262 µL), and centrifuge as above.
4. Discard the flowthrough, and transfer the spin column into a new 2-mL Collection Tube (supplied).
5. Pipet 750 µL of the cDNA Wash Buffer onto the spin column. Centrifuge for 1 min at ≥8000g (≥10,000 rpm). Discard flowthrough.
6. Open the cap of the spin column and centrifuge for 5 min at maximum speed (≤25,000g) to dry the membrane completely. Discard flowthrough and Collection Tube.
7. Transfer the spin column into a 1.5-mL Collection Tube, and pipet 14 µL of cDNA Elution Buffer directly onto the spin column membrane.
8. Incubate for 1 min at room temperature and centrifuge for 1 min at maximum speed (≤25,000g) to elute. The average volume of eluate is 12 µL (*see* **Note 11**).
9. After cleanup, proceed to **Subheading 3.5.4.** just below.

3.5.4. Synthesis of Biotin-Labeled cRNA by In Vitro Transcription

The GeneChip IVT Labeling Kit is used for this step.

1. Prepare a mastermix for all the samples. The following recipe is for a single reaction: 12 µL template cDNA from **Subheading 3.5.3.** (*see* **Note 12**), 4 µL 10X IVT Labeling Buffer, 12 µL IVT Labeling NTP Mix, 4 µL IVT Labeling Enzyme Mix, 8 µL RNase-free water. The total volume is 40 µL.
2. Mix well and incubate at 37°C for 16 h in a thermal cycler with heated lid to avoid condensation.
3. Labeled cRNA can be stored at −20°C for short-term storage or at −70°C for longer periods until further purified. However, it is recommended to proceed

directly to **Subheading 3.5.5.** just below after the 16-h incubation to avoid freeze/thaw cycles.

3.5.5. Cleanup and Quantification of Biotin-Labeled cRNA

To determine the concentration of the cRNA accurately, it is necessary to remove unincorporated NTPs. The Sample Cleanup Module is used for this purpose (*see* **Note 13**). Before using the IVT cRNA Wash Buffer for the first time, add 20 mL of ethanol (96–100%), as indicated on the bottle, to obtain a working solution.

1. Add 60 µL of RNase-free water to the IVT reaction, and mix by vortexing. The total volume is now 100 µL.
2. Add 350 µL IVT cRNA Binding Buffer to the sample and mix by vortexing.
3. Add 250 µL 96 to 100% ethanol to the lysate, and mix well by pipetting. Do not centrifuge.
4. Apply the sample (700 µL) to the IVT cRNA Cleanup Spin Column sitting in a 2-mL Collection Tube. Centrifuge for 15 s at ≥8000g (≥10,000 rpm).
5. Discard flowthrough and collection tube. Transfer the spin column into a new 2-mL Collection Tube (supplied).
6. Pipet 500 µL IVT cRNA Wash Buffer onto the spin column. Centrifuge for 15 s at ≥8000g (≥10,000 rpm) to wash. Discard flowthrough.
7. Pipet 500 µL 80% (v/v) ethanol onto the spin column and centrifuge for 15 s at ≥8000g (≥10,000 rpm). Discard flowthrough.
8. Open the cap of the spin column and centrifuge for 5 min at maximum speed (≤25,000g) to dry the membrane completely. Discard flowthrough and Collection Tube.
9. To elute the cRNA, transfer the spin column into a new 1.5-mL Collection Tube (supplied), and pipet 11 µL of RNase-free water directly onto the spin column membrane. Centrifuge for 1 min at maximum speed (≤25,000g).
10. Determine the cRNA yield by measuring absorbance at 260 nm, and determine the purity by measuring the 260:280 nm ratio using a spectrophotometer or the Nanodrop ND-1000.
11. To estimate the size distribution, run a 1 ng aliquot of the purified cRNA on the Agilent Bioanalyzer using the RNA labchip (*see* **Subheading 3.4.**) (*see* **Note 14**).

3.5.6. Fragmentation of the Biotin-Labeled cRNA

To achieve optimal binding sensitivity on Affymetrix GeneChip probe arrays, the biotin-labeled cRNA is fragmented by metal-induced hydrolysis into 35 to 200 base fragments (*see* **Note 15**). The Sample Cleanup Module is used for this step. It is usually best to perform the fragmentation on the same day as hybridization whenever possible to minimize freeze/thawing effects on the cRNA. The degree of fragmentation can vary substantially. Therefore, exact timing and accurate pipeting techniques are crucial.

1. For the fragmentation, program a thermal cycler with heated lid as follows: 94°C for 35 min, 4°C hold.
2. Prepare the following fragmentation mix: 20 μg cRNA (adjust the concentration from **Subheading 3.5.5.** to reflect carryover of the unlabeled total RNA input; the volume of cRNA should not exceed 21 μL), 8 μL 5X Fragmentation Buffer. Add RNase-free water to 40 μL final volume.
3. Incubate at 94°C for exactly 35 min. Put the sample on ice immediately after incubation.
4. Check the cRNA by the Agilent Bioanalyzer using the mRNA Smear Nano Assay (*see* **Subheading 3.4.**). Run 100 ng of unfragmented cRNA side by side with 1 μL fragmented undiluted cRNA directly from the fragmentation reaction. The assay should reveal long unfragmented cRNAs greater than 400 bp, whereas the average size of the fragmented cRNA should be around 100 nucleotides.
5. The remaining fragmented cRNA should immediately be used to assemble the hybridization cocktail. Do not leave reactions at 4°C for long periods. However, it is possible to store the assembled hybridization mix at –80°C until the day of hybridization.

3.6. Controls

Affymetrix provides two kinds of controls. The poly-A RNA controls correspond to probe sets on the GeneChip arrays for *Bacillus subtilis* genes (*lys*, *phe*, *thr*, and *dap*) that are not present in eukaryotic samples. These spike-in controls are added to the isolated RNA samples and then amplified and labeled together with the samples. Comparing the hybridization intensities of these controls on all GeneChip arrays in an experiment allows one to monitor the entire labeling process. However, the hybridization signals from poly-A spike-in controls do not allow one to assess the initial quantity and quality of the isolated RNA samples.

The other type is hybridization controls (*bioB*, *bioC*, *bioD*, *cre*) that are added to the hybridization mix. These controls only allow an evaluation of the efficiency of hybridization.

3.6.1. Poly-A RNA Spike-In Controls

Use the Eukaryotic Poly-A RNA Control Kit for this step. The in vitro synthesized transcripts in the kit are already premixed at different concentrations. The concentrated Poly-A Control Stock is diluted using Poly-A Control Dilution Buffer (provided in the kit) and spiked directly into RNA samples to achieve the following final concentrations, given as the ratio of copy numbers: *lys*, 1:100,000; *phe*, 1:50,000; *thr*, 1:25,000; *dap*, 1:6,667.

1. For a starting amount of 5 μg total RNA, the Poly-A Control Stock is diluted 1:10,000 in three serial dilutions of 1:20, 1:50, and 1:10.

2. Then 2 µL of the diluted stock are added to the 5 µg total RNA prior to labeling (*see* **Note 16**).

3.6.2. Hybridization Controls

The Eukaryotic Hybridization Control Kit contains a mixture of biotin-labeled cRNA transcripts of *bioB*, *bioC*, *bioD*, and *cre*, prepared in staggered concentrations (1.5, 5, 25, and 100 pM, respectively). *bioB*, *bioC*, and *bioD* represent genes in the biotin synthesis pathway of *E. coli*, and *cre* is a recombinase gene of bacteriophage P1. The Eukaryotic Hybridization Controls are spiked into the hybridization cocktail as described in **Subheading 3.7.** just below and are thus used to evaluate sample hybridization efficiency of the probe array.

3.7. Preparation of the Hybridization Cocktail and Hybridization

It is usually best to make all cocktails for a set of arrays fresh to minimize freeze/thawing effects on cRNA and at the same time minimize day-to-day variability.

The final concentration of the cRNA in the hybridization cocktail must be no less than 0.05 µg/µL. If the recommended fragmentation reaction (20 µg of cRNA fragmented in 40 µL) was followed, the sample meets this requirement.

3.7.1. Equilibration of Reagents

1. Equilibrate arrays to room temperature approx 15–20 min before use. Failure to do so may cause the microarray septa to crack and damage the microarray irreversibly.
2. Preheat the hybridization oven to 45°C (takes about 5 min).
3. Equilibrate the Hybridization Buffer to room temperature for 5–10 min before use.
4. Heat the 20X Eukaryotic Hybridization Controls and OligoB2 to 65°C in 1.7-mL microcentrifuge tubes for 5–10 min to ensure that the cRNA is completely resuspended. Briefly vortex to mix and spin down to collect on the bottom of the tube.
5. Thaw the Herring Sperm DNA and BSA at 37°C for 5–10 min. Briefly vortex to mix and spin down to collect on the bottom of the tube.
6. Keep the fragmented biotin-labeled cRNA from **Subheading 3.5.6.** on ice until ready to use.

3.7.2. Preparing the Affymetrix GeneChip Array

1. Remove the array from its packaging and inspect it for manufacturing defects (i.e., scratches on glass or wafer). Record the array type, lot number, and expiration date.
2. Wet the array by filling it through one of the septa with 200 µL of 1X Hybridization Buffer. It is necessary to use two pipet tips when filling the probe array

Table 1
Hybridization Cocktail for Standard Affymetrix GeneChip Probe Arrays

Component	Volume (μL)	Final concentration
Control oligonucleotide B2 (3 nM)	5	0.05 nM
20X Eukaryotic hybridization controls	15	1X (1.5, 5, 25, and 100 pM, respectively)
Herring Sperm DNA (10 mg/mL)	3	0.1 mg/mL
BSA (50 mg/mL)	3	0.5 mg/mL
2X Hybridization buffer	150	1X
100% Dimethyl sulfoxide (DMSO)	30	10%
Water	56	
Fragmented biotin-labeled cRNA (20 μg)	38	
Final volume	300	

cartridge: one for filling and the second to allow venting of air from the hybridization chamber. Use 100 μL for Test3 arrays and 300 μL for a standard GeneChip format.
3. Insert the array into the GeneChip cartridge carrier. Prepare a balanced carrier.
4. Insert the carriers into the preheated 45°C oven. Make sure that the carriers are properly secured.
5. Set the rotation speed to 60 rpm and prehybridize the microarrays for a minimum of 10 min but no longer than 25 min. Do not allow the microarrays to dry.

3.7.3. Preparation of the Hybridization Cocktail and Hybridization

1. Prepare the hybridization cocktail for a single probe array hybridization as shown in **Table 1**. The recipe takes into account that it is necessary to make extra hybridization cocktail owing to a small loss in volume during each hybridization. Scale up the volumes for hybridization to multiple arrays. Add the cRNA last. If an error is made, you will not have to redo the fragmentation step.
2. Heat the hybridization cocktail to 99°C for 5 min prior to use.
3. Transfer the hybridization cocktail to a 45°C heat block for 5 min to equilibrate the cocktail to the microarray temperature.
4. Spin the hybridization cocktail at maximum speed in a microcentrifuge for 5 min to remove any insoluble material from the hybridization mixture.
5. Remove the buffer solution from the probe array cartridge and fill with 200 μL of the clarified hybridization cocktail, avoiding any insoluble matter at the bottom of the tube. Upon filling, a small meniscus should be visible through the glass window, with few to no bubbles.
6. Seal both septa with lab tape or Tough-Spots.
7. Place the probe array into the prewarmed Hybridization Oven set to 45°C, and rotate at 60 rpm.
8. Hybridize for 16 h. Do not allow the microarrays to hybridize longer or dry.

3.8. Washing, Staining, and Scanning of Affymetrix GeneChip Probe Arrays

Washing and staining of Affymetrix GeneChip probe arrays is an automated process and requires the Affymetrix Fluidics station. The staining protocol includes a signal amplification step that employs a fluorescent molecule (streptavidin phycoerythrin) that binds to biotin, an antistreptavidin antibody (goat), and a biotinylated goat IgG antibody. A series of washes and stains with these reagents provides an amplified flour that emits light when the probe array is scanned with a confocal laser. Both the Fluidics Station and the scanner are controlled by the Affymetrix Microarray Suite (MAS) (older version) or the Affymetrix GeneChip® Operating Software (GCOS) (new version). GCOS also acquires data, manages sample and experimental information, and performs some gene expression data analysis.

The following protocols outline the general procedure for washing, staining, and scanning Affymetrix GeneChip arrays. Since there are several versions of the Fluidics Station (400 and 450/250), GeneArray and GeneChip scanner (2500 and 3000), and operating software (MAS and GCOS) in use, please refer to the specific Affymetrix User's Guides *(2)* for details on how to use the instrumentation and software.

3.8.1. Affymetrix GeneChip Probe Array Washing and Staining

Follow **steps 1** to **4** before taking the probe arrays out of the hybridization oven.

1. Launch the operating software and enter the experiment information according to the appropriate GCOS or Microarray Suite user's guides.
2. Set up and prime the Fluidics Station according to the Affymetrix User Manual for *Eukaryotic Sample and Array Processing* or the User's Guides of the instrument.
3. Prepare the streptavidin phycoerythrin (SAPE) stain solution (**Table 2**). The table lists the components needed for one to five probe arrays, adding a small volume for pipeting losses. Always prepare the stain solution fresh on the day of use, and prepare enough solution for all arrays to be processed on that day. SAPE should be stored in the dark at 4°C. Mix well and divide into two aliquots of 600 µL each per sample to be used for stains 1 and 3, respectively.
4. Prepare the antibody solution (**Table 3**). The table lists the components needed for one to five probe arrays, adding a small volume for pipeting losses. Mix well and use in aliquots of 600 µL to be used for stain 2. Prepare sufficient solution for all arrays to be processed on a given day.
5. After hybridization, recover the hybridization cocktail from the array and place it in the original tube. Store at −20°C for short-term storage or −80°C for long-term storage.
6. Fill the probe array completely with nonstringent Wash Buffer A (*see* **Note 17**).

Table 2
Streptavidin Phycoerythrin (SAPE) Stain Solution

Component	Final concentration	1	2.2	3.2	4.2	5.2
Water	—	540	1188	1728	2268	2808
2X Stain Buffer	1X	600	1320	1920	2520	3120
BSA, 50 mg/mL	2 mg/mL	48	105.6	153.6	201.6	249.6
SAPE, 1 mg/mL	10 µg/mL	12	26.4	38.4	50.4	62.4
Total (µL)		1200	2640	3840	5040	6240

Table 3
Antibody Solution

Component	Final concentration	1	2.2	3.2	4.2	5.2
Water	—	266.4	586.08	852.48	1118.88	1385.28
2X Stain Buffer	1X	300	660	960	1260	1560
BSA, 50 mg/mL	2 mg/mL	24	52.8	76.8	100.8	124.8
Goat IgG, 10 mg/mL	0.1 mg/mL	6	13.2	19.2	25.2	31.2
Biotinylated antibody, 0.5 mg/mL	3 µg/mL	3.6	7.92	11.52	15.12	18.72
Total (µL)		600	1320	1920	2520	3120

7. Start the Fluidics Station wash and stain run choosing the Fluidics scripts appropriate for your probe array. Insert the probe arrays into the designated module of the Fluidics Station; load the vial with the SAPE stain solution (Fluidics Station 400) or the two vials with the SAPE stain solution and the vial with the antibody solution (Fluidics Station 450/250) according to the Affymetrix User Manual for *Eukaryotic Sample and Array Processing*.
8. Proceed with the run and follow the prompts according to the protocol in the Affymetrix User Manual for *Eukaryotic Sample and Array Processing* for your instrument.
9. At the end of the run, remove the probe arrays from the Fluidics Station, and proceed with the probe array scan (*see* **Subheading 3.8.2.**).
10. If you do not scan the arrays right away, keep the probe arrays at 4°C and in the dark until ready for scanning. However, do not store arrays for a prolonged period, since this might lead to variability in the resulting data.
11. If there are no more samples to run for the day, shut down the Fluidics station following the procedure outlined in the Affymetrix User Manual for *Eukaryotic Sample and Array Processing*.

Expression Profiling Using Affymetrix GeneChip Probe Arrays 33

3.8.2. Scanning of Affymetrix GeneChip Probe Arrays

The scanner is also controlled by the Affymetrix Microarray Suite or GCOS.

1. Turn on the scanner at least 10 min before use to make sure that the laser is warmed up.
2. If the probe array was stored at 4°C, let it warm to room temperature before scanning.
3. Check that there are no bubbles visible in the glass window of the probe array. If bubbles are present, carefully remove the buffer and manually refill the array with nonstringent Wash Buffer A using a micropipet (*see* **Subheading 3.7.2., step 2**).
4. Remove any dirt or dust from the glass surface of the probe array using a non-abrasive tissue (do not use alcohol or Kimwipe tissues). Apply Tough Spots to each of the two septa to prevent any leaking of fluids during the scanning. Ensure that the tape or the Tough Spots remain completely flat.
5. Follow the Microarray Suite or GCOS or the appropriate scanner's user's manual for inserting the probe arrays into the scanner and starting the scanning process.
6. After scanning, check the image for the presence of image artifacts (i.e., high/low intensity spots, scratches, high regional or overall background, and so on) on the array. Right-click the mouse while the cursor is placed over the image, choose *Image Settings*, and uncheck *AutoScale*. View the image at a setting of 0 to 1000 to detect areas with high background.
7. Save the Affymetrix GeneChip probe arrays after scanning until you have checked the image and made sure that there are no manufacturing defects. Store the array wrapped in aluminum foil at 4°C. If there are defects, consult with your local Affymetrix representative for a possible replacement.

3.8.3. File Types

The Affymetrix operating software (MAS or GCOS) generates a number of files during processing of an Affymetrix GeneChip experiment.

The **Experiment Information File** (*.exp) contains information about the experiment name, sample, and probe array type. This file is not used for analysis but may be required to open other files for the designated probe array.

The **Data File** (*.dat) is the image of the scanned probe array.

The **Cell Intensity File** (*.cel) is derived from a *.dat file and is automatically created when a *.dat file is opened. It contains a single-intensity value for each probe cell delineated by the grid (calculated by the Cell Analysis algorithm).

The **Chip File** (*.chp) is the output file generated from the analysis of a probe array. It contains qualitative and quantitative analysis for every probe set. The file can be converted to a tab-delimited text file for easy uploading into data analysis software.

The **Report File** (*.rpt) is a text file summarizing data quality information for a single experiment. The report is generated from the analysis output file (*.chp).

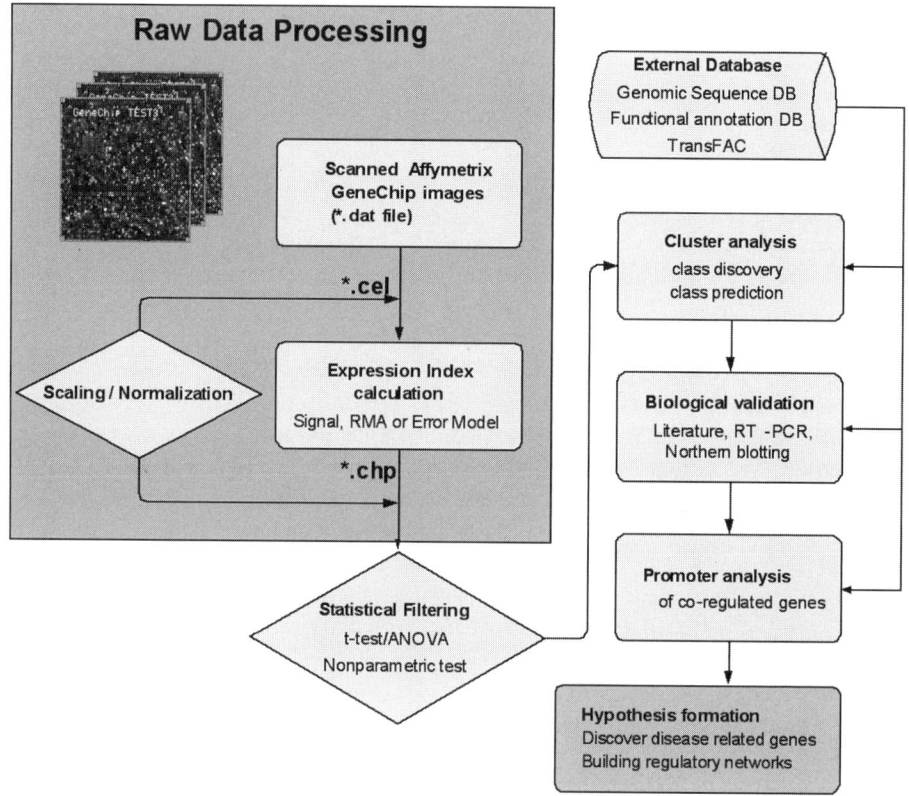

Fig. 1. Analysis steps involved in the first-order data analysis.

3.9. First-Order Data Analysis

Before the data can be used for statistical analysis, the raw values are scaled and normalized, and a gene expression index is calculated (**Fig. 1**).

Each transcript is measured by 11 probe pairs, which consist of a perfect match 25mer oligonucleotide (PM) and a 25mer mismatch oligonucleotide (MM) that contains a single base pair mismatch in the central position. The PM/MM design allows identification and subtraction of nonspecific hybridization and background signals, at the cost of introducing more variability in the gene expression indexes. Affymetrix GCOS uses a One-Step Tukey's Biweight Estimate to calculate the Signal, a quantitative metric that represents the relative level of expression of a transcript. It uses the MM intensity to estimate stray signal, which is subtracted from the PM. Further details about the analysis can be found in the Affymetrix User's Guides for MAS and GCOS, and the manual for *Data Analysis Fundamentals*.

Alternative approaches that use the PM intensity and other adjustment for background have been developed, such as Robust Multichip Average (RMA) *(3)*, and Model-Based Expression Index (MBEI) implemented in dChip *(4)*. It has been shown that the choice of the PM adjustment method can strongly influence the accuracy of the results *(5)*. However, no method seems to be superior to the others in all aspects.

The Affymetrix software also calculates a Detection *p*-value, which is evaluated against user-definable cutoffs to determine the Detection call. The Detection call indicates whether a transcript is reliably detected (Present) or not detected (Absent).

In GCOS, the average Signal intensity of the array is set to a default Target Signal of 500. The key assumption of this global scaling strategy is that there are few changes in gene expression among all the genes on the arrays that are being analyzed. Although this is a widely used approach, it might not be suitable for all experiments.

3.10. Quality Control Criteria

Generating high-quality microarray data requires vigorous quality control measures at each individual step of the process, starting with the experimental design of the study, the generation of samples, extraction of RNA, labeling of the probe, and microarray hybridization.

RNA quality control measures are outlined in **Subheading 3.4.** Briefly, RNA purity and yield are determined by optical density (OD) measurements at wavelengths of 260 and 280 nm. The OD 260:280 ratio should lie between 1.7 and 2.0. Otherwise, the RNA should be repurified. Further evaluation of the RNA quality is done using the Agilent Bioanalyzer and Lab-on-a-Chip. Electropherograms are created that detect degradation and measure the ribosomal 5S, 18S, and 28S bands. Ideally, the ratio of 28S:18S bands should be close to 2, but samples that show clear 18S and 28S peaks are acceptable. An additional index for measuring RNA quality is the yield (mass conversion rate) and average size of the biotin-labeled cRNA after RNA labeling.

3.10.1. Array Hybridization Quality Control

1. A general *visual inspection* of the entire GeneChip probe array should be performed after scanning, as outlined in **Subheading 3.8.2.** Probe arrays that show white speckling, holes, smudges, areas of saturation, or uneven hybridization should be repeated, or the affected probe sets should be manually masked. Alternatively, the image can be evaluated and outliers detected using dCHIP software *(4)*.
2. The boundaries of the probe area (viewed upon opening the *.dat/*.cel file) are easily identified by the hybridization of the *B2 Oligo*, which is spiked into each hybridization cocktail as part of the Eukaryotic Hybridization Control kit.

Hybridization of B2 is highlighted on the image by the alternating pattern of intensities on the border, the checkerboard pattern at each corner, and the array name, located in the upper left or upper middle of the array. B2 Oligo serves as a positive hybridization control and is used by the software to place a grid over the image. Some variation in B2 hybridization intensities across the array is normal. However, if the B2 intensities at the checkerboard corners are either too low or high, or are skewed owing to image artifacts, the grid will not align automatically. The user must align the grid manually using the mouse to click and drag each grid corner to its appropriate checkerboard corner.

3. The *Noise value (Raw Q)* can be found either in the Analysis Info tab of the Data Analysis (*.chp) file, or in the Expression Report (*.rpt) file. Noise (Raw Q) is a measure of the pixel-to-pixel variation of probe cells on a GeneChip array. RawQ values should remain consistent across the probe arrays of an experiment.

4. *Scaling factors (SF)* can be found in the Analysis Info tab of the .chp file output and in the Expression Report (.rpt) file. SF values should remain consistent across the experiment. The scaling factor for each given experiment should be within a two- to threefold range.

5. The number of probe sets called "*Present*" relative to the total number of probe sets on the array is displayed as a percentage in the Expression Report (.rpt) file. Percent Present (%P) values depend on multiple factors including cell/tissue type, biological or environmental stimuli, probe array type, and overall quality of RNA. Replicate samples should have similar %P values. Extremely low %P values are a possible indication of poor sample quality.

6. Most GeneChip expression arrays use β-*actin and GAPDH* as internal control genes. Transcripts of these genes are represented by 3' and 5' probe sets that are used to assess RNA sample and assay quality. The Signal values of the 3' probe sets are compared with the Signal values of the corresponding 5' probe sets. The ratio of the 3' probe set to the 5' probe set should be not more than 2 for the one-cycle assay. Additional rounds of amplification increase this number. A high 3' to 5' ratio may indicate degraded RNA or inefficient transcription of double-stranded cDNA. The 3' to 5' ratios for internal controls are displayed in the Expression Report (.rpt) file.

7. The *Eukaryotic Hybridization Controls bioB, bioC, bioD,* and *cre* are spiked in staggered concentrations (1.5, 5, 25, and 100 pM final concentrations for *bioB, bioC, bioD,* and *cre*, respectively) into the hybridization cocktail (*see* **Subheading 3.6.2.**). They should maintain a maximum 1:2 ratio of signal intensities of the 5' and 3' probe sets. *bioB* is at the level of assay sensitivity and should be called "Present" at least 50% of the time. *bioC, bioD,* and *cre* should always be called "Present" with increasing Signal values, reflecting their relative concentrations.

8. The *PolyA spike-in controls* (*see* **Subheading 3.6.1.**) should be called "Present" with increasing Signal values in the order of *lys, phe, thr,* and *dap*.

3.10.2. Graphical Quality Control

A number of graphical tools are available that allow one to visualize the quality of microarray data from several perspectives. These diagnostic plots include:

- A color image plot of the entire array to evaluate expression across the probe array's special layout.
- M versus A plots of two probe arrays display the log intensity ratio M versus the mean log intensity A. MvA plots make it easy to identify intensity-dependent biases in the data (i.e., "banana shape").
- Intensity boxplots are another good visualization tool for comparing the overall intensities of all probes across the arrays in an experiment.
- Intensity histograms are a good visualization tool for the distribution of intensities and for identifying saturation, which is revealed as an overrepresentation of high-intensity values.
- RNA degradation plots show expression as a function of the 3' to 5' position of probes from the *.cel file. The slope of these lines indicates potential RNA degradation in the samples.
- Principal components plots can help to evaluate inner group and between-group variability.

4. Notes

1. It is important that the tissue be snap-frozen immediately after tissue harvest. If a more time-intensive tissue dissection is required, adult heart tissue should be dissected and stored in RNA-stabilizing agents such as RNAlater (Qiagen) to preserve the RNA. RNAlater is not recommended for storage of embryonic tissue for subsequent dissection because it makes the tissue brittle.
2. For adult heart tissue, we found that phenol-based RNA isolation methods give higher yields than solid phase/glass filter-based methodologies alone. An additional purification step is necessary to yield RNA of high purity suited for microarray experiments. If sufficient starting tissue material is available, the use of a combination of phenol-based and solid phase/glass filter-based methods yields about 1 µg of high-purity RNA per mg of tissue.
3. If you are working with very small amounts of tissue, such as needle biopsy material, skip the weighing step and proceed directly to **step 3** using 1 mL of TRIZOL reagent.
4. For fibrous tissues, such as rat and mouse heart and skeletal muscle, the most difficult step in the isolation process is the complete disruption of all cells when one is preparing tissue homogenates. Owing to low cell density and the polynucleated nature of cardiomyocytes, the yield of total RNA is typically low. In addition, fibrous tissue is difficult to homogenize completely, which can result in degraded RNA and very low yield. For complete lysis of the cells, keeping the tissue completely frozen until homogenization and using a powerful homogenizer

is critical to isolating intact total RNA. Embryonic heart tissue can be easily homogenized using a hand-held glass homogenizer.
5. Samples should not be thawed before contact with TRIZOL reagent to avoid degradation of the RNA.
6. The RNA should not be stored in water for a longer period to avoid degradation. For long-term storage, keep the RNA in an alcohol solution at –80°C. For this purpose, add 2 vol of 100% ethanol and 1/10 vol of 3 M sodium acetate (NaOAc), pH 5.2, to the RNA in water. Before use, spin down the RNA in a microcentrifuge at 15,000g for at least 15 min at 4°C, decant the supernatant, and wash the pellet with 1 mL 80% ethanol. Centrifuge again for 5 min, remove the supernatant, and dry the pellet for 10 min at RT.
7. At this point, we frequently do a "prerun" of the chip to make sure that the chip does not have bubbles. Give the chip a quick vortex (10 s, 512g), and proceed with the assay (running the chip) as usual. Allow the assay to run until it *focuses* for 10 s. If this is successful, abort the run, and proceed with adding the samples.
8. Agilent has recently introduced the RNA Integrity Number (RIN) as part of a software extension. The RIN is a tool to grade RNA quality on a quantitative scale of 1 (worst) to 10 (best). Sample integrity is determined by the entire electrophoretic trace of the RNA sample, including the presence or absence of degradation products.
9. Total RNA versus PolyA + RNA: good-quality microarray data have been obtained using total RNA or PolyA+ RNA as starting material for sample labeling. The results obtained from both types of sample are similar but not identical. Therefore, only samples prepared using the same sample preparation protocol should be compared. It is advisable to isolate total RNA first and check the initial RNA quality before proceeding with PolyA+ RNA isolation. PolyA+ RNA that has been purified once using oligo-dT columns might still contain significant amounts (up to 50%) of ribosomal and other nonpolyadenylated RNAs. Two rounds of purification over oligo-dT columns usually yield up to 95% pure PolyA+ RNA but result in significant loss of material. Therefore, PolyA+ RNA isolation depends on the availability of a sufficient amount of starting material.
10. The Affymetrix GeneChip One-cycle target labeling protocol is recommended for starting amounts of 1 to 15 μg total RNA or 0.2 to 2 μg PolyA RNA. Usually, 5 to 8 μg of total RNA yield around 80 to 100 μg of biotin-labeled cRNA. For hybridization, 15 to 20 μg of biotin-labeled cRNA is needed for a standard GeneChip. One microgram of RNA will yield just sufficient amounts for preparation of one hybridization cocktail. Here, the quality of RNA has to be excellent to ensure that enough biotin-labeled cRNA is being made. If less than 1 μg of total RNA is available, an additional round of amplification is required to generate the 15 μg amount of biotin-labeled cRNA needed for hybridization. Although most amplification methods claim that they faithfully maintain relative RNA abundance, amplification does shorten the length of the resulting transcripts and introduces artifacts, especially if the detection probes on the microarray are not sufficiently 3' biased. Hence, it is recommended to keep amplification to a minimum.

11. The volume of the eluate should be carefully measured by pipeting. Recovery of 11 µL or less may be a signal that the cDNA has not been completely eluted from the column.
    a. Check whether any liquid is present above the rim of the membrane. If liquid is seen along the rim of the membrane, use a P10 or P2 pipetor with a filtered, pointed (nonbeveled) pipet tip to aspirate the liquid gently from the rim and then eject the liquid directly onto the membrane itself. Centrifuge at maximum speed for 1 min.
    b. If there is still volume missing, perform an additional centrifugation step at maximum speed for 3 to 5 min.
    c. If a. and b. fail to recover the missing volume, add the missing volume (up to 5 µL) of cDNA Elution Buffer directly onto the membrane and repeat **step 8**.
12. If you are starting with more than 8 µg of total RNA, use no more than 6 µL of the double-stranded cDNA from **Subheading 3.5.3.** per IVT reaction.
13. Do not extract biotin-labeled RNA with phenol-chloroform. The biotin will cause some of the RNA to partition into the organic phase. This will result in low yields.
14. The average size distribution after one round of amplification should center around 1500 nucleotides. Additional rounds of amplification result in smaller fragment sizes, which can lead to signal loss if the microarray is not sufficiently 3'-biased.
15. Fragmentation is a critical step in the target preparation. Overfragmentation can cause high background, low signal, and high 3':5' ratios. If this is the case, fragmenting a new aliquot of cRNA for hybridization may solve the problem.
16. For different starting amounts, adjust the dilution accordingly to arrive at the same proportion of the spike-in controls to the RNA sample. Do not pipet less than 2 µL volumes, as this might increase the pipeting error and lead to inconsistencies.
17. The Affymetrix user guide for *Eukaryotic Sample and Array Processing* states that, if necessary, at this point, the probe array can be stored at 4°C for up to 3 h. However, it has been found that probe array processing at different time points after hybridization introduces variability. Hence, we recommend processing the probe array immediately after hybridization, if possible.

References

1. Downey, T., Pevsner, J., Jeon, O.-H., Murillo, F. M., and Spitznagel, E. L. (2004) Estimating and removing batch effects from microarray data, in *7th International Meeting of the Microarray Gene Expression Data Society*, September 8–10, 2004, Toronto, ON, Canada.
2. Affymetrix Manuals and User's Guides, available for download from www.Affymetrix.com:
    a. Expression Analysis Technical Manual, which contains the following sections:
        Section 1: GeneChip Expression Analysis Overview
        Section 2: Eukaryotic Sample and Array Processing
        Section 3: Prokaryotic Sample and Array Processing
        Section 4: Fluidics Station Maintenance Procedures
        Section 5: Appendices

b. Affymetrix Fluidics Station 400 Manual
c. Affymetrix Fluidics Station 450 Manual
d. Hybridization Oven Manual
e. Data Analysis Fundamentals
f. GeneChip® Operating Software Manual
g. Microarray Suite User's Guide, Version 5.0

3. Irizarry, R. A., Bolstad, B. M., Collin, F., Cope, L. M., Hobbs, B., and Speed, T. P. (2003) Summaries of Affymetrix GeneChip probe level data. *Nucleic Acids Res.* **31,** e15.
4. Li, C. and Wong, W. H. (2001) Model-based analysis of oligonucleotide arrays: expression index computation and outlier detection. *Proc. Natl. Acad. Sci. USA* **98,** 31–36.
5. Choe, S. E., Boutros, M., Michelson, A. M., Church, G. M., and Halfon, M. S. (2005) Preferred analysis methods for Affymetrix GeneChips revealed by a wholly defined control dataset. *Genome Biol.* **6,** R16.
6. Knudtson, K. L., Griffin, C., Brooks, A., et al. (2002) Factors contributing to variability in DNA microarray results: the ABRF Microarray Research Group 2002 study. www.abrf.org.

3

Serial Analysis of Gene Expression (SAGE)
A Useful Tool to Analyze the Cardiac Transcriptome

Kirill V. Tarasov, Sheryl A. Brugh, Yelena S. Tarasova, and Kenneth R. Boheler

Summary

Serial analysis of gene expression (SAGE), a functional genomics technique, can be used for global profiling of gene transcripts. It relies on the preparation and sequencing of cDNA concatemers, but it does not require prior knowledge of the genes to be assayed (as with microarrays). Once analyzed, SAGE data provide both a qualitative and quantitative assessment of potentially every transcript present in a particular cell or tissue type. In this chapter, we describe the fundamental principles of SAGE, describe a complete protocol for the generation of SAGE libraries, and show how it has been employed to generate the first SAGE reference data set of the mouse myocardium. Following the protocols described here, investigators should be able to generate unique mouse heart SAGE libraries, which can be directly compared with our reference library. This permits the identification of transcripts that are differentially expressed as a function of time, age, genetic background or transgenic state, among other factors. SAGE is thus a powerful technique that permits a comprehensive analysis of changes in mRNA abundance. The results provide a snapshot of altered patterns of gene expression in response to any genetic or environmental stimulus that can be used to generate new biological hypotheses or test existing paradigms.

Key Words: SAGE; transcriptome; mRNA; gene expression; sequencing; heart; functional genomics.

1. Introduction

Serial analysis of gene expression (SAGE) permits the simultaneous evaluation of thousands of transcripts in a single sequencing-based assay. SAGE relies on two major principles. First, short DNA sequences are sufficient to identify individual gene products (transcripts or mRNAs), and second, concatenation (linking together) of short DNA sequences or tags increases the efficiency of identification of expressed transcripts in a sequencing reaction. The tag (transcript) profile generated by SAGE relies primarily on 14 to 21 base nucleotide sequences (SAGE tags) for gene identification *(1,2)*. The technique generates large numbers of short tags, originating from the last (most 3') unique location of an enzyme recognition site in a single transcript. When the tags are sequenced, the technique can theoretically identify up to 4^{10} (1,048,576) unique transcripts. The newer LongSAGE method (17 base tags) can distinguish 4^{17} different transcripts, a number sufficient to be virtually unique even within the whole genome *(2)*.

A general scheme for the generation of SAGE libraries is given in **Fig. 1**. The method involves seven major steps: (1) mRNA preparation and cDNA synthesis; (2) cleavage of biotinylated cDNA with an anchoring enzyme and isolation of these fragments by binding to magnetic beads; (3) linker ligation to the bound cDNA; (4) release of cDNA tags using a tagging (type IIs) restriction enzyme (*see* **Note 1**), followed by blunt ending; (5) ligation of individual tags to form ditags, followed by a polymerase chain reaction (PCR) amplification; (6) isolation of ditags, followed by ditag ligation to form concatemers; and (7) concatamer cloning and plasmid amplification, followed by sequencing. Once completed, the sequence files are subjected to a data analysis process that identifies the recognition site of the anchoring enzyme (AE). The ditags are extracted and can be used to identify the transcripts of origin. Each of these steps will be described in detail below.

One major advantage of SAGE is its comparative power. SAGE libraries can be compared both within a laboratory and with other publicly available libraries (http:// www.ncbi.nlm.nih.gov/SAGE). Ultimately, the comparative power of SAGE increases as a function of the number of publicly available libraries, to confirm or refute the authenticity of putative transcripts with differential gene expression either within a tissue or among tissues. The latter can be used to identify novel tissue-restricted transcripts.

2. Materials

2.1. Reagents

1. Dimethyl sulfoxide (DMSO).
2. Ammonium sulfate [$(NH_4)_2SO_4$].
3. Magnesium chloride ($MgCl_2$).
4. Sodium perchlorate ($NaClO_4$).
5. β-Mercaptoethanol.
6. Agarose.
7. Adenosine triphosphate (ATP).
8. Sodium chloride (NaCl).

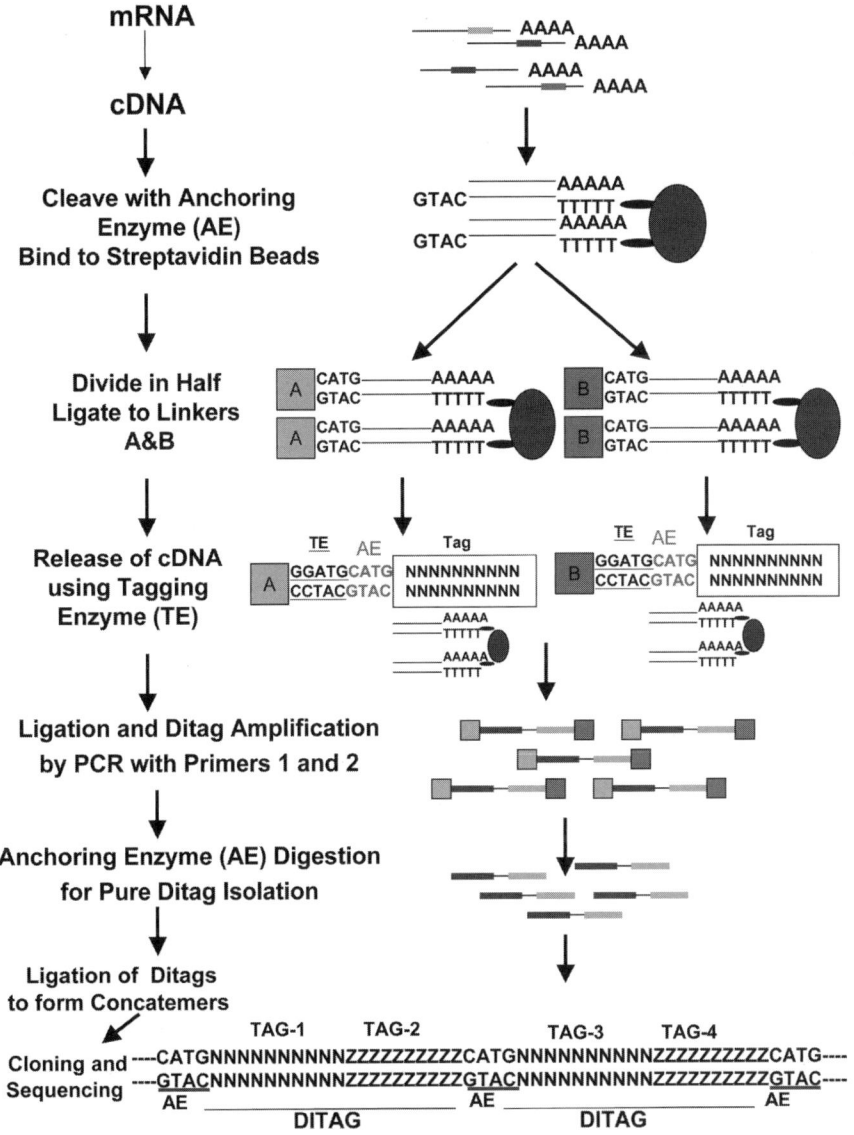

Fig. 1. General scheme for the generation of SAGE libraries. The details of each step are given in the text. Two assumptions are critical for SAGE analyses: (1) short DNA sequences (10–14 bp) are sufficient to identify individual gene products, and (2) concatenation (linking together) of short DNA sequences or tags increases the efficiency of identifying expressed mRNAs in a sequence-based assay. Once a SAGE library is completed, each individual tag sequence can be run against GenBank databases to identify the corresponding gene product.

2.2. Electrophoresis Reagents and DNA Ladders

1. Polyacrylamide solution 19:1 (Bio-Rad, cat. no. 161-0144).
2. Polyacrylamide solution 37.5:1 (Bio-Rad, cat. no. 161-0148).
3. N,N,N',N'-tetramethylenediamine (TEMED; Bio-Rad, cat. no. 161-0800).
4. Ammonium persulfate (Bio-Rad, cat. no. 161-0700).
5. Polyacrylamide Criterion Precast Gel (15% TBE, 20% TBE Bio-Rad, cat. no. 345-0056) or similar.
6. 20- or 25-bp DNA ladder.
7. 100-bp DNA ladder.
8. 1-kb DNA ladder.

2.3. Premade (Purchased) Stock Solutions

1. 0.5 M Ethylenediaminetetraacetic acid, disodium (EDTA), pH 8.0.
2. Phenolchloroform/isoamyl alcohol mix, 25:24:1, pH 8.0.
3. Tris/acetate/EDTA (TAE), 50X.
4. 1 M Tris-HCl, pH 7.5 and 1 M, pH 8.0 (PC8).
5. 10 M Ammonium acetate (NH_4OAc).
6. 5 M NaCl; 0.25 M EDTA, pH 7.0.
7. Diethylpyrocarbonate (DEPC, 0.1%)-treated water.
8. Tris/borate/EDTA buffer (TBE), 10X, pH 8.3.
9. SOC media.
10. Isopropanol.
11. Absolute ethanol.
12. Glycerol.

2.4. Prepared Solutions

1. LoTE: 3 mM Tris-HCl (pH 7.5), 0.2 mM EDTA (pH 7.5). Store at 4°C.
2. 2X Binding and washing buffer (B+W) (*see* **Note 2**): 10 mM Tris-HCl (pH 7.5), 1 mM EDTA, 2.0 M NaCl; store at room temperature (RT).
3. 12% Polyacrylamide solution (44 mL final): for a 20 × 16 × 0.1-cm gel, mix 29.4 mL of dH_2O, 13.2 mL of 40% PAAG mix (19:1), 875 µL of 50X TAE, 437.5 µL of 10% ammonium persulfate, and 37.5 µL of TEMED. Stir well.
4. 8% Polyacrylamide solution (44 mL final): for a 20 × 16 × 0.1-cm gel, mix 33.8 mL of dH_2O, 8.8 mL of 40% PAAG mix (37.5:1), 875 µL of 50X TAE, 438 µL of 10% ammonium persulfate, and 37.5 µL of TEMED. Stir well.
5. LB with 10% glycerol (*see* **Note 3**): for 1 L, add 10 g tryptone, 5 g yeast extract, 10 g NaCl; add 100 mL anhydrous glycerol to a 1-L bottle. Fill to 1 L with dH_2O and autoclave, cool to 55°C, and add Zeocin to a final concentration of 50 µg/mL. Store in the dark at 4°C for up to 1 mo.
6. Low-salt LB agar plates or lawns with Zeocin (*see* **Note 4**): 1% tryptone, 0.5% yeast extract, 0.5% NaCl, 1.5% Bacto-agar, 50 µg/mL Zeocin. For preparation, add tryptone, yeast extract, and NaCl, stir, adjust pH of the solution to 7.5 with NaOH, add Bacto-Agar, and bring water to the final volume. Autoclave, cool to 55°C, add 250 µL of Zeocin 100 µg/µL stock, pour into the plates, and seal with parafilm. Store in the dark at 4°C for up to 1 mo.

2.5. Enzymes

1. T4 DNA ligase (1 and 5 U/μL) (*see* **Note 2**).
2. T4 polynucleotide kinase.
3. DNA polymerase I, Klenow fragment.
4. Taq DNA polymerase.
5. AmpliTaq Gold polymerase (Applied Biosystems, cat. no. 4311814).
6. Restriction enzymes: *Nla*III (*see* **Note 5**); *Bsm*FI, *Sph*I.

2.6. Other Materials

1. Bacto-tryptone.
2. Bacto-yeast extract.
3. Bacto-agar.
4. ElectroMAX DH10B cells (Invitrogen, cat. no. 18290-015).
5. pZeRO-1 (Invitrogen, cat. no. 2500-01).
6. Zeocin (Invitrogen, cat. no. R250-01).
7. SYBR Green I (Molecular Probes, cat. no. S-7567).
8. Glycogen (Boehringer, cat. no. 901 393).
9. Dynabeads M-280 (Dynal, cat. no. 112.05).

2.7. Kits

1. FastTrack 2.0 kit for preparation of mRNA (Invitrogen, cat. no. 1593-02).
2. cDNA Synthesis kit (Invitrogen, cat. no. 18267-013).
3. dNTP kit (Amersham, cat. no. US77100).

2.8. Oligo-DNA Linkers and Primers (see Note 6)

1. Linker A1 (gel-purified): 5'-TTT GGA TTT GCT GGT GCA GTA CAA CTA GGC TTA ATA GGG ACA TG-3'.
2. Linker A2 (gel-purified): TCC CTA TTA AGC CTA GTT GTA CTG CAC CAG CAA ATC C [amino mod. C7]-3'.
3. Linker B1 (gel-purified): TTT CTG CTC GAA TTC AAG CTT CTA ACG ATG TAC GGG GAC ATG.
4. Linker B2 (gel-purified): TCC CCG TAC ATC GTT AGA AGC TTG AAT TCG AGC AG [amino mod. C7]-3'.
5. Primer 1: GGA TTT GCT GGT GCA GTA CA.
6. Primer 2: CTG CTC GAA TTC AAG CTT CT.
7. Biotinylated oligo dT (gel-purified): 5'-[biotin]T_{18}.
8. M13 forward: GTA AAA CGA CGG CCA GT.
9. M13 reverse: GGA AAC AGC TAT GAC CAT G.

2.9. Equipment

1. Thermocycler (*see* **Note 7**).
2. Electrophoresis units for polyacrylamide and agarose gel electrophoresis.
3. Electroporation system Gene-Pulser II/Pulse Controller II (Bio-Rad) or equivalent.

4. 0.2-cm Electroporation cuvets (Bio-Rad, cat. no. 165-2086).
5. Magnetic particle concentrator (MPC-E type, Dynal).

2.10. Plastics

1. SpinX microcentrifuge tubes (Costar, cat. no. 8160).
2. 96-Well Costar Flat Bottom Plates (VWR, cat. no. 29442-070).
3. Thermo-Fast® 96-well Detection plate and adhesive sealing sheets (Marsh, cat. no. AB-1100 and cat. no. AB-0558).
4. MicroAmp Reaction tubes (Applied Biosystems, cat. no. 801-0580).
5. MicroAmp Caps (Applied Biosystems, cat. no. 801-0535).
6. 25 × 25-cm Bio Assay Trays (Marsh Bio, cat. no. QH2216).

2.11. Software

1. SAGE data preprocessing and processing: SAGE2000 V4.12 or better is recommended (available at http://www.sagenet.org).
2. WinZip (available at www.winzip.com) for Windows; MS Access, MS Excel (Microsoft).

3. Methods
3.1. Preliminary and Routine Procedures

All incubations are performed in a water bath. All centrifugation steps, unless otherwise noted, are performed in an Eppendorf microcentrifuge at room temperature (RT).

1. Kinasing (phosphorylation) reaction for linkers.
    a. In two Eppendorf tubes, add 6 μL of LoTE, 2 μL of 10X kinase buffer, 2 μL of 10 mM ATP, and 1 μL of T4 polynucleotide kinase (10 U/μL).
    b. Add 9 μL of Linker A2 (350 ng/μL) to tube marked "Linker A" and 9 μL of Linker B2 (350 ng/μL) to tube marked "Linker B."
    c. Incubate both tubes at 37°C for 30 min and then at 65°C for 10 min.
    d. Add 9 μL of Linker A1 to "Linker A" tube and 9 μL of Linker B1 to "Linker B" tube.
    e. Incubate both at 95°C for 2 min, then at 65°C for 10 min, at 37°C for 10 min, and at RT for 20 min.
    f. Kinased linkers are stable at –20°C for up to a year.
2. Control reaction for linkers.
    a. Prepare four Eppendorf tubes marked "1+," "1–," "2+" and "2–."
    b. Add 21 μL of LoTE, 6 μL of 5X Ligation buffer, and 1 μL of selected linker (Linker A to the tubes marked "1" and Linker B to the tubes marked "2") in each tube.
    c. Incubate all four tubes at 50°C for 2 min and then at RT for 15 min.

Analysis of the Cardiac Transcriptome by SAGE

    d. Add 2 µL of T4 Ligase (1 U/µL) to the tubes marked "+" and 2 µL of LoTE to the tubes marked "–."
    e. Incubate both at 16°C for 2 h.
    f. Run half of each sample on 20% Precast TBE polyacrylamide gels at 130 V for 3 h, and stain with SYBR Green I (20 µL/200 mL of 1X TBE buffer) for 15 min on a shaker.
    g. Kinased linkers should form linker-linker dimers (80–100 bp) after ligation, whereas unkinased linkers will not self-ligate. Only linkers dimerized with >70% efficiency are acceptable.

3. Prepare pZeRO-1.
    a. Add 14 µL of dH$_2$O, 2 µL of Buffer 2, 2 µL of pZeRO-1 (1 µg/µL), 2 µL of *Sph*I (5 U/µL).
    b. Incubate at 37°C for 25 min, then add 60 µL of TE buffer (final concentration 25 ng/µL), and heat-inactivate at 68°C for 15 min.
    c. Store on ice until used, ideally the same day. Cleaved pZeRO-1 plasmid can be stored at –20°C for few weeks, but its cloning efficiency decreases with time.

4. Phenol-chloroform extraction.
    a. Bring sample volume to 200 µL with LoTE and add an equal volume (200 µL) of phenol/chloroform/isoamyl alcohol (PC8, 25:24:1, pH 8.0).
    b. Separate the aqueous phase, which contains the DNA, from the organic phase by centrifugation at maximum speed for 2 min.
    c. Transfer the aqueous phase into a fresh Eppendorf tube, while ensuring that no organic solution is included.

5. Ethanol (EtOH) precipitation. DNA in aqueous solution is precipitated by the addition of ethanol in the presence of ammonium acetate followed by cooling. Because of the small amount of DNA and small fragment sizes, glycogen is added as a carrier. For a 200 µL vol of sample/LoTE:
    a. Add 3 µL of glycogen (20 mg/mL), one-half vol 10 *M* ammonium acetate, and 2.5 vol ice-cold (–20°C) 100% ethanol and vortex.
    b. Pellet the precipitated DNA by centrifugation at maximum speed for 20 min.
    c. Remove the ethanol with care and wash with 75% ethanol to remove salt, which is critical prior to ligation reactions.
    d. Centrifuge at 10,000*g* for 10 min, dry the pellet, and resuspend in LoTE.
    e. In certain instances, an increased volume of ethanol or 15 min dry ice/ethanol incubation may be required at the precipitation step, as indicated in the text.

6. Isopropanol precipitation.
    a. Bring sample volume to 450 µL with LoTE, and add 3 µL of glycogen, 150 µL of 2 *M* sodium perchlorate, and 330 µL of isopropanol.
    b. Vortex, and spin at maximum speed for 10 min at RT.
    c. Aspirate the supernatant, dry the pellet, and resuspend in specified volume of LoTE.

3.2. SAGE Protocol

This protocol is a modified version based on those described by Velculescu et al. *(1)*, Kenzelmann and Muhlemann *(3)*, and Anisimov et al. *(4–6)* (*see* **Note 8**). The whole protocol (except sequencing and data analysis) takes about 5 to 6 d; however, this time will vary depending on the experience of the investigator.

3.2.1. mRNA Preparation and cDNA Synthesis

1. We recommend preparing total RNA from cells or tissues using a guanidine isothiocyonate method *(7)* (*see* **Note 9**). The RNA must be at good quality before proceeding with the protocol.
2. We have had good results with the FastTrack 2.0 kit (Invitrogen) for the preparation of Poly(A)$^+$ (mRNA) from total RNA. About 1 mg of total RNA is usually required for mRNA preparation (if you have lesser amounts of total RNA, *see* **Note 10**). Ideally, you should have 5 µL/µg of mRNA for the preparation of a SAGE library.
3. Starting with 5 µg of mRNA, use the cDNA Synthesis kit according to manufacturer's protocol. Substitute the 5'-biotinylated Oligo(dT)$_{18}$ primer (500 ng/µL) for the oligo d(T) primer (supplied with this kit) in the first-strand synthesis reaction. After the second-strand reaction is terminated with 25 µL of 0.25 *M* Na$_2$EDTA (pH 7.5), extract the cDNA sample with PC8, EtOH-precipitate, and resuspend the pellet in 20 µL of LoTE (*see* **Note 11**).

3.2.2. Cleavage of Biotinylated cDNA with Anchoring Enzyme (AE) and Binding to Magnetic Beads

1. Take 10 µL of cDNA sample, add 163 µL of LoTE, 2 µL of bovine serum albumin (BSA; 100X), 20 µL of 10X restriction enzyme Buffer 4, and 5 µL of *Nla*III (anchoring enzyme, 10 U/µL). Incubate at 37°C for 1 h, extract with PC8, EtOH-precipitate, and resuspend in 20 µL of LoTE.
2. Prepare two Eppendorf tubes marked "1" and "2." Add 100 µL of resuspended M-280 streptavidin magnetic beads to each of two Eppendorf tubes. Use the magnetic particle concentrator to immobilize beads and remove storage buffer.
3. Wash beads once with 200 µL 1X B+W buffer.
4. Add 100 µL of 2X B+W buffer, resuspend, and add 90 µL of dH$_2$O and 10 µL of cDNA digestion products to each Eppendorf tube. Incubate at RT for 15 min, carefully resuspending beads intermittently.
5. Wash beads three times with 200 µL 1X B+W and then once with 200 µL LoTE, removing the wash each time. Proceed immediately to the next step.

3.2.3. Ligating Linkers to Bound cDNA

1. Resuspend the magnetic beads in 25 µL of LoTE and add 8 µL of 5X Ligase buffer.

Analysis of the Cardiac Transcriptome by SAGE

2. Add 5 μL of annealed Linker A (A1, A2) to the tube marked "1" and 5 μL of annealed Linker B (B1, B2) to the tube marked "2." Incubate both tubes at 50°C for 2 min and then at room temperature for 15 min.
3. Add 2 μL of T4 Ligase (high concentration, 5 U/μL) to each tube, and incubate at 16°C for 2 h.
4. Wash the beads four times with 200 μL 1X B+W and then two times with 1X Buffer for *Bsm*FI. Proceed immediately to the next step.

3.2.4. Release of cDNA Tags by Tagging Enzyme (TE) and Blunt Ending

1. Resuspend the magnetic beads from each tube in 86 μL of LoTE, add 10 μL of 10X Buffer 4, 2 μL of 100X BSA, and 2 μL of *Bsm*FI (2 U/μL, tagging enzyme). Incubate at 65°C for 1 h, mixing intermittently.
2. Collect the supernatants, extract with PC8, EtOH-precipitate, and resuspend the pellets in 10 μL of LoTE.
3. To each of two tubes (1 and 2) with 10 μL of tagging enzyme digestion products, add 32.4 μL of dH$_2$O, 5 μL of 10X second-strand buffer (from cDNA synthesis kit), 1 μL of 100X BSA, 1 μL of dNTPs (25 m*M*), and 0.6 μL of Klenow (5 U/μL). Incubate at 37°C for 30 min.
4. Add 150 μL of LoTE, extract samples with PC8, run ethanol precipitation with 1 mL of ethanol and ammonium acetate instead of 500 μL, and resuspend pellets in 6 μL of LoTE.

3.2.5. Ligating Tags to Form Ditags Followed by PCR Amplification of Isolated Ditags

1. For ditag ligation reaction, take 2 μL of blunt-ended samples 1 and 2 from the previous step, add 1.2 μL of 5X Ligase buffer and 0.8 μL of high-concentration T4 Ligase (5 U/μL). For negative control reactions, add H$_2$O, do not add enzyme. Incubate at 16°C overnight.
2. Add 14 μL of LoTE to the samples, and proceed directly to the next step. Otherwise, samples can be saved for up to 8 h at 4°C or frozen at –20°C indefinitely.
3. Use 1-μL aliquots of the samples to make serial LoTE dilutions (*see* **Note 12**): 1:10, 1:25, 1:50, and 1:100 for ditag ligation reaction, 1:10 and 1:25 for negative control reactions, and the pure LoTE sample as an additional negative control.
4. Prepare a PCR mix using 30.5 μL of sterile water, 1 μL of selected ligation reaction or control dilution, 5 μL of 10X PCR Gold Buffer, 4 μL of MgCl$_2$ (25 m*M*), 3 μL of DMSO, 4 μL of dNTPs (25 m*M*), 1 μL of Primers 1 and 2 (350 ng/μL each), and 0.5 μL of AmpliTaq Gold polymerase (5 U/μL). Use one strip (eight tubes) of MicroAmp Reaction tubes for PCR. Cycling conditions (*see* **Note 7**) were: the initial hold; 10 min at 95°C, followed by 28 cycles of 30 s at 95°C, 1 min at 55°C, 1 min at 72°C, final extension 72°C for 5 min, hold at 4°C.
5. Load 20 μL of individual PCR products mixed with 4 μL of any 6X Loading buffer (containing bromphenol blue) into the wells of 10-well 20% preready Novex TBE polyacrylamide gel. Load 100-bp DNA ladder to distal wells of the

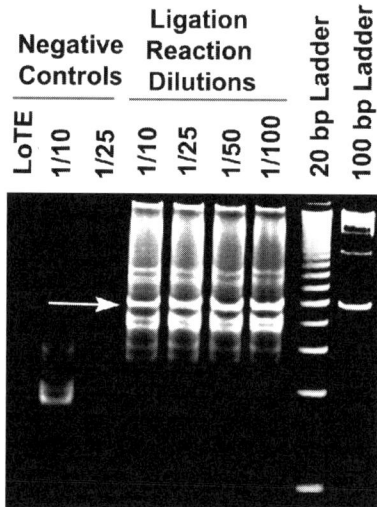

Fig. 2. Electrophoresis of PCR-amplified ditags in a 20% TBE polyacrylamide gel. The major band of 102 base pairs (arrow) represents amplification products from ligated ditags. Other bands represent nonspecific amplification products or linker-linker ligation products (80 bp).

gel. Use 1X TBE as a running buffer. Set voltage to 30 V until all the samples have entered the gel, and then reset voltage to 120 V. Run until bromphenol blue dye reaches the end of the gel (approx 5 h).

6. Stain gel with SYBR Green I (20 μL/200 mL of 1X TBE buffer) for 15 min while shaking. Ditag ligation reaction PCR amplification products should be seen as a major 102-bp band with another bright band at approx 80 bp (linker-linker ligation). Any other bands should be considerably less bright. Negative control reactions should lack the 102-bp band, and LoTE (negative control) should not produce any bands (**Fig. 2**). The optimal dilution of the ditag ligation reaction products for large-scale PCR is determined based on the brightness of the 102-bp band relative to background. The highest dilution that gives the greatest brightness to background ratio is considered optimal. Usually, 1:50 to 1:100 dilutions produce good results.

7. Large-scale PCR amplifications are performed similarly to test PCR conditions, but with the selected LoTE dilutions as a template. Use a Thermo-Fast 96-well Detection plate; cover the plate with adhesive sealing sheets and spin (*see* **Note 13**) to eliminate bubbles in the wells. Prepare 120 reactions in 50 μL in 96-well plates.

8. Collect PCR products in 12 Eppendorf tubes (500 μL in each), extract samples with PC8, ethanol-precipitate (using 1 mL of absolute ethanol), and resuspend pellets in 18 μL of LoTE (216 μL total).

9. Add 54 μL of 6X Loading buffer (270 μL total) and load PCR products onto eight wells of 12% polyacrylamide gel (19:1). Load 100-bp DNA ladder to distal wells of the gel. Use 1X TAE as a running buffer. Set voltage to 30 V until the entire sample has entered the gel, and then reset voltage to 130 V.
10. At the end of the electrophoresis, stain the gel with SYBR Green I (20 μL/ 200 mL of 1X TAE buffer) for 15 min while shaking.
11. Pierce the bottoms of eight 0.5-mL Eppendorf tubes with a needle or surgical blade to form a hole (diameter of about 0.8–1.0 mm), and place them in regular Eppendorf tubes. Using a surgical blade, cut 102-bp bands off the gel, and put them into the individual 0.5-mL tubes.
12. Spin tubes at maximum speed for 2 min at room temperature, and then discard 0.5-mL tubes.
13. Add 300 μL of LoTE to each Eppendorf tube, and vortex. Incubate at 65°C for 15 min, mixing intermittently.
14. Transfer the contents of each Eppendorf tube to individual SpinX tubes, and spin at maximum speed for 2 min.
15. Discard SpinX cartridges, ethanol-precipitate the samples (liquid phase) using 940 μL of absolute ethanol, and resuspend pellets in 14.8 μL vol of LoTE (118.5 μL total).

3.2.6. Isolation of Ditags
Followed by Ligation of Ditags to Form Concatamers

1. Collect all samples in one Eppendorf tube, and add 15 μL of 10X restriction Buffer 4, 1.5 μL of 100X BSA, and 15 μL of *Nla*III (10 U/μL) (*see* **Note 14**). Incubate at 37°C for 1 h 15 min.
2. Extract sample with PC8, and then add 50 μL of LoTE to bring sample volume to 200 μL. EtOH-precipitate on dry ice/ethanol bath, spin at full speed at 4°C for 20 min, and resuspend the pellet in 30 μL of LoTE.
3. Add 6 μL of 6X Loading buffer and load onto two wells of 12% polyacrylamide gel (19:1). Two lanes are reserved for 100- and 25-bp DNA ladders. Run and stain the gel as described in **Subheading 3.2.5., step 6**).
4. Using a surgical blade, cut 24- to 26-bp bands off the gel (**Fig. 3**), and place them in individual bottom-pierced 0.5-mL tubes.
5. Spin tubes at 12,000*g* for 2 min, and then discard 0.5-mL tubes. Add 300 μL of LoTE to each Eppendorf tube, and vortex. Incubate at 37°C for 15 min, mixing intermittently. Transfer the contents of each Eppendorf tube to individual SpinX tubes, and spin at 12,000*g* for 2 min.
6. Discard SpinX cartridges, and distribute liquid phases from two SpinX tubes into three Eppendorf tubes; EtOH-precipitate on dry ice/ethanol bath. Resuspend and combine the pellets in 7 μL of LoTE.
7. To the pooled purified ditags (7 μL total), add 2 μL of 5X Ligation buffer and 1 μL of high-concentration T4 Ligase (5 U/μL). Ligate ditags to form concatemers by incubation at 16°C for 3 h (*see* **Note 15**), add 90 μL of LoTE, incu-

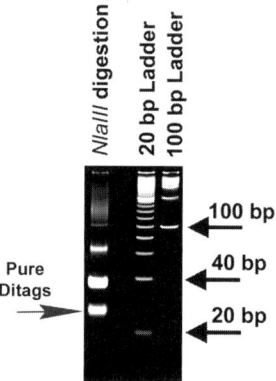

Fig. 3. Electrophoresis of NlaIII digestion products in a 12% TAE polyacrylamide gel. Pure ditags can be seen as 24- to 26-bp products (arrow, left). Other bands represent products of incomplete anchoring enzyme digestion (linker-ditag structures) or undigested linker-ditag-linkers.

bate at 60°C for 5 min, extract sample with PC8, EtOH-precipitate, and resuspend pellet in 10 µL of LoTE.
8. Incubate ditag concatemer sample (see **Note 16**) at 60°C for 5 min, add 190 µL of LoTE, extract sample with PC8, EtOH-precipitate, and resuspend pellet in 10 µL of dH$_2$O. Incubate at 60°C for 5 min, and immediately load sample onto one well of 8% polyacrylamide gel (37.5:1). Load 100-bp DNA ladder and 1-kb DNA ladder to distal wells of the gel. Run and stain the gel as described above.
9. Using a surgical blade, cut concatemers with lengths of 600 to 2500 bp out of the gel and divide into two parts (**Fig. 4**). Place each portion into two labeled bottom-pierced 0.5-mL tubes. Spin tubes at maximum speed in microfuge for 2 min, and then discard 0.5-mL tubes.
10. Add 300 µL of LoTE to each Eppendorf, and vortex. Incubate at 65°C for 15 min, mixing intermittently.
11. Transfer the contents of each Eppendorf tube to two SpinX tubes, and spin at maximum speed for 2 min.
12. Discard SpinX cartridges; transfer liquid phases from two SpinX tubes to one Eppendorf tube. Ethanol-precipitate using 1000 µL of absolute ethanol, and resuspend pellets in 6 µL total volume of LoTE.

3.2.7. Cloning Concatemers and Sequencing
1. To the 6 µL of purified concatemers, add 2 µL of 5X Ligase buffer, 1 µL of pZeRO-1, previously linearized with SphI (25 ng/µL), and 1 µL of T4 Ligase (1 U/µL). Incubate at 16°C overnight.
2. Add 190 µL of LoTE to the samples, extract with PC8, EtOH-precipitate, wash pellets three times with 70% ethanol, and resuspend pellets in 6 µL of LoTE. Use self-ligated pZeRO-1 vector as a negative control.

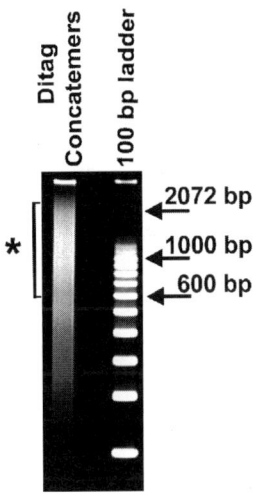

Fig. 4. Electrophoresis of concatemerized products in an 8% TAE polyacrylamide gel. Cloning should be performed with concatemers of >500 bp in length. The ideal cloning products (600–2500 bp) are shown by an asterisk.

3. Transform bacteria by adding 2 µL of each DNA sample to 40 µL of freshly thawed ElectroMAX DH10B cells on ice; mix gently with the pipet tip in a test tube, rather than pipeting.
4. Transfer the cold mixture to the bottom of 0.2-cm electroporation cuvets (avoid bubbles in the mix), and immediately pulse-electroporate using the following constants: voltage setting, 2.5 kV; capacitor, 25 µF; resistance, 200 Σ/Ω. After the pulse, immediately transfer cells to the 15-mL bacterial tube with 1 mL of prewarmed SOC media (no Zeocin), place in a shaker incubator, and let the bacteria recover at 37°C and 200 rpm rotation for 40 min.
5. Plate 250-µL aliquots of the cell/SOC solution onto the 25 × 25-cm Bio Assay Trays with low-salt LB agar plates supplemented with Zeocin (50 ng/µL) and 80 µL of cell/SOC solution onto the 10-cm Petri dishes (see **Note 17**). Incubate plates for about 20 h at 37°C.
6. Use about 15 individual colonies from the Petri dishes for each sample to check the cloning efficiency and ditag concatemer structure in the clones by PCR. For this, run 25-µL PCR reactions with 18 µL of dH$_2$O, 2.5 µL of 10X PCR buffer, 1.5 µL of DMSO, 1.5 µL of dNTPs (25 mM), 0.5 µL of each M13 forward and reverse primers (both 350 ng/µL), and 0.1 µL of thermostable Taq DNA polymerase (0.5 U), inoculated with a single bacterial colony picked by a sterile toothpick. Cycling conditions are: the initial hold; 2 min at 95°C, followed by 25 cycles of 30 s at 95°C, 1 min 30 s at 56°C, 1 min 30 s at 70°C, final extension 70°C for 5 min, hold at 4°C.

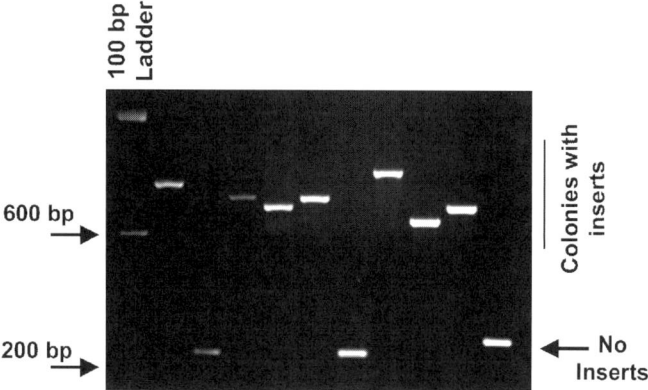

Fig. 5. PCR products of fragments from test clones. Clones with inserts demonstrate a variety of insert sizes that depend on the concatemer length. Bands of 226 bp represent amplification products from colonies that do not contain inserts.

7. To the 5-μL aliquots of control PCR reactions, add 1 μL of 6X loading buffer and load onto the wells of 1.2% agarose gels with 1 μg/mL ethidium bromide. Load 100-bp DNA ladder to distal wells of the gels, and use 0.5X TBE as a running buffer. Run electrophoresis reactions at a constant voltage setting of 120 V.
8. Estimate the cloning efficiency by the ratio of clones with large inserts to the total number of clones amplified with PCR. Background clones with no inserts will appear on the gel as 226-bp bands (**Fig. 5**).
9. To the remaining 20 μL of PCR products for 6 to 10 clones with inserts, add 180 μL of LoTE, extract sample with PC8, add 250 run μL of LoTE (to the total volume of 450 μL for each sample), precipitate samples with isopropanol, and resuspend pellet in 10 μL of LoTE.
10. For sequencing, use forward and/or reverse M13 primers (*see* **Note 18**). If inserts of the tested clones demonstrate correct concatemer structure (**Fig. 6**) and the cloning efficiency is satisfactory, then the SAGE library should be sequenced.
11. Currently some commercial sequencing companies provide automated picking of colonies from agar lawns on 25 × 25-cm Bio Assay Trays. Otherwise proceed to **step 12**.
12. For submission of libraries in glycerol stock format, inoculate wells of 96-well plates filled with 250 μL of LB with 10% glycerol/Zeocin media by colonies from Petri dishes using sterile toothpicks. It is recommended that one well in each plate be filled with sterile LoTE to serve as a negative control in sequencing reactions.
13. Incubate plates for about 12 h at 37°C in the dark. Seal plates with adhesive SealPlate film, freeze at –80°C, and ship for sequencing on dry ice.
14. We recommend that you duplicate each plate and store duplicates until the library has been successfully sequenced.

```
CCCAATTTTTCTTATGGGGGCGAATTGGGGCCCTCTAGATGCATGTTGTTTTTGTATTCGTGAGCCACC
ATGTGCTGCCCTCCACTGGAATGGTCCATGATTAGTTACCCGCCCGTCGTCTCCGTGGGTGCCCAGTTT
GCTCCCTCGCCATGCTGGGCTCTGGAGCCAGATGAACATGTGAAACCAGCCGATCTCCAAGGCATGTAA
AGAGGCCGTAACAATGAGCCATGCAAATGCTGTGAAAATCACAACCATGAGCAGTCCCCTCTGTCCTCC
TTCTATGATACTGACATTCTGGGACACCCCATGTAGGCAAGACAAATTCCTCGTTCATGAGGTACACTT
CGCCGCGTCCAGACATGTGGACATAGCCGGGGTATCCTCATGGGAGCCATTGGAATGTCAATTCCATGC
AAGAGCATCACTCCGCATTCCATGATGTAGTAGTGCCTGGGCCACCCATGGTAGATTTGATATTACAC
AGAACATGTTATTACTGTGCATCTCTCCAATCATGAGACACTCCTTACCCATTTCCTTCATGACAGGTG
ATGCCTGTCACCGTGCATGCGCCGCCGGCTTTGTGAGCCACCATGACTGTGTCAGACCGATAGATCTCC
ATGCTCGAGCGGCCGCCAGTGTGATGGATATCTGCATAATTCCAGCACACTGGCGGCCGTTACTAGTGG
ATCCGAGCTCGGTACCAATCTTGATGCATTGCTTGATATTCCATAGTGTCCCTAGATAGCTTGGCGTAA
TCAAGGTCTAACTGTTTCTTGGGTGAATTGTTTCCCCTCAAATTCCCCCAACCACGACCGGAACTTAAG
```

Fig. 6. Correct concatemer structure of sequenced plasmid DNA. Fragments of 24 to 26 bp, which are flanked by *Nla*III sites (in bold, underlined), represent authentic ditags. These ditags can be extracted from the raw data files and used to identify transcripts, both quantitatively and qualitatively. Comparisons among SAGE libraries (http://www.ncbi.nlm.nih.gov/sage/) facilitate the identification of factors present or differentially expressed in the cell/tissue of interest.

3.3. SAGE DATA Analysis (see Note 19)

Data analysis should be performed exactly as described in the guide "SAGE Analysis Software" Version E (July 14, 2003), available at http://invitrogen.com/Content/sfs/manuals/SAGEsoftware_man.pdf (*see* **Note 20**). The SAGE analysis software is useful for the extraction of tag sequences from raw sequence data, the quantification of tag frequency, the identification of genes encoded by the tag sequences (using a reference sequence database), and the comparison of tag abundances between SAGE libraries. For comparisons among SAGE libraries, we use MS Access© (*see* **Subheading 3.4.**).

Other useful SAGE-related weblinks:
SAGE Homepage: http://www.sagenet.org/;
SAGEmap Homepage: http://www.ncbi.nlm.nih.gov/SAGE/;
UniLib: http://www.ncbi.nlm.nih.gov/UniLib/index.cgi;
SAGE Genie: http://cgap.nci.nih.gov/SAGE.

3.4. SAGE Reference Library for AMH

We originally employed SAGE to sequence a total of 88,860 tags or 23,941 unique tags from adult female C57BL/6 mouse hearts (AMH) *(6)*. The catalog generated in this study gave an accurate and quantitative reflection of mRNA transcripts present in AMH (based on Q-PCR and microarray comparisons). This library can be used for meaningful comparisons with other cardiac SAGE libraries present online or prepared in an investigator's laboratory. The AMH transcriptomes that we published demonstrated many unique features. A high proportion of genes were involved in energy metabolism, similar to that seen in skeletal muscle *(8–10)*. Mitochondrial transcripts accounted for 18.7% of total transcripts, whereas sarcomeric proteins accounted for ≥3.2% of all tags. In fact, we found that cardiac tissues contained the highest percentage of non-tRNA mitochondrial genome-derived transcripts described thus far (18.7% vs 17.7% in skeletal muscle). At that time, only 10% of the unique tags matched known genes, but this percentage has increased substantially over the past 2 yr as more mouse sequencing data have become publicly available *(11–14)*. These published SAGE results thus represent the first quantitative expression profile of AMH and serve as a reliable transcriptome reference to identify dynamic changes in cardiac gene expression in mice.

As an example of how analyses can be performed among libraries, we have compared our AMH SAGE library (GSM1681) with 23 other SAGE libraries available at the Gene Expression Omnibus Database (GEO; December, 2003, http://www.ncbi.nlm.nih.gov/geo/; *see* Table 1 in http://www.grc.nia.nih.gov/branches/lcs/sage2004.htm). We excluded SAGE libraries derived from animals/tissues/cell cultures that were treated with drugs or originated from transgenic animals. For this comparative analysis, we first imported 23 SAGE catalogs into MS Access. Second, we created a "query," whereby the field "SAGE tags" in AMH were linked to "UniGene Cluster ID," "Gene Description," and tag frequencies (tag per million [TPM]) from all these SAGE libraries. We were able to identify 18 known genes that were uniquely expressed in heart (*see* Table 2 in http://www.grc.nia.nih.gov/branches/lcs/sage2004.htm). Additionally, we identified numerous tags that were present at least sixfold higher in AMH relative to any noncardiac catalog. We went on to compare some of these genes with microarray data available at the Genomics Institute of the Novartis Research Foundation (http://expression.gnf.org/cgi-bin/index.cgi), excluding redundant tags (i.e., tags associated with more than one gene) from this analysis. These data indicate that 10 genes were absent from the Affymetrix chip, whereas six tags were unknown and might be of interest for further investigation of cardiac-restricted transcripts (*see* Table 3 in http://www.grc.nia.nih.gov/branches/lcs/sage2004.htm).

4. Notes

1. SAGE uses two major enzymes: *Bsm*FI *and Nla*III. The tagging enzyme *Bsm*FI is a type IIs restriction enzyme that cuts about 10 to 14 bp downstream of the GGAC recognition sequence, whereas the anchoring enzyme, *Nla*III, is a type II restriction enzyme that yields a 3' overhang.
2. It is easy to get confused by B+W solutions (2X or 1X) or T4 ligase (5 U/µL and 1 U/µL) concentrations for each particular step. Please pay attention to the concentrations of each reagent.

3. Previously we used TB Medium to amplify the library in glycerol stock format, but currently we use LB media with 10% glycerol, according to the recommendations of Agencourt. TB Medium Base: to make 900 mL, add 12 g of tryptone, 24 g of yeast extract, 4 mL of glycerol, and autoclave. Phosphate buffer: to make 100 mL, add 2.3 g of KH_2PO_4 and 12.5 g of K_2HPO_4 and filter-sterilize. TB Medium/Zeocin: to make 1 L, add 900 mL of TB Medium Base, 100 mL of phosphate buffer, and 500 µL of Zeocin 100 µg/µL stock. Store in the dark at 4° for up to 2 wk.
4. Zeocin is light sensitive and comparatively unstable. Zeocin-containing media should be kept in the dark for no longer than 2 wk. Low-salt LB agar/Zeocin plates should be protected from light immediately after pouring and until bacterial colonies have grown to an appropriate size.
5. The most commonly used anchoring enzyme (*Nla*III) has a half-life of only about 3 mo at –20°C. Be sure to use fresh batches of enzyme for each SAGE project. If you wish to use a different anchoring enzyme, see **Note 8**.
6. Oligonucleotides (Linkers A1, B1, A2, B2, and biotinylated oligo dT) should be gel-purified.
7. We have successfully employed several thermocyclers (GeneAmp PCR Systems 9600, 9700 from Applied Biosystems and Master Gradient from Eppendorf) for the construction of SAGE libraries. If you use other thermocyclers, then we recommend that you test the system before large-scale amplification.
8. This protocol is optimized for "standard" SAGE using the *Nla*III/*Bsm*FI enzyme pair. Other anchoring and restriction enzymes could be used to generate longer tags (LongSAGE) *(2)* or generate SAGE tags from genes lacking *Nla*III recognition site (*Sau*3A is suggested and recognized by SAGEmap) *(15,16)*.
9. RNAse/DNAse-free conditions are important for the successful construction of SAGE libraries. Ensure that all pipets are accurate and solutions are prepared properly, and use DEPC-treated glassware and plasticware during the early steps of library construction and aerosol-barrier pipet tips throughout the procedure.
10. If you have small volumes of starting material, you can use these modifications to the SAGE protocol: microSAGE *(17)*; miniSAGE *(18)*.
11. At the cDNA synthesis steps, we recommend using $\alpha^{32}P$ labeling only for test purposes, i.e., ensure first- and second-strand synthesis. We have done this with a small portion of the original mix for cDNA synthesis, but often we skip this protocol. It is most useful for those who have never prepared a cDNA library. If it is used, the incorporation can be employed to calculate cDNA yield, and isotope labeling allows cDNA tracing up to the tagging enzyme digestion step.
12. We recommend always preparing fresh dilutions of the ligation reactions for the PCR amplification step. Do not use dilutions that have been frozen.
13. To spin 96-well plates, we use an RT6000 centrifuge with an H1000B rotor (Sorvall) at 600 g for 3 min.
14. We use modified conditions to reduce potential problems associated with incomplete digestion during the step of ditag isolation *(19)*. In the original protocol, this step was described as follows: for PCR products in 90 µL of LoTE,

add 226 μL of LoTE, 40 μL of 10X restriction Buffer 4, 4 μL of BSA (100X), and 40 μL of *Nla*III (anchoring enzyme, 10 U/μL). Incubate at 37°C for 1 h. You can also perform an additional procedure versus the original protocol by adding a single purification step before *Nla*III digestion of the ditags *(20)*.
15. Although concatenation times can vary from 45 min to overnight, depending on the volume of ditags purified, a 3-h concatenation generally produces good results.
16. A modification of the original SAGE protocol allows better resolution of concatemers on polyacrylamide gel, which improves the overall yield of ditags in a SAGE library *(3)*: 65°C for 15 min, chill on ice for 10 min, and then load onto the gel. We recommend using the "hot" variation of the same modification: 60°C for 5 min, PC8-extract, EtOH-precipitate, 60°C for 5 min, and load immediately.
17. A selection based on X-Gal/IPTG blue/white coloring of colonies with or without inserts has been suggested *(21)* to increase the cloning efficiency. Although this modification can enhance the overall yield of clones with inserts, the use of pZeRO vectors obviates this requirement.
18. For large-scale sequencing, we suggest the use of M13 forward primers. For clones with long inserts, an additional M13 reverse-primed sequencing could be beneficial, yielding more tags, but it also leads to a rapid accumulation of duplicate dimers owing to the partial overlapping of the sequences.
19. When you perform your data analyses, keep in mind three major limitations to SAGE: (1) multiple tag hits (it is possible that one tag can be associated with several genes or several tags with one gene); (2) there are sequences that lack *Nla*III digestion sites *(16)*; and (3) some tags are not informative, e.g., when the CATG site is located immediately upstream of the poly(A+) tail.
20. We have previously detailed procedures for downloading GenBank databases from the National Center of Biotechnology Information (NCBI) web site and step-by-step instructions for tag extraction from GenBank sequences for identification of "tag to gene matches." However, since the NCBI web site has been significantly improved, it is now preferable to use the SAGEmap_tag_ug-full or SAGEmap_tag_ug-rel files as described in the Chapter "Downloading a UniGene Reference Database," found in the "SAGE Analysis Software."

References

1. Velculescu, V. E., Zhang, L., Vogelstein, B., and Kinzler, K. W. (1995) Serial analysis of gene expression. *Science* **270,** 484–487.
2. Saha, S., Sparks, A. B., Rago, C., et al. (2002) Using the transcriptome to annotate the genome. *Nat. Biotechnol.* **20,** 508–512.
3. Kenzelmann, M. and Muhlemann, K. (1999) Substantially enhanced cloning efficiency of SAGE (serial analysis of gene expression) by adding a heating step to the original protocol. *Nucleic Acids Res.* **27,** 917–918.
4. Anisimov, S. V., Tarasov, K. V., Tweedie, D., Stern, M. D., Wobus, A. M., and Boheler, K. R. (2002) SAGE identification of gene transcripts with profiles unique to pluripotent mouse R1 embryonic stem cells. *Genomics* **79,** 169–176.

5. Anisimov, S. V., Tarasov, K. V., Riordon, D., Wobus, A. M., and Boheler, K. R. (2002) SAGE identification of differentiation responsive genes in P19 embryonic cells induced to form cardiomyocytes in vitro. *Mech. Dev.* **202,** 25–74.
6. Anisimov, S. V., Tarasov, K. V., Stern, M. D., Lakatta, E. G., and Boheler, K. R. (2002) A quantitative and validated SAGE transcriptome reference for adult mouse heart. *Genomics* **80,** 213–222.
7. Chirgwin, J. M., Przybyla, A. E., MacDonald, R. J., and Rutter, W. J. (1979) Isolation of biologically active ribonucleic acid from sources enriched in ribonuclease. *Biochemistry* **18,** 5294–5299.
8. Welle, S., Bhatt, K., and Thornton, C. A. (1999) Inventory of high-abundance mRNAs in skeletal muscle of normal men. *Genome Res.* **9,** 506–513.
9. Welle, S., Bhatt, K., and Thornton, C. A. (2000) High-abundance mRNAs in human muscle: comparison between young and old. *J. Appl. Physiol.* **89,** 297–304.
10. St-Amand, J., Okamura, K., Matsumoto, K., Shimizu, S., and Sogawa, Y. (2001) Characterization of control and immobilized skeletal muscle: an overview from genetic engineering. *FASEB J.* **15,** 684–692.
11. Marra, M., Hillier, L., Kucaba, T., et al. (1999) An encyclopedia of mouse genes. *Nat. Genet.* **21,** 191–194.
12. Kawai, J., Shinagawa, A., Shibata, K., et al. (2001) Functional annotation of a full-length mouse cDNA collection. *Nature* **409,** 685–690.
13. Rogers, J. and Bradley, A. (2001) The mouse genome sequence: status and prospects. *Genomics* **77,** 117–118.
14. Tanaka, T. S., Jaradat, S. A., Lim, M. K., et al. (2000) Genome-wide expression profiling of mid-gestation placenta and embryo using a 15,000 mouse developmental cDNA microarray. *Proc. Natl. Acad. Sci. USA* **97,** 9127–9132.
15. Lash, A. E., Tolstoshev, C. M., Wagner, L., et al. (2000) SAGEmap: a public gene expression resource. *Genome Res.* **10,** 1051–1060.
16. Pleasance, E. D., Marra, M. A., and Jones, S. J. (2003) Assessment of SAGE in transcript identification. *Genome Res.* **13,** 1203–1215.
17. Datson, N. A., van der Perk-de Jong, J., van den Berg, M. P., de Kloet, E. R., and Vreugdenhil, E. (1999) MicroSAGE: a modified procedure for serial analysis of gene expression in limited amounts of tissue. *Nucleic Acids Res.* **27,** 1300–1307.
18. Ye, S. Q., Zhang, L. Q., Zheng, F., Virgil, D., and Kwiterovich, P. O. (2000) MiniSAGE: gene expression profiling using serial analysis of gene expression from 1 microg total RNA. *Anal. Biochem.* **287,** 144–152.
19. Damgaard Nielsen, M., Millichip, M., and Josefsen, K. (2003) High-performance liquid chromatography purification of 26-bp serial analysis of gene expression ditags results in higher yields, longer concatemers, and substantial time savings. *Anal. Biochem.* **313,** 128–132.
20. Angelastro, J. M., Klimaschewski, L. P., and Vitolo, O. V. (2000) Improved NlaIII digestion of PAGE-purified 102 bp ditags by addition of a single purification step in both the SAGE and microSAGE protocols. *Nucleic Acids Res.* **28,** E62.
21. Angelastro, J. M., Ryu, E. J., Torocsik, B., Fiske, B. K., and Greene, L. A. (2002) Blue-white selection step enhances the yield of SAGE concatemers. *Biotechniques* **32,** 484–486.

4

Functional Genomics by cDNA Subtractive Hybridization

Christophe Depre

Summary

The regulation of myocardial gene expression is highly sensitive to any extracellular or intracellular stimulus that affects contractile function. Subtractive suppression hybridization represents a large-scale, unbiased method for detecting transcriptionally and posttranscriptionally regulated genes, both known and unknown, independently of the prevalence of these transcripts. The strength of subtractive hybridization relies on its unbiased nature and its power to extract even low-abundance transcripts. Therefore, the subtraction experiments can reveal "unexpected" gene profiles and can represent a starting point for the cloning and characterization of novel genes. However, the procedure and the subsequent sequencing of the subtracted genes are labor intensive, and, because the method is purely qualitative, it also requires alternative methods of validation.

Key Words: Gene expression; genomics; heart; ischemia; novel genes; RNA; sequencing; subtractive hybridization.

1. Introduction

The myocardium can develop a remarkable plasticity at the gene level *(1–9)*, which is the direct consequence of variations in physiological conditions. Analyzing the large-scale regulation of myocardial gene expression in response to changes in physiological parameters is the goal of cardiac functional genomics. Depending on the experimental hypothesis, two broad approaches can be used. The horizontal approach determines how a change in the function of the organ results in a change of gene expression, which leads to the determination of genomic profiles. The vertical approach is used to determine the function of a specific gene, usually novel (never annotated) or unknown (never characterized in the tissue of interest), combining various experimental methods (from mRNA expression to the generation of genetically modified animals), which

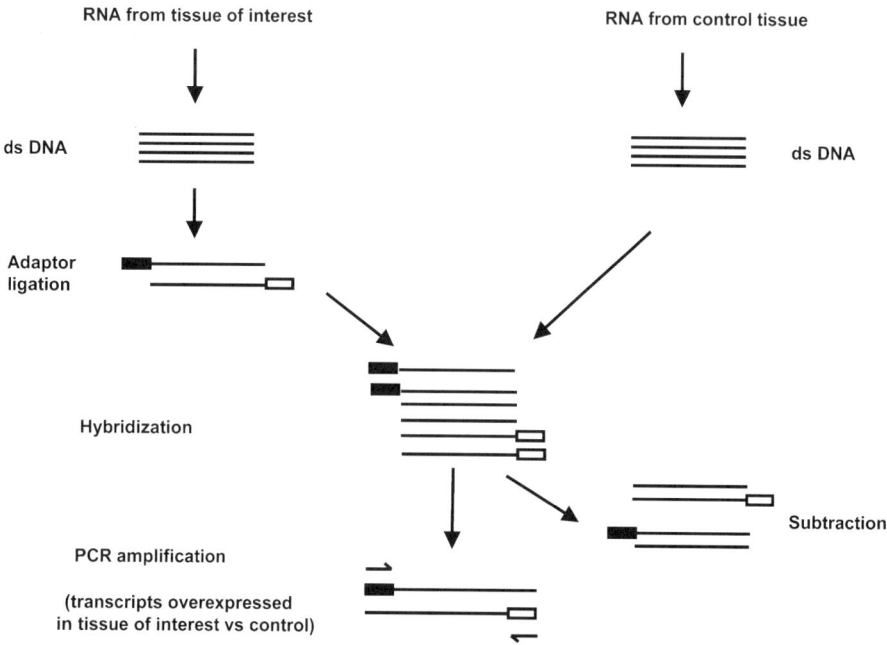

Fig. 1. Overview of cDNA subtractive hybridization.

leads to the discovery of new cellular mechanisms. The subtractive hybridization methodology *(10)* can be used as a discovery tool in both approaches to investigate a large array of genes in a specific condition *(11)* and to detect novel or unknown genes *(12,13)*.

cDNA subtractive suppression hybridization represents a large-scale, unbiased method for detecting transcriptionally and posttranscriptionally regulated genes, both known and unknown, independently of the prevalence of these transcripts *(14)*. The procedure is summarized in **Fig. 1**. The subtraction is typically performed between two biological samples, the sample of interest and the control, from which total RNA and, subsequently, poly-A messenger RNA is extracted. After reverse transcription and double-stranded DNA synthesis, one of the two samples is ligated to specific adaptors. The two preparations are denatured and mixed together to let complementary strands hybridize at random between the adaptor-ligated preparation (called the tester) and the unligated preparation (called the driver), which represents the subtraction in itself. If one specific transcript is overexpressed in the tester, two adaptor-ligated complementary strands rehybridize together. This DNA can be further amplified by polymerase chain reaction (PCR) using primers matching the sequence of the adaptors. Therefore, a single pair of primers can amplify at

Functional Genomics by cDNA Hybridization

once all the potentially regulated transcripts, both known and unknown. The PCR products are then subcloned for sequencing. Querying the sequences in different databases identifies the subtracted genes. The results need to be validated and quantified by alternative methods, such as Northern blot, quantitative PCR, and *in situ* hybridization *(11)*.

The main advantage of subtractive hybridization is its unbiased nature. Potentially, all the genes regulated by a specific condition can be found. Subtractive hybridization is therefore an excellent tool for discovery. It can show "unexpected" gene profiles or lead to the discovery and characterization of novel genes. Another advantage is that the experiment does not depend on the species studied. The different genome projects have substantially increased the number of sequences available in public databases, so that querying sequences for "unusual" species is not a problem in practice. This methodology is therefore particularly useful for the genomic exploration of species for which no microarrays are available *(11,12)*. Other advantages of this methodology compared with microarrays include the possibility of detecting low-copy transcripts. In addition, the technique generates subcloned gene products, which can be used for printing custom-made microarrays.

Despite these considerable advantages, several limitations, which are mainly methodological, must be considered. Subtractive hybridization is a random procedure, which means that interesting clones can be "missed," and therefore the technique is not as comprehensive as a high-density microarray. Especially in the heart, it is impossible to avoid "contamination" of the library by non-nuclear genes from the mitochondrial DNA. Because of the large amount of mitochondria in the cardiomyocyte, such contamination can represent 20 to 40% of the overall library *(11,15)*. This contamination can be limited by performing additional rounds of hybridization, at the expense of losing low-abundance transcripts from nuclear-encoded genes. Another limitation is that the subtraction experiments require 2 to 3 µg of poly-A messenger RNA as starting material, which, after correct purification and cleaning, corresponds to about 100 µg of total RNA in the heart. The procedure of subtractive hybridization is time consuming (at least 4 full days) and recapitulates most of the basic techniques of molecular biology, including RNA extraction, poly-A mRNA purification, reverse transcription, double-stranded DNA synthesis, digestion, adaptor ligation, denaturation, hybridization, PCR amplification, subcloning, sequencing, and Northern blotting.

2. Materials

Although all the reagents needed to perform a cDNA subtractive hybridization can be found in commercially available kits, it might be wise to obtain separately specific products for key experiments, such as the digestion enzyme

*Rsa*I or the T4 DNA ligase, from alternative sources. However, considering the complexity of the procedure, it is a major advantage to have all the other reagents gathered in one single kit, such as the PCR-Select cDNA Subtraction Kit (BD Bioscience).

2.1. RNA Extraction and Poly-A RNA Purification

1. Triazol reagent (Sigma). The RNA can be further purified with the RNeasy Mini kit (Qiagen).
2. Chloroform.
3. Isopropanol.
4. 80% Ethanol.
5. NucTrap Midi-kit for poly-A mRNA purification (BD Bioscience).

2.2. Reverse Transcription

1. 5X RT buffer: 250 mM Tris-HCl, pH 8.5, 40 mM MgCl$_2$, 150 mM KCl, 5 mM dithiothreitol (DTT).
2. 10 μM Oligo-dT primer with *Rsa*I restriction site: 5'-TTTT<u>GTAC</u>AAGCT$_{31}$-3'.
3. Reverse transcriptase.

2.3. Double-Stranded DNA Synthesis

1. 5X buffer: 100 mM Tris-HCl, pH 7.5, 500 mM KCl, 50 mM ammonium sulfate, 25 mM MgCl$_2$, 0.25 mg/mL bovine serum albumin (BSA).
2. Enzymes: DNA polymerase I, RNAse H, DNA ligase, T4 DNA polymerase.
3. Stop mix: 0.2 M EDTA, 1 mg/mL glycogen.
4. 4 M Ammonium acetate.
5. Phenol/chloroform/isoamyl alcohol solution (25:24:1). The phenol should be equilibrated in TNE buffer (50 mM Tris-HCl, pH 7.5, 150 mM NaCl, 1 mM EDTA).
6. Chloroform/isoamyl alcohol solution (24:1).
6. Ethanol (both 100% and 80%).

2.4. DNA Digestion

1. 10 U/μL *Rsa*I restriction enzyme.
2. 10X digestion buffer: 100 mM Tris-HCl, pH 7.5, 100 mM MgCl$_2$, 10 mM DTE.

2.5. Adaptor Ligation

1. 5X buffer: 250 mM Tris-HCl, pH 7.8, 50 mM MgCl$_2$, 10 mM DTT, 0.25 mg/mL BSA.
2. 400 U/μL T4 DNA ligase.
3. Adaptors.

   Adaptor 1:
   5'-CTAATACGACTCACTATAGGGCTCGAGCGGCCGCCCGGGCAGGT-3'

   Adaptor 2:
   5'-CTAATACGACTCACTATAGGGCAGCGTGGTCGCGGCCCAGGT-3'.

2.6. Hybridization

1. 4X hybridization buffer: 50 mM HEPES-HCl, pH 8.3, 500 mM NaCl, 0.02 mM EDTA, 10% (wt/vol) PEG 8000.
2. Dilution buffer: 20 mM HEPES-HCl, pH 8.3, 50 mM NaCl, 0.2 mM EDTA.

2.7. PCR Amplification

1. 10X buffer: 100 mM Tris-HCl, pH 8.3, 500 mM KCl, 15 mM MgCl$_2$, 10 mM DTT, 0.25 mg/mL BSA.
2. 5 U/μL Taq polymerase.
3. 10 mM dNTP mix.
4. Primers:

    Primer 1: 5'-CTAATACGACTCACTATAGGGC-3'.
    Nested primer 1: 5'-TCGAGCGGCCGCCCGGGCAGGT-3'.
    Nested primer 2: 5'-AGCGTGGTCGCGGCCCAGGT-3'.

5. Purification of PCR product: QIAquick PCR purification kit (Qiagen)

3. Methods

Before you start, take into consideration the following recommendations.

1. It is critical to keep in mind that a successful completion of each step of the procedure relies on the efficiency with which the previous step was performed. The kits contain control samples that must *absolutely* be run simultaneously with the experimental samples to verify the correct achievement of each experimental step. Considering that cDNA subtractive hybridization is time consuming, requires a large amount of starting material, and can provide a source of data for several years of research, this precaution is imperative and should never be overlooked.
2. Because of the multiple steps of the procedure, it is also very important to adopt from the beginning a logical labeling of the tubes that will remain consistent throughout the experiment.

3.1. RNA Preparation

Experiments are usually started from frozen tissue. About 200 mg of tissue is usually sufficient to collect the amount of RNA needed for the experiment. RNA purification can be done by classical phenol extraction *(16)* using commercially available solutions, followed by isopropanol precipitation and ethanol wash. The RNA pellet can be resuspended in DEPC water or in a buffer of sodium citrate, pH 6.7. It is recommended but not mandatory to clean up the RNA after extraction (Rneasy Mini kit, Qiagen). Compared with other tissues, such as kidney or liver, the mRNA content per gram of cardiac tissue is relatively low. It is recommended to pool several samples from each group, to ensure that the subtraction is quite representative of the condition tested.

Remember that all the information that will be collected from the subtraction closely depends on the quality of the starting material, which is the poly-A RNA. RNAse-free conditions are indispensable to preserve the structure and the full length of the RNA as well as possible. Different techniques are available for poly-A mRNA purification, some insisting more on the exclusion of the contaminating ribosomal RNA, and some insisting more on the recovery of rare transcripts. The latter choice is probably the best in this case, to ensure that the library to be subtracted contains as many different transcripts as possible. The quality of the RNA must be verified by the 260/280 absorbance ratio or on a denaturing 1.2% agarose gel. The absorbance ratio of the purified poly-A RNA must be as close as possible to 2.1 and should not be lower than 1.95.

3.2. Reverse Transcription (RT)

Subtraction between two libraries begins by reverse transcription of the poly-A mRNA into a cDNA. Reverse transcription is usually performed with an oligo-dT primer, rather than random hexamers, to increase the length of the products. Because of this priming method, the need for good-quality poly-A RNA is self-explanatory. The primer must contain an *Rsa*I digestion site that will be cut in the following steps to allow the ligation of the adaptors. Although the final volume of the RT is very small (10 µL), the reaction should be started in a 0.5-mL RNAse-free tube because the same tube will be kept for the following steps. It is recommended to start this reaction early in the morning because there is a busy day ahead!

1. The reaction is set as follows: 2 µL of 2 µg poly-A RNA, 1 µL of 10 µ*M* primer, 3 µL of DEPC water, 2 µL of 5X buffer, 1 µL of 10 m*M* dNTP mix, and 1 µL of 20 U reverse transcriptase. If the poly-A RNA concentration is lower than 1 µg/µL, the volume of DEPC water can be decreased as needed.
2. Incubate the primer first with the poly-A RNA at 70°C for 5 min, and then transfer the tubes in ice.
3. Add the other components of the RT reaction and incubate at 42°C for 2 h. To avoid excessive evaporation of this small volume, an air incubator is preferred to a thermocycler.
4. Stop the reaction by transferring the tubes at 4°C (*see* **Note 1**).

3.3. Double-Stranded DNA Synthesis

This step should be done immediately after the RT, using the same tube.

1. Set the reaction as follows: 10 µL RT reaction, 48.5 µL DEPC water, 16 µL 5X buffer, 1.5 µL 10 m*M* dNTP mix, and 4 µL enzyme mix.
    The enzyme mix contains: 6 U/µL DNA polymerase 1, 0.25 U/µL RNAse H, and 1.2 U/µL DNA ligase.

Functional Genomics by cDNA Hybridization 67

2. Incubate for 2 h at 16°C in a thermocycler; then add 2 µL of 3 U/µL T4 DNA polymerase and incubate for 30 more minutes at 16°C.
3. Stop the reaction with 4 µL of Stop mix (EDTA/glycogen).
4. Add 100 µL of phenol/chloroform/isoamyl alcohol, mix well, and centrifuge for 5 min at maximum speed in a bench centrifuge (*see* **Note 2**).
5. Recover the aqueous phase and add 100 µL of chloroform/isoamyl alcohol. Vortex and spin as in **step 4**.
6. Recover the aqueous phase, and then add 40 µL of 4 *M* ammonium acetate and 300 µL of 100% ethanol. Mix well and spin for 20 min in a bench centrifuge (15,000*g*). Make sure to keep the tubes at room temperature to avoid salt precipitation. The pellet is very tiny but should be visible.
7. Remove the supernatant and carefully add 500 µL of 80% ethanol. (The pellet might move at this point.) Centrifuge again at maximum speed for 5 min.
8. Remove the supernatant and make sure that there is no residual ethanol around the pellet. Do not overdry, however, or the pellet will be difficult to solubilize.
9. Dissolve the pellet in 45 µL of DEPC water. Take an aliquot of 5 µL and freeze it. This sample will be used later as a predigestion control.

3.4. RsaI Digestion

The DNA from both samples is subsequently digested by the four-base-cutter restriction enzyme *Rsa*I.

1. Set the reaction as follows: 40 µL DNA sample, 5 µL 10X buffer, 5 µL 10 U/µL *Rsa*I.
2. Incubate at 37°C. Although an incubation time of 1.5 to 2 h is suggested in some protocols, it is highly recommended to leave the incubation overnight in an air incubator. This ends your first day of experiments.
3. The following morning, stop the reaction with 4 µL Stop mix.
4. Take an aliquot of 5 µL and freeze it. This sample will be used later as a postdigestion control. The remaining of the reaction must follow exactly the same protocol of phenol/chloroform/isoamyl alcohol extraction (*see* **Note 2**) and ethanol precipitation as described in **Subheading 3.3.**
5. Dissolve the pellet in 5.5 µL DEPC water and store at −20°C. This sample is referred to below as the *Rsa*I-digested cDNA.
6. On an agarose gel, run the samples that were aliquoted as predigestion and postdigestion controls (*see* **Note 3**), together with molecular weight markers ranging from 3.0 to 0.5 kb. As a result of the digestion, the postdigestion control should be characterized by a shift of the cDNAs to lower molecular weight compared with the corresponding predigestion control. After correct digestion, most of the cDNAs should be in the 0.5- to 1.0-kb range.

3.5. Adaptor Ligation

The principle of subtractive hybridization consists of hybridizing an adaptor-ligated library (or tester) with a library without adaptors (or driver). All the transcripts expressed to a higher level in the adaptor-ligated, tester library can

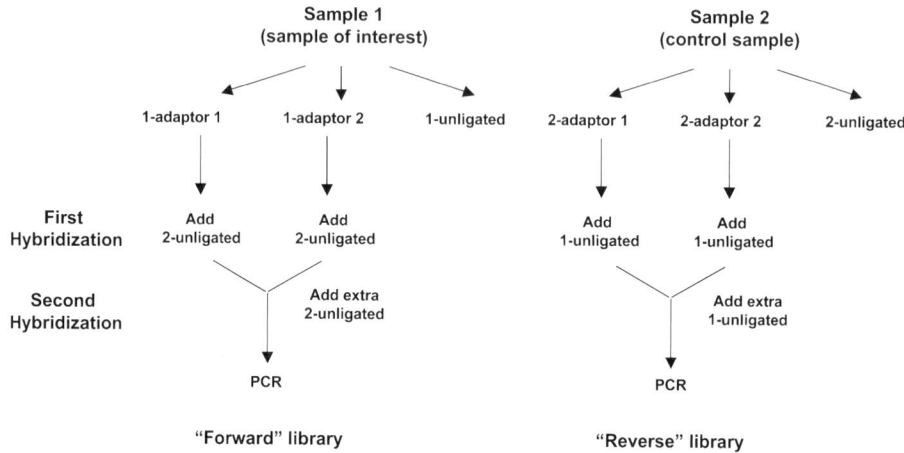

Fig. 2. Details of the procedure of ligation and hybridization.

subsequently be amplified by PCR using primers binding to the adaptors. Based on this principle, two hybridizations can be done for each pair of samples, depending on which library is ligated to the adaptors. The "forward" library will show all the transcripts that are overexpressed if the sample of interest is the tester, whereas the "reverse" library will show all the transcripts that are downregulated if the sample of interest is the driver. Therefore, at this stage each sample will be divided in two parts. One part will be ligated to the adaptors (tester) and the other will not (driver). To decrease the risk of false positives during hybridization (*see* **Subheading 3.6.**), two different adaptors will be ligated separately on each sample. This principle is illustrated in **Fig. 2**.

1. The digestion step described in **Subheading 3.4.** ended up with the *Rsa*I-digested cDNAs resuspended in 5.5 µL DEPC water. To prepare the ligation of the adapters on the tester, first dilute 1 µL of each *Rsa*I-digested cDNA with 5 µL DEPC water. For each sample (both the sample of interest and the control sample), set the reactions as follows:

	Adapter-1	Adapter-2
Diluted cDNA	2 µL	2 µL
10 µ*M* Adaptor 1	2 µL	—
10 µ*M* Adaptor 2	—	2 µL
5X Ligation buffer	2 µL	2 µL
DEPC water	3 µL	3 µL
400 U/µL T4 DNA ligase	1 µL	1 µL

2. Incubate at 16°C overnight. This ends your second day of experiments.
3. The following morning, stop the reaction with 1 µL Stop mix.

4. Denature the ligase by heating for 5 to 10 min at 75°C. (There is no need for a phenol/chloroform/isoamyl alcohol extraction at this stage.)
5. The ligation efficiency can be verified by PCR using the nested primers 1 and 2.
6. The samples are now ready for the first hybridization.

3.6. Hybridization

The hybridization will be performed between the adaptor-ligated testers from one sample and the driver of the other sample. The driver corresponds to the nonligated *Rsa*I-digested cDNA obtained in **Subheading 3.4.** The duration of the hybridization is a crucial aspect of the experiment for the following reason. The hybridization between two complementary strands of DNA is proportional to the concentration of the strands and the time of hybridization. Therefore, the driver will be added in excess compared with the tester to accelerate the hybridization between both preparations. (Remember that the cDNA digest used to make the tester was diluted in **Subheading 3.5.** before adaptor ligation.) Adjusting the time of hybridization is a normalization mechanism to ensure that all the transcripts (rare or abundant) will hybridize relatively equally. Because of this normalization, the chance to detect a transcript after subtraction is relatively independent of its original abundance. Two successive rounds of hybridization are usually performed to decrease the risk of false positives. In the first hybridization, the driver is added separately to the tester ligated with adapter 1 or adapter 2. In the second hybridization, more driver is added, and both adapter-ligated samples are combined (**Fig. 2**).

3.6.1. First Hybridization

At this point, we will hybridize the tester from one sample with the driver from the other sample. The experiment is performed in both directions to create both a "forward" and a "reverse" library, as mentioned in **Subheading 3.5.** If this seems confusing, refer to the diagram of **Fig. 2** to label your tubes correctly. As a reminder, we had 5.5 µL of *Rsa*I-digested cDNA at the end of **Subheading 3.4.**, from which 1 µL was taken and diluted to create the tester in **Subheading 3.5.** Therefore, we have 4.5 µL left to be used as driver in the two rounds of hybridization that follow.

1. For the first hybridization, set the reaction as follows:

	Reaction 1	Reaction 2	Reaction 3	Reaction 4
Driver cDNA (unligated) (sample 1)	1.5 µL	1.5 µL	—	—
Adaptor 1-ligated tester (sample 2)	1.5 µL	—	—	—

	Reaction 1	Reaction 2	Reaction 3	Reaction 4
Adaptor 2-ligated tester (sample 2)	—	1.5 µL		
Driver cDNA (unligated) (sample 2)	—	—	1.5 µL	1.5 µL
Adaptor 1-ligated tester (sample 1)	—	—	1.5 µL	—
Adaptor 2-ligated tester (sample 1)	—	—	—	1.5 µL
4X hybridization buffer	1 µL	1 µL	1 µL	1 µL

2. Heat at 98°C for 5 min to denature the samples, and then incubate at 68°C to allow the hybridization between the driver and the testers. Cover the samples with mineral oil to avoid evaporation.
3. For the reasons detailed above, the incubation time for the first hybridization *must* be 8 ± 1 h.
4. Immediately follow with the second hybridization.

3.6.2. Second Hybridization

In the second hybridization, we will mix together the testers ligated on adaptor 1 and adaptor 2 in the presence of an excess of driver from the other sample. After the first hybridization, there is still 1.5 µL left from the *Rsa*I-digested cDNA of **Subheading 3.4.** that will be used as driver for the second hybridization.

1. Prepare the supplemental driver as follows: 1 µL *Rsa*I-digested cDNA (unligated), 1 µL 4X hybridization buffer, and 2 µL DEPC water.
2. Cover with mineral oil and heat at 98°C for 5 min.
3. Draw separately (*see* **Note 4**) the contents of the two hybridized testers of the first hybridization into a pipet tip and mix them with the supplement of denatured driver from the other sample, according to **Fig. 2**.
4. Set the reaction as follows:

	Reaction 5	Reaction 6
Supplemental denatured driver (sample 1)	4 µL	—
Reaction 1	4 µL	—
Reaction 2	4 µL	—
Supplemental denatured driver (sample 2)	—	4 µL
Reaction 3	—	4 µL
Reaction 4	—	4 µL

5. Incubate overnight at 68°C. This ends your third day of experiments.
6. The following morning, add 200 µL of dilution buffer.

Functional Genomics by cDNA Hybridization 71

7. These two tubes (labeled Reaction 5 and Reaction 6) represent the final subtraction, one tube containing the "forward" library and the other the "reverse" library. Both tubes should be conserved at –20°C. If the procedure was executed properly, these tubes might contain enough information to keep a whole laboratory busy for several years. One microliter from each tube is sufficient to conduct the amplification by PCR (*see* **Subheading 3.7.**), which may reveal hundreds of regulated genes, known or unknown.

3.7. PCR Amplification of Subtracted cDNAs

It is recommended to follow the subtraction immediately with a PCR amplification on d 4. Two rounds of PCR with nesting will be performed to improve the specificity of the amplified products. The first round of PCR is executed with a single primer that binds to a sequence shared by both adaptors 1 and 2. The nested PCR is performed with two primers, each binding one of the two adaptors.

1. Set the reaction as follows: 1 µL subtracted sample, 17.5 µL DEPC water, 2.5 µL 10X buffer, 2.5 µL 10 mM dNTP mix, 1 µL 10 µM primer 1, and 0.5 µL Taq polymerase.
2. Start the incubation with 5 min at 72°C to fill in the adaptor ends.
3. Start the PCR using the following parameters: 95°C for 1 min and then 30 cycles of 95°C for 15 s, 60°C for 30 s, and 72°C for 2 min.
4. Take 3 µL of the PCR product and dilute with 27 µL of water. Set the nested PCR as follows: 1 µL diluted PCR product, 16.5 µL DEPC water, 2.5 µL 10X buffer, 2.5 µL 10 mM dNTP mix, 1 µL 10 µM nested primer 1, 1 µL 10 µM nested primer 2, and 0.5 µL Taq polymerase.
5. Start the PCR using the following parameters: 95°C for 1 min and then 12 cycles of 95°C for 10 s, 68°C for 30 s, and 72°C for 2 min.
6. This reaction can be scaled up if more PCR product is needed.
7. The PCR products should be cleaned up with the QIAquick PCR purification kit (Qiagen) before subcloning in a vector.
8. The pGEM Teasy kit (Promega) is particularly convenient for easy ligation of the cDNAs, identification of positive colonies with X-Gal, and sequencing from either the T7 or the Sp6 promoter.

3.8. Sequencing

Although the hybridization in itself can be achieved in a week, identification of the subcloned genes is more labor intensive. Nowadays, the sequencing work is greatly facilitated by the use of automated sequencers using a 96-well or 384-well format. The sequencing also determines the quality of the library by the extent of the different gene products found and by the redundancy of the same products. Finding new products by sequencing a library follows an exponential curve. It is reasonable to interrupt the sequencing when the last batch of clones does not bring more than 5% of new information (for instance, if less than five new gene products are found in a 96-well sequencing plate). There will always be a high interference of mitochondrial DNA in cardiac libraries. Querying the sequences in public databases will determine the identity of the regulated gene and will reveal whether the product is known or novel (*see* **Note 5**).

Some investigators prefer to verify the quality of the library first by microarrays, which represent a reliable technique to verify at a glance which clones from the library are truly regulated. It consists of spotting all the clones isolated on nylon membranes or glass chips and hybridizing them with labeled RNA from the different experimental groups tested. The labeling can be radioactive (using ^{33}P isotope instead of ^{32}P for a better resolution) or nonradioactive (using fluorescent dyes or chemiluminescence). This technique therefore offers some information about the magnitude of changes in expression for each gene product. These chips can also be hybridized with labeled fragments of the mitochondrial genome to define which clones correspond to mitochondrial DNA instead of nuclear-encoded genes.

3.9. Validation

Whatever the results of the hybridization and the identity of the isolated clones, a method of validation is always required because the subtractive hybridization is a qualitative method by nature. It is necessary to rely on methods that truly quantify the changes in expression of one specific gene of interest. One of the most popular methods is the Northern blot, because the probe is already available in the form of the clone sequenced from the library. The average size of the cDNA fragments obtained during the subtraction (0.6–1.2 kb) corresponds to the recommended size of a usual probe for Northern blot. In addition, the tissue localization of the transcript can be validated further by *in situ* hybridization. Again, the subcloned DNA fragment can be used as a probe, although some investigators prefer to use oligo probes. Finally, quantitative PCR (using specific fluorogenic probes or as a screening technique with the SybrGreen) will be the preferred choice when one is analyzing a large number of samples or making a time-course, or when the RNA supply is limited.

4. Notes

1. Some protocols recommend the addition of ^{32}P-labeled dCTP to the reaction in order to follow the incorporation of nucleotides. In practice, however, this alternative does not make a major difference and requires additional precautions against radioactive contamination.
2. Because of the small volumes and the crucial need to avoid any phenol contamination for the following steps, the phenol/chloroform extraction is greatly facili-

tated by the use of Phase-Lock Gel columns (Eppendorf) that automatically separate the aqueous and the organic phases.
3. When assessing the efficiency of digestion by *Rsa*I on an agarose gel, it is recommended to label the samples with the more sensitive SybrGreen dye (1:10,000 dilution) instead of ethidium bromide because the signal will be very faint. The gel does not show specific bands but rather a smear that moves to lower molecular weight in the postdigestion control. If the result is not conclusive, it is recommended to repeat the digestion before moving to the next step because a poor digestion will automatically induce a poor ligation. If the digestion needs to be repeated, the sample can be recovered by phenol/chloroform/isoamyl alcohol extraction followed by ethanol precipitation, as explained in **Subheading 3.3.**
4. The second hybridization is performed with the adaptor-1 tester, the adaptor-2 tester, and an excess of denatured driver. Each of these three reactions has a volume of 4 µL and should be covered with a drop of mineral oil. To achieve a correct hybridization, it is important that the three mixtures enter into contact simultaneously. To achieve that, set a pipetor at a volume between 20 and 40 µL. Draw the adaptor-1 tester in the tip, then draw some air, and then draw the adaptor-2 tester. (Do not worry if you take some mineral oil.) If done correctly, both testers are in the tip but they are separated by air. Then, transfer all the contents of the tip into the tube containing the excess of denatured driver and mix well by pipeting up and down. The three samples came in contact simultaneously and the second hybridization can now start.
5. Theoretically, novel genes are gene products that do not recall any known sequence in public databases. However, because the cDNA synthesis starting the subtraction experiments is primed by oligo-dT, an "unknown" sequence can also correspond to the 3' untranslated region of a known gene. Different criteria determine which gene products are those with the highest priority for further investigation. The first rule of thumb is to start with a long sequence of at least 0.5 kb. A second rule is to focus on gene products that show a tissue specificity, which can be assessed easily with a multitissue Northern blot. A tissue-specific gene is likely to be more original than a ubiquitous gene product. The Northern blot will also determine the size of the transcript. The third rule is to concentrate on true positives, i.e., genes showing a differential expression between the two experimental groups, as validated by quantitative PCR or Northern blot.

References

1. Nadal-Ginard, B. and Mahdavi, V. (1989) Molecular basis of cardiac performance: plasticity of the myocardium generated through protein isoform switches. *J. Clin. Invest.* **84,** 1693–1700.
2. Schneider, M., Roberts, R., and Parker, T. (1991) Modulation of cardiac genes by mechanical stress. The oncogene signalling hypothesis. *Mol. Biol. Med.* **8,** 167–183.
3. Komuro, I. and Yazaki, Y. (1993) Control of cardiac gene expression by mechanical stress. *Annu. Rev. Physiol.* **55,** 55–75.

4. Komuro, I., Kaida, T., Shibazaki, Y., Kurabayashi, M., Katoh, Y., and Yazaki, Y. (1990) Stretching cardiac myocyte stimulates proto-oncogene expression. *J. Biol. Chem.* **265,** 5391–5398.
5. Chien, K. R., Zhu, H., Knowlton, K. U., et al. (1993) Transcriptional regulation during cardiac growth and development. *Annu. Rev. Physiol.* **55,** 77–95.
6. Chien, K., Knowlton, K., Zhu, H., and Chien, S. (1992) Regulation of cardiac gene expression during myocardial growth and hypertrophy: Molecular studies of an adaptive physiologic response. *FASEB J.* **5,** 3037–3046.
7. Robbins, J. (1996) Regulation of cardiac gene expression during development. *Cardiovasc. Res.* **31,** E2–E16.
8. Sadoshima, S. and Izumo, S. (1997) The cellular and molecular response of cardiac myocytes to mechanical stress. *Annu. Rev. Physiol.* **59,** 551–571.
9. Schoenfeld, J., Vasser, M., Jhurani, P., et al. (1998) Distinct molecular phenotypes in murine muscle development, growth and hypertrophy. *J. Mol. Cell Cardiol.* **30,** 2269–2280.
10. Diatchenko, L., Lau, Y.-F. C., Campbell, A. P., et al. (1996) Suppression subtractive hybridization: A method for generating differentially regulated or tissue-specific cDNA probes and libraries. *Proc. Natl. Acad. Sci. USA* **93,** 6025–6030.
11. Depre, C., Tomlinson, J. E., Kudej, R. K., et al. (2001) Gene program for cardiac cell survival induced by transient ischemia in conscious pig. *Proc. Natl. Acad. Sci. USA* **98,** 9336–9341.
12. Depre, C., Hase, M., Gaussin, V., et al. (2002) H11 Kinase is a novel mediator of myocardial hypertrophy in vivo. *Circ. Res.* **91,** 1007–1014.
13. Depre, C., Wang, L., Tomlinson, J., et al. (2003) Characterization of pDJA1, a cardiac-specific chaperone found by genomic profiling of the post-ischemic swine heart. *Cardiovasc. Res.* **58,** 126–135.
14. Hubank, M. and Schatz, D. (1994) Identifying differences in mRNA expression by representational difference analysis of cDNA. *Nucleic Acids Res.* **22,** 5640–5648.
15. Sehl, P., Tai, J., Hillan, K., et al. (2000) Application of cDNA microarrays in determining molecular phenotype in cardiac growth, development, and response to injury. *Circulation* **101,** 1990–1999.
16. Chomczynski, P. and Sacchi, N. (1987) Single-step method of RNA isolation by acid guanidium thiocyanate-phenol-chloroform extraction. *Anal. Biochem.* **162,** 159–169.

5

Statistical Methods in Cardiac Gene Expression Profiling
From Image to Function

Sek Won Kong

Summary

By providing genome-scale information on gene expression, microarray technology has gained popularity in diverse areas including clinical medicine. However, the analysis and interpretation of microarray data are often complicated. This chapter describes various strategies for microarray data analysis. The analysis starts with the scanned image of a microarray. The image information is processed and summarized to numerical values that represent the abundance of transcripts. Technical variability and systematic biases can be minimized with the proper procedures of background correction and normalization. Considerable numbers of genes are not expressed or not detected by microarray technology. Those genes can be filtered out before further statistical comparison to reduce the dimensionality of the problem. The next step in analysis involves statistical comparison, cluster analysis, and visualization. Genes from the same cluster are considered to be coexpressed and/or coregulated. Also, we can group coexpressed genes into categories by their biological function and cellular location. By combining prior knowledge and statistical results, we can make an inference based on the gene expression profiles.

Key Words: Microarray; quality assessment; low-level analysis; functional annotation.

1. Introduction

By providing a genome-wide view of gene expression, microarray has become a common exploratory technique in many areas of biological and clinical studies. An enormous amount of data has been produced during the last decade with this technology, and data on the order of terabytes will have accumulated in the next few years. However, there are several limitations of this technology. A profile of mRNA expression is a snapshot of a dynamic process, and the kinetics of RNA synthesis has been measured in the order of minutes

(1). In the extreme case, normal tissues show diurnal variations in their gene expression profile *(2)*. Thus the profiles of a given condition at a given time point do not provide enough information for deciphering the interaction among genes and gene products. These limitations should be considered and reflected in the initial step of experimental design.

Several points should be considered in the designing step. First, most microarray experiments have a limitation in terms of sample size, and hence a complex design with many covariates makes analysis less powerful and unnecessarily complicated *(3)*. The sample preparation and hybridization protocol should be the same for a study. For example, the comparability between two different amplification protocols has been reported to be limited since systematic bias is introduced during the amplification step *(4)*. The biological variance is in general regarded as greater than the technical one; thus biological replicates are more important than technical replicates within a limited number samples in a group *(5)*.

Analysis methods for microarray data have been continuously refined during the last decade. However, the statistical and mathematical algorithms are not easy to understand for most biological and clinical researchers. Much software has been developed to help the research community analyze and interpret microarray data, but this software often gives conflicting results for the same input data. In some cases, different software applications give different results for an algorithm by the same name, presumably owing to small differences in implementation not disclosed to the user. For instance, the dendrograms from average linkage hierarchical clustering using Euclidian distance are not always the same among different software applications. To improve the reproducibility from the same data set and to have more control over each step of analysis, R/Bioconductor libraries have been used in this chapter (*see* **Note 1**). **Figure 1** shows the flow of steps in microarray analysis.

2. Materials

2.1. Microarray Data Sets in Public Repository

A microarray data set in a public repository enables a researcher to explore a variety of data analyses. Gene Expression Omnibus (GEO) and ArrayExpress are two of the largest public data repositories. In particular, the data from large-scale projects for specific disease like cancer and cardiac disease have been made public to facilitate research. CardioGenomics (http://www.cardiogenomics.org) of Programs for Genomic Application (PGA) has been producing benchmark data sets for the research community using the Affymetrix platform. It provides more than 350 profiles from animal models of heart disease and 150 profiles from human cardiac tissues. In this chapter, we use a data set from CardioGenomics for demonstration purposes (*see* **Note 2**).

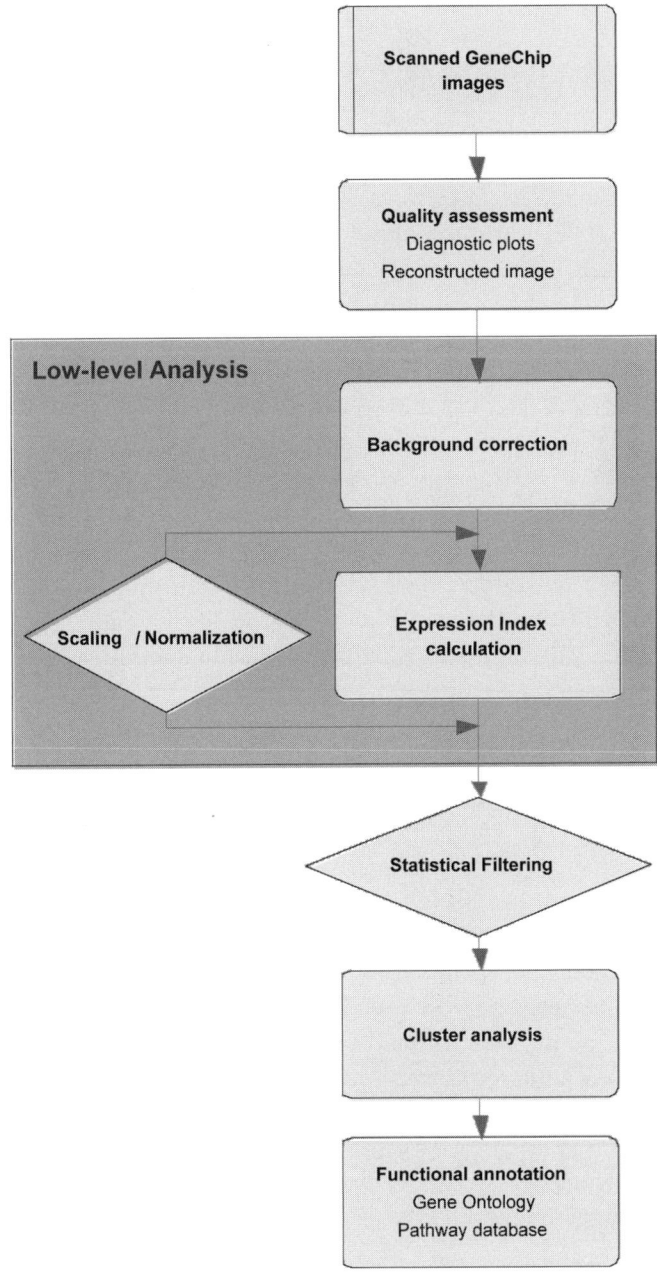

Fig. 1. A flow chart for microarray data analysis.

2.2. Analysis Software Tools

Analysis tools can be categorized into three groups *(42)*. The first category is the enterprise-level server/client environment (*see* **Note 3**). The second category is specialized for specific purposes and usually comes with a graphical user interface (GUI) for stand-alone workstations. The last category is actually an environment, which provides the libraries and toolboxes for microarray analysis. Software in first category is very expensive and provides data storage and a management environment like the Oracle™ database. Examples in this category are Rosetta Resolver (Rosetta Biosoftware) and GeneData Expressionist (GeneData). They provide a certain level of customization for specific situation, and the analysis pipeline can be built for data processing and results management. Tools belonging to the second category come with a GUI for a stand-alone desktop environment so that the user can analyze the data according to the procedures suggested by the software. They are developed with predefined steps for analysis-specific function in mind. Most academically available as well as commercial tools belong to this category. Although the steps in the analysis are not as flexible as those in the third category, the learning curve is less stiff than it is for the third category. The last category is a command line-based programmable environment. The software in this category requires a certain level of programming skill; thus the learning curve is stiff.

Most software has already been used in the statistics and engineering fields. With various libraries and packages specialized for microarray data analysis, they provide a flexible environment so that each procedure in the analysis can be customized and optimized. S-plus/ArrayAnalyzer, R/Bioconductor, MatLab/Bioinformatics toolbox, and SAS/Microarray belong to this category. Finally, the choice of tools largely depends on the number of arrays in an experiment, the array platform, and scalability. In this chapter, most of the examples and graphs were created using R/Bioconductor and MEV (*see* **Subheading 3.6.**). The latter belongs to the second category and was especially developed for cluster analysis. R/Bioconductor provides a highly flexible environment for microarray data analysis. Both software applications are freely available to academia, and they support multiple platforms, i.e., Microsoft Windows, Linux, and Mac OS.

3. Methods

3.1. Quality Control of RNA

The first step in analysis should include assessment of RNA quality and integrity. The Bioanalyzer from Agilent can be used to check the integrity of RNA in a specific case. The degradation can be detected by visual inspection of the electropherogram. Moderately degraded samples are not suitable for microarray experiments, but they can still be used for reverse transcriptase-polymerase chain reaction (RT-PCR). An Affymetrix TestChip can be used for checking hybridization efficiency of labeled RNA and degradation of RNA. There are probes designed to aid in quality control. These control probe sets have sequences derived from the different part of the coding region, i.e., the 3', mid, and 5' sides of transcripts. By calculating the ratio of the 5' (or mid) to 3' side probe set, the degree of degradation can be assessed.

3.2. Image Processing

Once samples are hybridized and scanned, a scanner produces the digital image data. The computational aspect of microarray analysis starts with image processing and feature extraction. It is important to store the raw image data uncompressed because the raw data can be damaged during compress/decompress. Most of the scanner has the dynamic range of 16 bits ($2^{16} = 65,536$) (*see* **Note 4**) and produces large image files (*see* **Note 5**). The first step in image processing is called feature extraction. The spot center and borders are defined by applying a grid, and then foreground and background intensities are calculated from pixel intensities. For the GeneChip, the scanned image file has DAT as the suffix. The Microarray Analysis Suite (MAS) and GeneChip Operating System (GCOS) from Affymetrix typically import a DAT file with an EXP file (*see* **Note 6**). The CEL file is created from the DAT file by feature extraction (*see* **Note 7**).

This procedure has been improved over the years, and the probe intensities from previous generations of Affymetrix software might be different from those generated using the latest one *(6)*. The quality of the probe can be assessed, and a bad probe can be masked to exclude it from further analysis. The reconstructed image of the CEL file should be inspected visually for gross artifacts.

3.3. Quality Assessment

GCOS provides several parameters of quality measurement. Details of these parameters are described in the Statistical Analysis Manual. The 3'/5' ratios of the control probe sets, Scaling Factor (SF), and RawQ value can be used to address the sample quality, hybridization efficiency, and background noise, respectively. There are reference values for these parameters *(7)*.

Whereas a single spot in a cDNA microarray represents a single target transcript, between 11 and 16 probe pairs represent one target transcript in GeneChip. In most cases one probe outlier in the probe set does not make big difference in expression level, for at least two reasons: first, a robust statistical algorithm (one-step Tukey's Bi-weight in the case of MAS 5 Signal intensity)

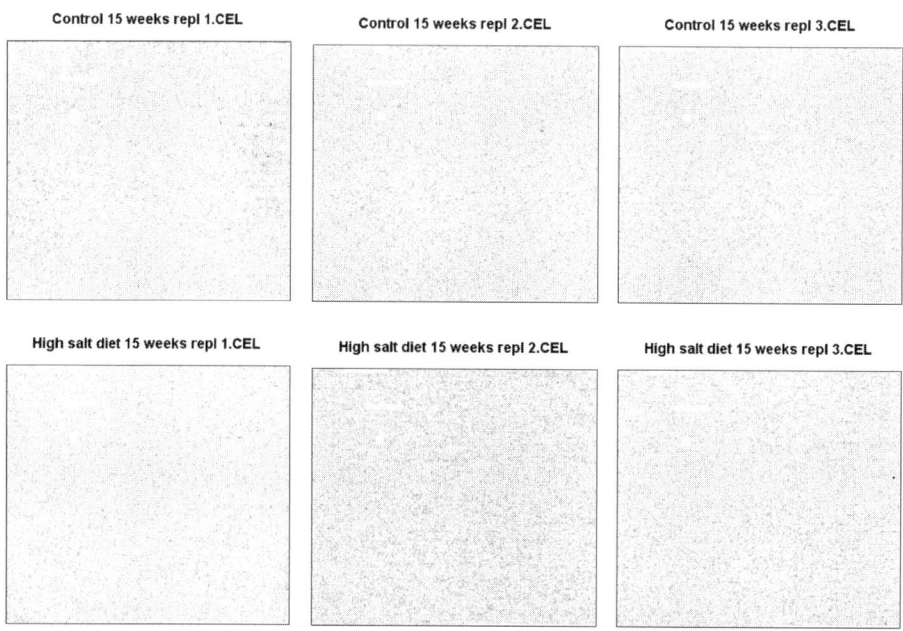

Fig. 2. Reconstructed image of the CEL files.

is used for expression index calculation, and second, the probe pairs that make up the probe set are randomly located on the GeneChip, and small artifacts in one area do not strongly affect the expression index in general. In practice, if the artifact area is smaller than 10% of the total area, the array can be used for further analysis *(43)*.

Bioconductor *affy* and *affyPLM* libraries provide several measures to address the quality issue (*see* **Note 8**). **Figure 2** shows the reconstructed images from a CEL file (*see* **Note 9**). By visually inspecting these, we can identify some technical artifacts. **Figure 3A** shows the histograms of the log-transformed intensity. For the arrays that were scanned with a 100% PMT setting, a hump can be observed in the highest intensity region (**Fig. 3B**) (*see* **Note 10**). The box and whisker plot in **Fig. 3C** can be used to compare the distribution of the interquartile range among arrays. **Figure 3D** shows a so-called *RNA digestion plot*. This plot was initially devised for detecting RNA degradation by comparing the mean intensity of the same position in the probe set, but it is also useful for

Fig. 3. *(opposite page)* Diagnostic plots. (**A**) Histogram of the log-transformed intensities. The images are with the scanner set to 10% PMT. (**B**) The images were scanned with the scanner set to 100% PMT. A hump is observed in the higher intensity area.

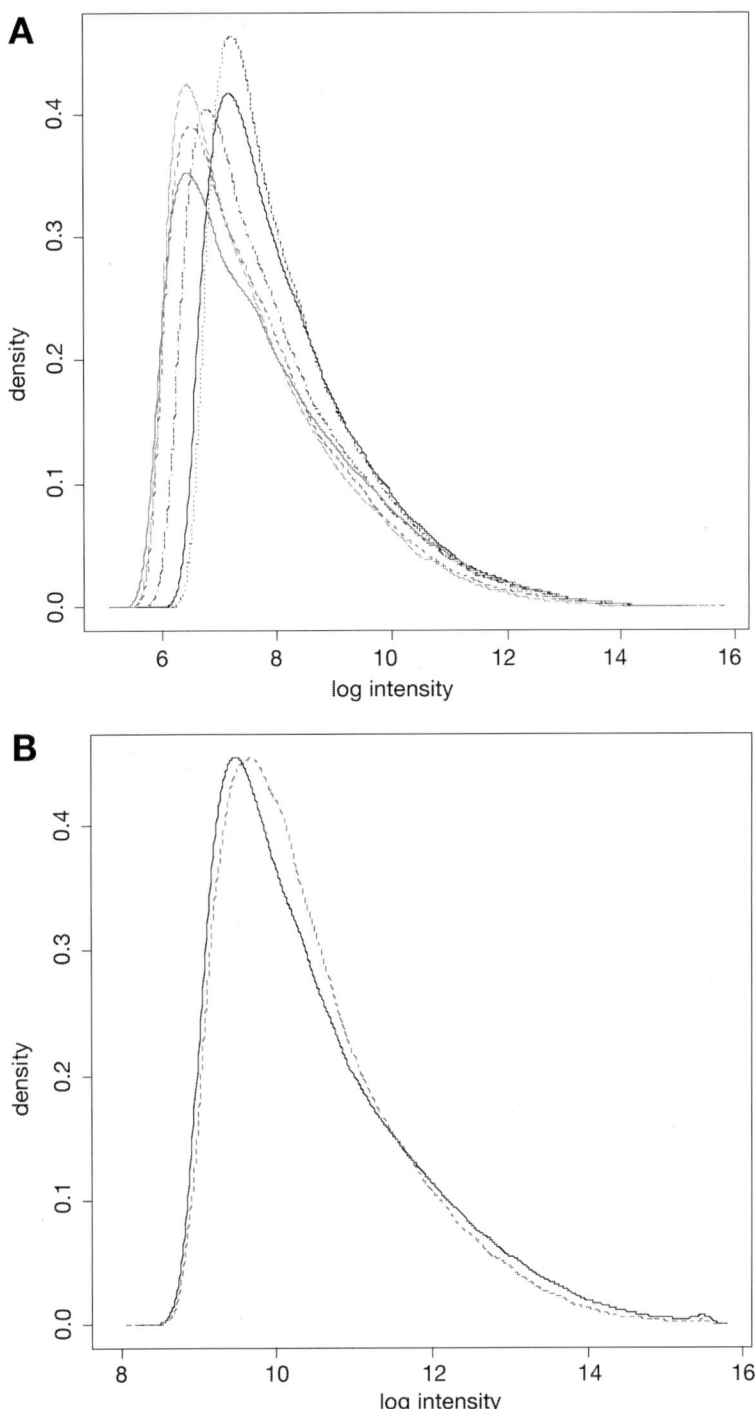

Fig. 3. *(continued on next page)*

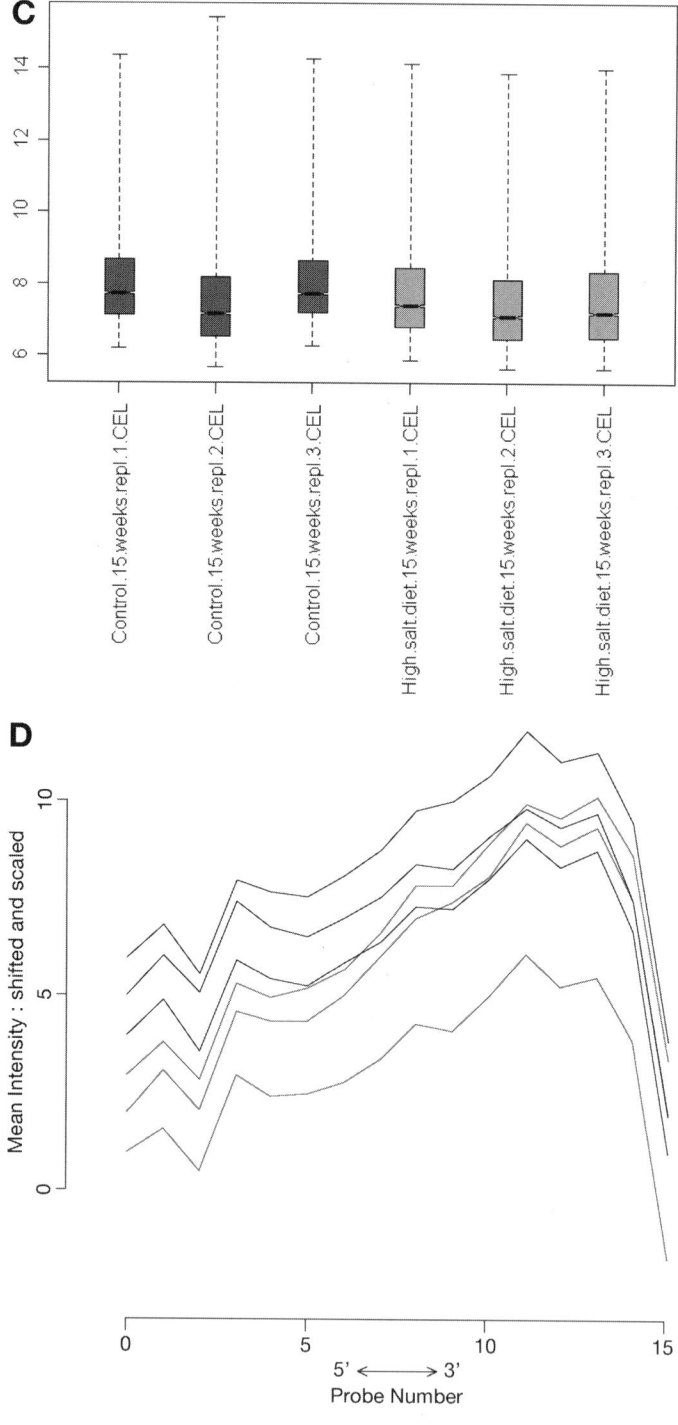

Fig. 3.

comparing the samples prepared with the standard protocol and the small sample protocol. The probe outliers can be excluded for expression index calculation, and the sample outliers can be excluded for further statistical comparison. Biological and/or technical replicates can also be used for detecting outliers by comparing probe intensity across arrays (*see* **Note 11**).

3.3. Preprocessing: Background Correction and Normalization

The information in the CEL file should be further processed to generate the expression level of each transcript. The step involved in this procedure is called low-level analysis, which consists of at least three procedures: background correction (*see* **Note 12**), normalization (*see* **Note 13**), and expression index calculation (*see* **Note 14**). The purpose of low-level analysis is to reduce systematic bias and random noise from nonspecific hybridization and artifacts. To quantify and compensate for the amount of inhomogeneity, several statistical models have been employed *(8)*. It has been noted that the choice of normalization algorithm greatly affects the result, and this is also true for the background correction algorithm *(9)*. Among such background correction algorithms, GCRMA is unique *(10)*. It considers three types of background signal, i.e., the optical noise of the scanner, nonspecific hybridization, and GC content bias. These three parameters are modeled and calculated for each GeneChip type to assess stray background signal (*see* **Note 15**). The background correction might increase the variance, especially for low-abundance transcripts, and can be identified as so-called *fish tail* in intensity scatter plot (**Fig. 4A**) (*see* **Note 16**).

The second step is normalization. This has been an issue in low-level analysis from the very beginning of microarray experimentation, and diverse linear and nonlinear algorithms have been implemented for probe and/or probe set level normalization. The underlying assumption of normalization is that most of the genes are not differentially expressed and that among those differently expressed genes, the numbers of upregulated and downregulated genes are similar.

For the GeneChip, probe level normalization has been recommended after the observation that the shapes of probe intensity distributions across different microarrays are not similar *(11)*. Quantile normalization and invariant set normalization have been widely used for probe level normalization. Some caution should be used when the aggressive normalization method is applied. For example, quantile normalization makes the distribution of probe intensity

---

Fig. 3. *(previous page)* Diagnostic plots. (**C**) Box and whisker plot of the log-transformed intensities. (**D**) RNA digestion plot.

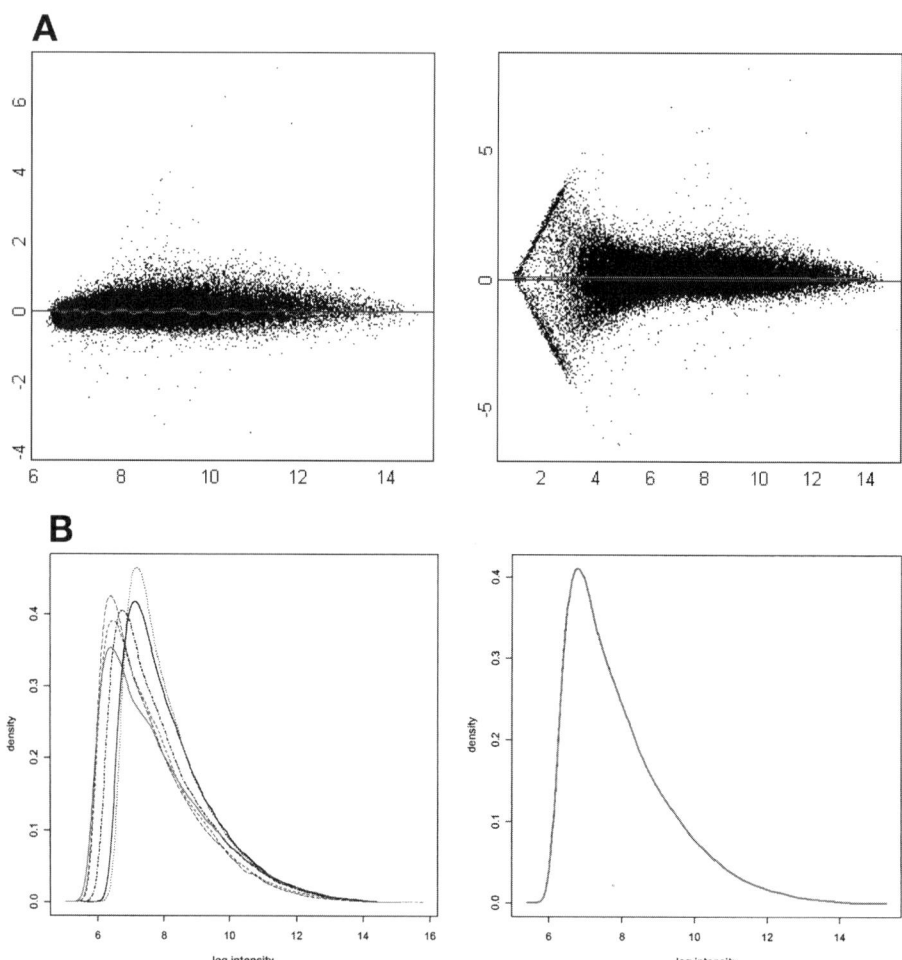

Fig. 4. Background correction and normalization. (**A**) The fishtail effect of background correction (right panel). The MvA plots were created with the log-transformed intensities of the Perfect Match probes. (**B**) Histogram of the log-transformed probe intensities before (left panel) and after (right panel) quantile normalization.

exactly the same (**Fig. 4B**) (*see* **Note 17**); thus biological differences could be washed out during between-group normalization.

In the case of cDNA microarray, two normalization steps are needed, within-array and between-array normalization. In addition, several other factors contribute a measurement bias. For example, two fluorescent dyes, Cy3 and Cy5, have different labeling efficiency and different quantum efficiency. Yang and colleagues investigated these issues carefully and implemented the normalization algorithms in Bioconductor library *manorm* (*12*).

3.4. Expression Index Calculation

For the GeneChip, 11 to 16 probes represent a single transcript. The summarization procedure of 11 to 16 probes into a single number that represents the level of expression is called the expression index calculation. This step has been improved over time since *Average Difference* was implemented. A statistical algorithm i.e., *Signal*, has been used so far; however, a novel algorithm called PLIER (Probe Logarithmic Intensity ERror) was recently introduced by Affymetrix (*6*). There has been some disagreement on how to extract the best information from the GeneChip. Several researchers have developed custom processing algorithms; MBEI (*see* **Note 18**), RMA (*see* **Note 19**), and GCRMA (*see* **Note 20**) are just a few of them (*see* **Note 21**). To aid in choosing the algorithm, Choe and colleagues compared the different preprocessing and expression index calculation methods using a spiked-in data set (*13*). In their study, 3860 RNA species were used to build a spike-in data set, and 2551 RNA species were used as background either present or absent. One hundred to 200 RNAs are spiked in at each fold-change level of interest, ranging from 1.2-fold to 4-fold. They compared current Affymetrix expression summary procedures using the Bioconductor *affy* package, dChip, and MAS 5 software. It was evident that there was variation in identifying differentially expressed genes, but it is still not clear whether any single algorithm is superior to others.

As described, many different methods are available for low-level analysis of GeneChips. In general, it is recommended to start with the expression level reported by Affymetrix GCOS software. If necessary, the background correction strategy can be changed based on the local and global background inhomogeneity of the data, and the normalization strategy can be changed based on the degree of perturbation in the experiment. For the several genes of interests, it is recommended that the results from the microarray be compared with a more confirmatory technique such as quantitative RT-PCR (*see* **Subheading 4.3.**).

3.5. Statistical Filtering

The purpose of using statistical methods in microarray data analysis is to identify the differentially expressed genes as well as coexpressed genes. To use inferential statistics, the underlying distribution should be assumed. For microarray data, however, it is often difficult to justify the distributional assumptions required for these methods, primarily owing to small sample size. In practice, once the data are transformed to the log scale, the Gaussian normal distribution can be used as an approximation.

The gene expression profile is a special case of multivariate condition in which the number of variables is far greater than the number of samples. The general analysis strategy for multivariate data is as follows *(14)*:

1. Data reduction or structure simplification.
2. Sorting and grouping.
3. Investigating of the dependence among variables.
4. Prediction.
5. Hypothesis construction and testing.

The statistical filtering and clustering that have been widely used for gene expression data correspond to data reduction and sorting and grouping. The simplest procedure is to calculate a statistic for one gene at a time. This reduces the multivariate problem to a univariate one so that one can use, for example, t- or F-statistics and p-values. However, each gene is not independent of the others, and correlation among genes is ignored in this approach. The other approach is a bivariate analysis, which considers gene pairs instead of a single gene *(15)*.

In both cases, multiple comparison problems (MCPs) arise. In microarray analysis, two popular approaches examine the family wise error rate (FWER) and the false discovery rate (FDR) *(16)*. The FWER can be interpreted as the probability of at least one type I error, i.e., false positives, and the FDR is the expected proportion of type I errors among rejected hypotheses. The number of genes found to be differentially expressed can be different by the choice of threshold as well as the MCP method. Of note, Storey and Tibshirani introduced the q-value, a method that can translate p-values into q-values once some assumptions are met (*see* **Note 22**) *(17)*.

Practically, there is still no guidance on what statistical test to choose. A conservative approach to this issue is to use a combination of several methods. For instance, Differential Expression via Distance Summary (DEDS) can be used in this context (*see* **Note 23**) *(18)*. The t-score, F-score, B-score, fold change, and SAM statistic score can be calculated using this library (*see* **Note 24**), and the new score of distance summary (DS) can be calculated by combining several statistical scores.

One of the key differences in the methods is the way in which the variance of each gene is estimated with the small number of samples. The simplest case is two-group comparison. We can calculate the t-statistics and adjust for multiple testing to generate the corrected p-values for each gene. In this case, the variance estimate necessary for the t-statistic was calculated with a small number of samples and may be incorrect. To address this issue, several variations of the t-test have been proposed. Significance analysis of microarray (SAM) added the intensity-dependent variance to the variance estimated from samples

(19), and the Cyber-T test modeled the intensity-dependent variance using a Bayesian approach *(20)*. Kim and Park investigated the possibility of using public microarray data sets to improve the estimation of variance of each gene *(21)*. They reported that the variance estimation using public databases could aid in a precise estimation of the variance. This is helpful in the sense that the variance estimation using a small number of samples can be highly unstable.

3.6. Cluster Analysis

Clustering methods generally use either a hierarchical or a partitioning approach. Heat maps and dendrograms from hierarchical clustering have been used to find groups of genes and samples with similar profiles *(22)*. Clustering can be applied to samples, to genes, or to samples and genes simultaneously. The latter is called two-way clustering. Clustering starts with the calculation of the similarity (or dissimilarity) matrix, which contains the "distance" between every pair of genes. The expression measure is a quantitative variable in most cases, and the most commonly used distance metrics are Euclidean distance and the correlation coefficient, which emphasize different aspects of the data. For binary data, i.e., when each gene is determined to be either present or absent, the proportion of co-occurrence can be calculated as a measure of similarity.

Hierarchical methods are divided into the divisive (top-down) and the agglomerative (bottom-up) methods. The more common is the bottom-up approach, which starts with each object (genes or samples) as its own cluster and combines the most similar objects to form a bigger cluster. These steps will continue until one big cluster is formed. This method can be classified further depending on how the distances between two clusters are calculated. Average linkage, complete linkage, and single linkage methods are commonly used. The number of clusters and the content of a cluster vary by the distance metric and joining method selected. **Figure 5** shows the difference among these clustering methods using normal tissue samples (*see* **Note 25**).

Partitioning methods are the clustering methods that address the problem of partitioning a set of n objects into c disjoint classes with the number of classes predefined. The K-means and K-medians algorithms are commonly used.

However, this raises an immediate question of "How many classes (clusters) are there in the data set?" Several methods have been proposed. For instance, the Gap statistics proposed by Tibshirani and colleagues find cluster number *(23)*. Yeung and colleagues suggested the Figure of Merit (FOM) as an estimate of the predictive power of a clustering method. This method was initially developed to compare different clustering methods; however, it also provides a guideline on the number of clusters in a given data set *(24)*. Self-Organizing

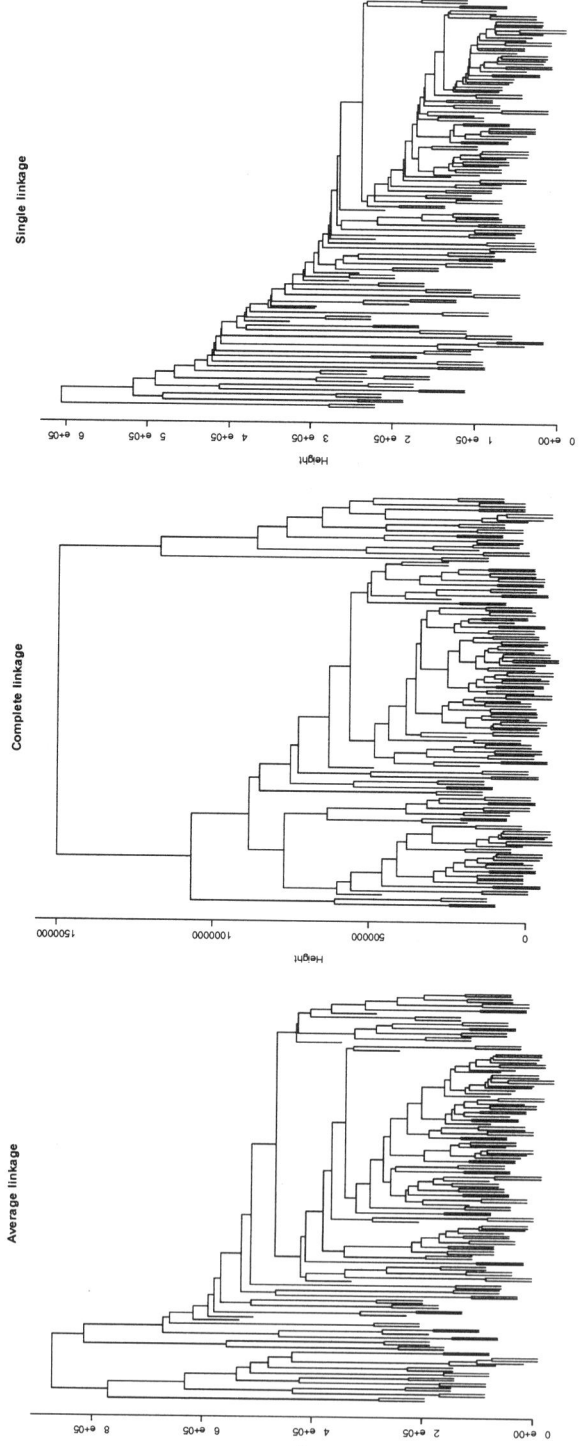

Fig. 5. Dendrograms from agglomerative hierarchical clustering of normal human tissue samples. Average linkage (left panel), complete linkage (center panel), and single linkage (right panel) methods were used.

Maps (SOMs) is a partitioning method that constrains the clusters to be represented in a regular, low-dimensional structure, such as a rectangular grid.

Heat maps and dendrograms from hierarchical methods and the line graphs from the partitioning methods are informative in terms of summarizing the data visually. However the visualization techniques of multivariate analysis such as principal component analysis (PCA) and multidimensional scaling (MDS) could give additional insights into the data structure. For example, Pomeroy and colleagues showed that the different types of brain tumors could be identified with a 3D PCA plot built on 50 marker genes *(25)*.

The Multi-Experimental Viewer (MEV) from the TIGR Institute is a useful open source tool that performs statistical and cluster analysis (*see* **Note 26**). More than 17 algorithms are implemented for the analysis and visualization of microarray data. For the purpose of demonstration, the normal human tissue data set (GSE803) was downloaded from the GEO database (*see* **Note 27**). The samples consist of 12 healthy human tissues, and each sample is typically composed of a pool of 10 to 25 individuals. The genes that received the absent call across all the samples and the control probe sets were excluded. The data were imported to MEV and transformed to log 2 intensities. The variance filter was used to identify the top 5% of the highly variable genes. The remaining 492 genes were used for cluster analysis. **Figure 6** shows the different methods of clustering and visualization. **Figure 7A** and **B** shows the PCA plot of 12 normal human tissues and 492 genes, respectively *(26)*.

3.7. Functional Annotation

The probe sequences are designed using the NCBI RefSeq database (http://www.ncbi.nlm.nih.gov/RefSeq/) at any given time (*see* **Note 28**). Netaffx™ provides the latest functional annotation for the GeneChip; it is updated every 3 mo *(27)*. Sometimes the annotation based on a single database such as LocusLink and UniGene is not correct; thus it is necessary to compare the annotations from several different databases. The Ubiquitous Bio-Information Computing (UBIC^2) from Wong Lab provides a method for integrating diverse biological information from different sources (http://bayes.fas.harvard.edu/UBIC2/). The upstream sequences as well as the functional annotation can be easily retrieved with a list of genes.

Another method to get the latest annotation is to compare the probe sequence with the current version of genomic sequences. For instance, GeneAnnot from the Weizmann Institute provides the latest and nonredundant annotation for GeneChip probe set IDs by direct sequence comparison of probes with GenBank, RefSeq, and Ensembl mRNA sequences (http://genecards.weizmann.ac.il/geneannot/). Finally, Bioconductor also provides annotation information in the form of Metadata (*see* **Note 29**). With this, the multiple

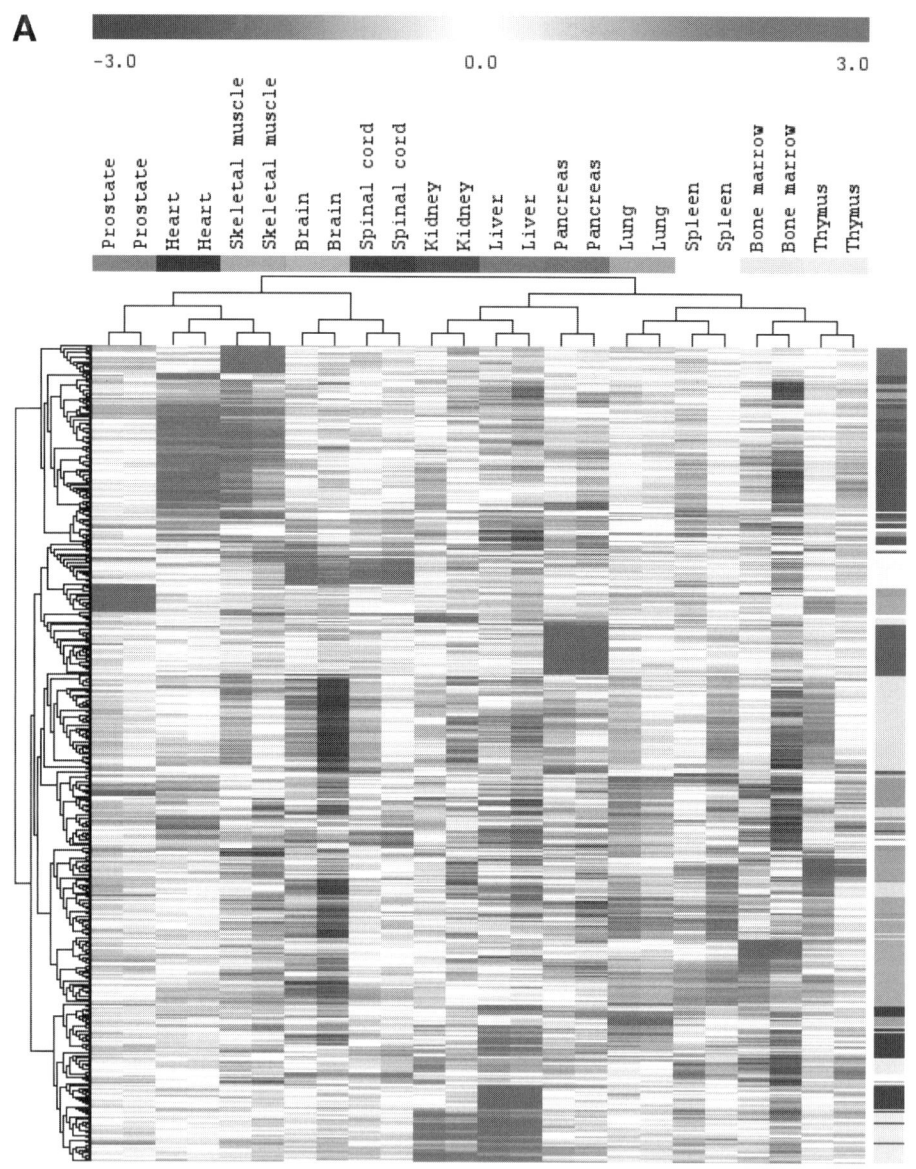

Fig. 6. Cluster analysis with normal human tissue samples. (**A**) Two-way clustering with the agglomerative hierarchical method. The color codes were used to identify the same tissue types, and the results of the K-means clustering were also color coded.

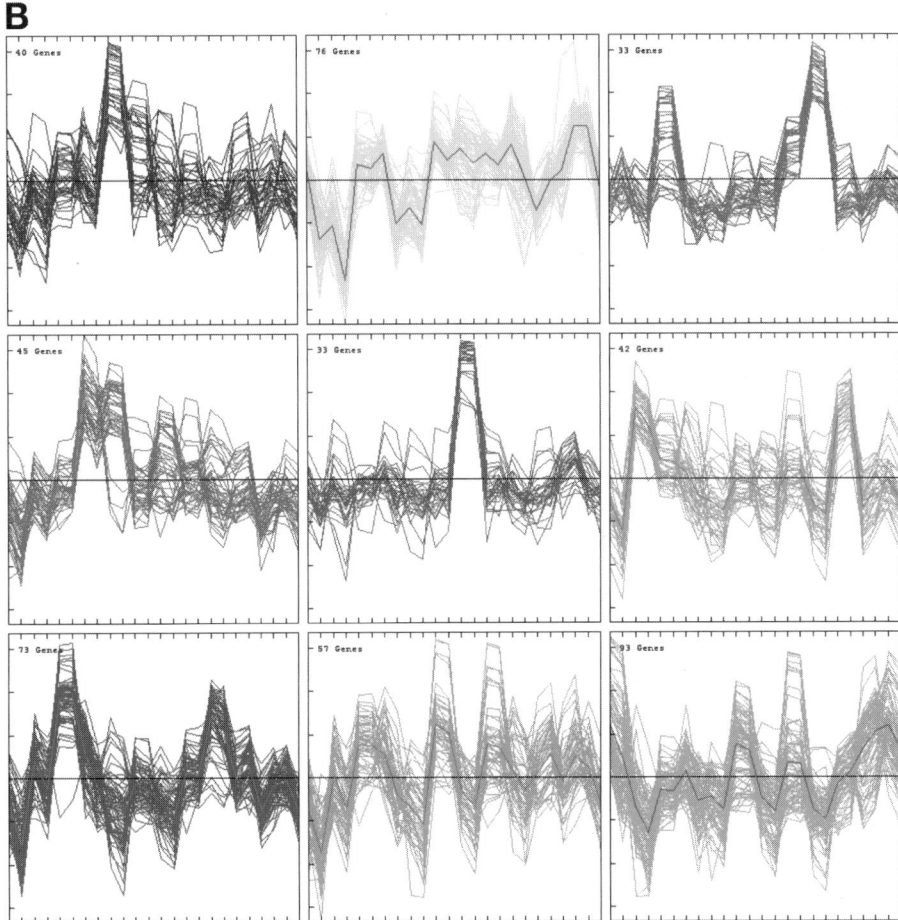

Fig. 6. Cluster analysis with normal human tissue samples. (**B**) A *K*-means clustering with nine clusters.

annotation information of differently expressed genes can be conveniently retrieved together.

3.8. Functional Analysis with Gene Ontology and Pathways

Once the list of target genes is available after the statistical filtering and cluster analysis just described, the next step is interpretation into a biologically relevant result. This step is considered the most challenging one in microarray experiments, and the interpretation can be especially beneficial if controlled terminologies have been used for describing the function of genes and proteins. In an effort to facilitate this process, the Gene Ontology (GO) Consortium

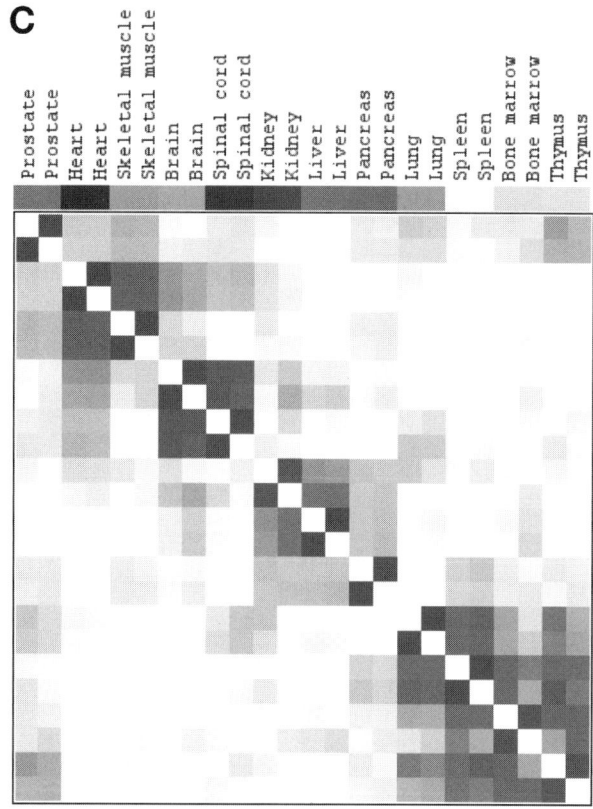

Fig. 6. Cluster analysis with normal human tissue samples. (**C**) A color correlation matrix of the samples. The result of a hierarchical clustering was imposed.

provides systematic and hierarchical nomenclatures of molecular functions, biological processes, and cellular components. A group of genes together builds a GO term that has the specific annotation; each GO term has the unique identifier, and the identifiers of genes and proteins from the different databases are mapped to the GO term. The contents of a GO term are equivalent not to a pathway but to a collection of genes (or gene products) that are functionally related, involved in the same processes, or localized together within a cellular compartment.

In analyzing the data, it would be meaningful if the majority of genes from a cluster belong to the same GO term. To test this statistically, the Chi-square and Fisher's exact tests based on the hypergeometric distribution have been used. GenMAPP (*see* **Note 30**), FatiGO (*see* **Note 31**), DAVID/EASE (*see* **Note 32**), and GeneMerge (*see* **Note 33**) are examples of the software that

Statistical Methods in Cardiac Gene Expression Profiling

Fig. 6. Cluster analysis with normal human tissue samples. (**D**) A color correlation matrix of 492 genes. The result of a *K*-means clustering was implied.

calculate the statistical scores for each GO term for given gene expression data (*see* **Note 34**).

Instead of considering a single cluster, we can perform the same analysis on the group of differentially expressed genes between conditions. A recent improvement in this approach is to consider all potentially differentially expressed genes rather than a small number. This is useful especially when only a very small number of genes or even no single gene is differently expressed. Mootha and colleagues identified the group of genes from the oxidative phosphorylation (OXPHOS) pathway that are significantly different between a type 2 diabetes group and a normal glucose tolerance group *(28)*. In their study, each member of the OXPHOS pathway showed no statistical significance when the expression levels between two groups were compared. However, as a group, referred to a *gene set*, significant differences were seen. Segal and colleagues also presented groups of genes that are iden-

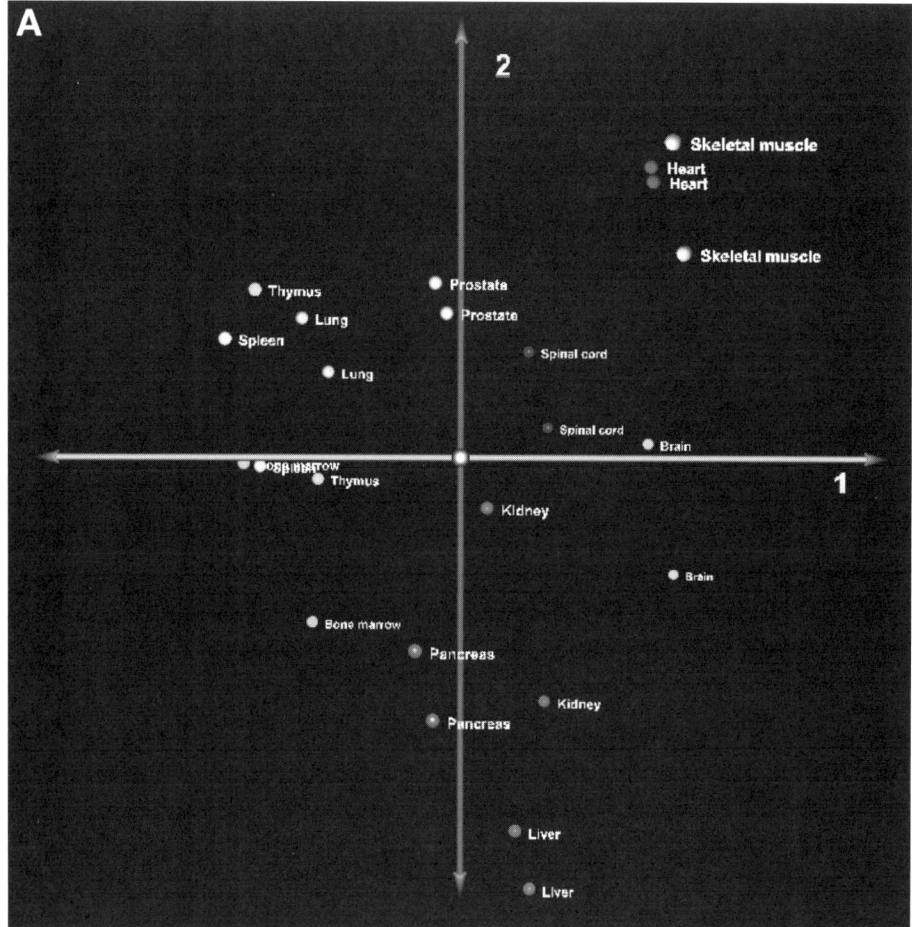

Fig. 7. Principal components analysis (PCA). **(A)** A PCA plot of 12 normal human tissues. The result of a *K*-means clustering was overlaid with the same color code used in **Fig. 6B**.

tified as significant by different types of cancer tissue using a gene set-based approach *(29)*.

4. Further Issues in Microarray Experiments

4.1. Comparison of the Data Prepared with the Standard Protocol and the Small Sample Protocol

When the amount of the starting material (total RNA) is not enough for labeling and hybridization, a second round of amplification is needed. For the GeneChip, the standard procedure requires 5 µg of total RNA as the amount of

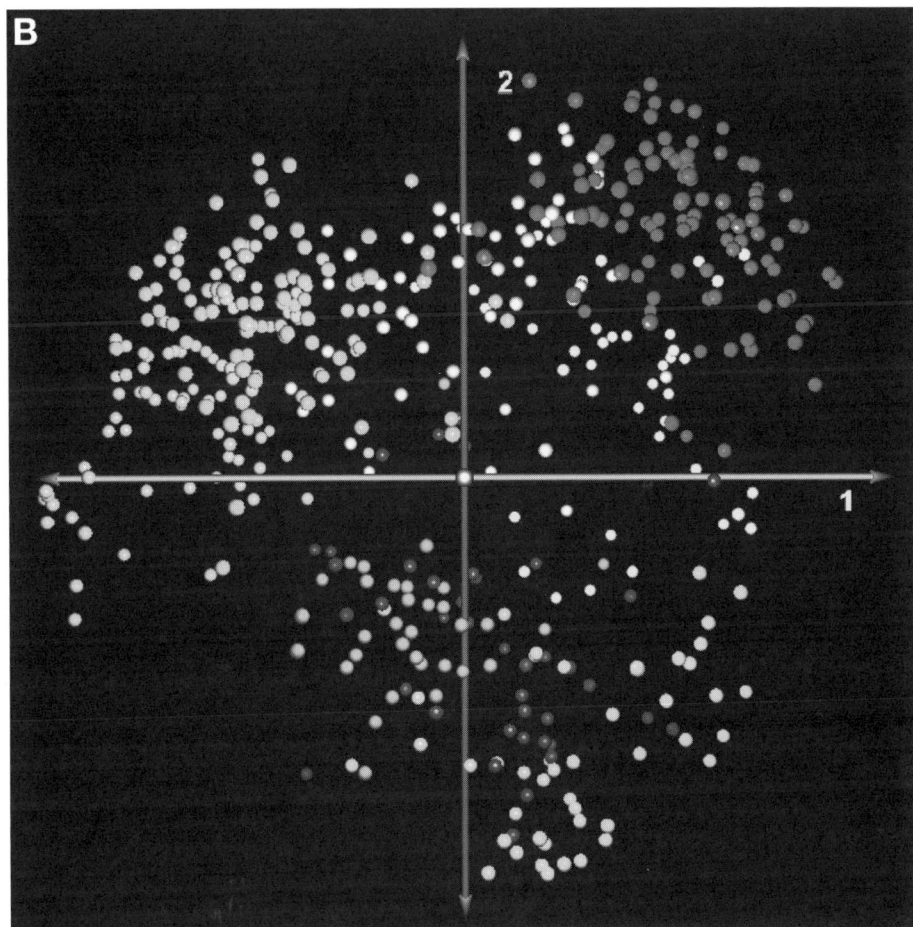

Fig. 7. Principal components analysis (PCA). **(B)** A PCA plot of 492 genes. The result of a *K*-means clustering was overlaid with the same color code used in **Fig. 6B**.

starting material. For a small sample case, Affymetrix provides a standardized protocol with second round of amplification using the random primers and T7-Oligo (dT) primers *(30)*. More technical details of small sample protocols are described in Baugh et al. *(31)*, Eberwine et al. *(32)*, and Dulac et al. *(33)*. It has been reported that data from the two different protocols are not directly comparable *(4)*.

4.2. Combining the Data from Different Platforms and Generations

Given the large amount of microarray data in public repositories, the need for combining and comparing data created using different microarray platforms is increasing. The probe sequences and corresponding intensities are two pieces of information that do not change over time. When one is comparing several different platforms or different generations of the same platform, it is important to have the exact probe sequence information and the original microarray image. Kong and colleagues suggested a method for comparing the different generations of Affymetrix GeneChips using probe sequence information *(34)*. However, there is a limitation to this approach in that a certain number of genes will be excluded from analysis because they do not satisfy the criteria for comparability between the platforms.

4.3. Validation of the Results from Microarray Experiments

Quantitative RT-PCR (qRT-PCR) has been used as a method of choice for validating the results from a microarray experiment. The correlation between microarray technology and qRT-PCR is good in general *(35)*. However, the dynamic range of the two technologies is different, and also the correlation of measurement is not linear *(36)*. Microarray technology is not as sensitive as qRT-PCR, and the detection sensitivity for low-abundance transcripts is limited. The fold change estimated from two technologies can be very different, especially when the denominator is small enough to receive the absent call in the case of GeneChip.

5. Notes

1. R is distributed through the Comprehensive R Archive Network (CRAN) (http://cran.r-project.org/). R binaries are available for most computing environments, including Microsoft Windows, Mac OS, Linux, and UNIX. Installation is straightforward for most operating systems. Examples in this chapter use several Bioconductor libraries, which can be installed from R. The reader can reproduce the same results by typing in the commands provided with each example. Throughout this chapter, the commands to R are set in a monospaced typewriter font like this, and the names of the library packages are in a font *like this* (italic).
2. The data set is available at http://cardiogenomics.med.harvard.edu/groups/proj1/pages/download_rat.html. This data set consists of three treatment groups: control group, exercise group, and high-salt diet group. In this chapter, three samples of a 15-wk control group and three samples of a 15-wk high-salt-diet group were used. The detailed description of the data set is available from Kong et al. *(37)*.
3. The server environment provides a central repository for the microarray data, and the database usually comes with a LIMS (Lab Information Management System).
4. For Affymetrix, a dynamic scan range of 15 and 1/2-bit was used until the GeneChip Scanner 3000 became available.
5. To scan the microarray, a scanner uses the photomultiplier tube (PMT) as a light source. The voltage setting of the PMT can be changed according to the conditions. For a cDNA microarray, the brightest pixel intensity can be adjusted to just below 16 bits so that scanner can make use of its full 16-bit dynamic range.

Statistical Methods in Cardiac Gene Expression Profiling 97

Among the public data sets, there are arrays that had been scanned with 100% PMT setting compared with the current standard setting of 10%. This is one of the obstacles one encounters when comparing data sets from different laboratories. The presence of saturated cells can be easily identified with a histogram of probe intensities. The CEL files from the 10% PMT setting cannot be compared directly with those from a 100% PMT setting because of the saturation in the highest intensity spots. **Figure 3B** shows a histogram of saturated arrays.

6. The EXP file has the description of the experiment and the sample annotation.
7. About 100 pixel intensities within a single cell are summarized in two measurements: the intensity and its standard error. If the standard error is large, the estimated intensity might not be valid. These probes can be masked as outliers.
8. The diagnostic plots shown in **Fig. 3** were created using the functions in the Bioconductor *affy* library. The script for **Figs. 2** and **3** is as follows:

```
# Change the working directory to where the CEL files are
located. The CEL files of three control samples and three
high-salt-diet samples are used from the data set described
in Note 2.
# Load affy and affyPLM libraries:
    >library(affy).
    >library(affyPLM).
# Read six CEL files to create the object named "rat.data."
    >rat.data = ReadAffy().
# Save the affybatch object named "rat.data." "rat.
data.Rdata" can be used later by loading ("rat.
data.Rdata").
    >save(rat.data, file = "rat.data.Rdata").
# Calculate the fitness of the model as described in the
affyPLM manual.
    >Pset.rat.data = fitPLM(rat.data).
# Create the reconstructed images of the CEL files (Fig. 2):
    >image(Pset.rat.data).
# Create a histogram of log-transformed probe intensity
(Fig. 3A):
    >hist(rat.data).
# Create a box and whisker plot of log-transformed probe
intensity (Fig. 3C).
#Adjust the graph margin for the sample names:
    >par(mar = c(16.4,4.1,4.1,4.1)).
    >boxplot(rat.data, notch = TRUE, boxwex = 0.3, boxfill
= c(2,2,2,3,3,3)).
# Calculate the mean intensity by its location within the
probe set:
    >AffyRNAdeg = AffyRNAdeg(rat.data).
```

```
# Plot the above results (Fig. 3D):
>plotAffyRNAdeg(AffyRNAdeg, cols = c(2,2,2,3,3,3)).
```

9. For a detailed description and the theory behind this, please refer to the *affyPLM* manual (http://www.bioconductor.org/repository/devel/vignette/QualityAssess.pdf).
10. The PMT is described in **Note 5**. Two CEL files that had been scanned with 100% PMT setting were downloaded from the CardioGenomics website (http://cardiogenomics.med.harvard.edu/groups/proj1/pages/download_band.html) for the purpose of demonstrating the histogram with saturated CEL files.
11. dChip (http://www.dchip.org) provides a detection algorithm for the probe, probe set, and array outliers *(38)* in a similar way.
12. The Affymetrix default algorithm for background correction regards the lowest 2% signal intensities in each subregion as a local background signal. The background value is computed as a weighted sum of the background values of the neighbor zones with the weight being inversely proportional to the square of the distance to a given zone. The current version (version 1.5) of the Bioconductor *affy* package offers four options for background correction: i.e., "mas," "none," "rma," and "rma2."
13. The Bioconductor *affy* library provides seven options for normalization; "constant," "contrasts," "invariantset," "loess," "qspline," "quantiles," and "quantiles.robust." The *vsn* library provides a variance-stabilizing normalization method that can be applied to the probe level and the probe set level *(39)*.
14. "Expression index calculation" refers to a procedure that summarizes 11 to 16 probe-pair intensities to one number, which corresponds to the abundance of a transcript.
15. The background correction of GCRMA is an integrated part of the function gcrma(). This function gcrma() is the same as rma() except for the background correction algorithm. The *gcrma* library must be loaded first to use this function.
16. **Figure 4A** was created with the following commands. PM refers to Perfect Match probes.

```
# Create an MvA plot using PM intensities of two control
samples:
    >mva.pairs(pm(rat.data[,c(1,3)])).
# Correct the background using the method "mas" that is
the implementation of Affymetrix method:
    >bgcorrected.rat.data = bg.correct(rat.data, method = "mas").
# Create an MvA plot with the background-corrected PM
intensities:
    >mva.pairs(pm(bgcorrected.rat.data[,c(1,3)])).
```

17. Quantile normalization transforms the distribution of the probe intensities so they are the same.

```
# Create the histogram of intensities:
    >hist(rat.data).
```

Statistical Methods in Cardiac Gene Expression Profiling 99

```
# Normalize the probe intensities using the quantile nor-
malization algorithm:
    >normalized.rat.data = normalize(rat.data, method =
    "quantiles").
# Create a histogram of the quantile normalized intensities:
    >hist(normalized.rat.data, lwd = 2).
```

18. The Bioconductor *affy* library provides the function for various expression methods.

    ```
    # Calculating MBEI as an expression index. The rat.data
    was created and saved from Note 8. Change the working direc-
    tory to where "rat.data.Rdata" is located:
        >library(affy).
        >load("rat.data.Rdata").
        >mbei.rat.data = expresso(rat.data, bg.correct = FALSE,
        pmcorrect.method = "pmonly," normalize.method = "invariantset,"
        summary.method = "liwong").
    # Write the output to a tab-delimited text file:
        >write.exprs(mbei.rat.data, "mbei.rat.data.txt").
    ```

19. To calculate the RMA as an expression index:

    ```
    # Change the working directory to where the CEL files are
    located:
        >library(affy).
        >rma.rat.data = justRMA().
    # Write the output to the tab-delimited text file:
        >write.exprs(rma.rat.data, "rma.rat.data.txt").
    ```

20. To calculate the GCRMA as an expression index:

    ```
    # Change the working directory to where the CEL files are
    located:
        >library(gcrma).
        >gcrma.rat.data = justGCRMA().
    # Write the output to a tab-delimited text file:
        >write.exprs(gcrma.rat.data, "gcrma.rat.data.txt").
    ```

21. These methods were developed with the spiked-in and dilution data set from Affymetrix and GeneLogic *(40)*. One of the limitations of these data sets is that only limited numbers of cRNAs are present in samples. For the Latin square data set of Affymetrix, 42 cRNAs of known concentration were spiked in; for the GeneLogic data set, only 11 cRNAs were spiked in.

22. This conversion of q-value from p-value can be achieved with the Bioconductor *qvalue* library. It also provides a graphical user interface.

23. The *DEDS* package is available at http://www.biostat.ucsf.edu/jean/DEDS.htm or can be installed using the script:

    ```
    >install.packages("DEDS",contriburl="http://www.biostat.
    ucsf.edu/jean/software").
    ```

24. With the *DEDS* library, various statistical scores can be calculated.

```
# Import data created in Note 19 into R.
  >rat.data = read.delim("rma.rat.data.txt", row.names = 1).
# Make the list of probe set IDs as row names:
  >rat.data.gnames = rownames(rat.data).
# Define the class assigned to each column, i.e., assign
each sample to a class. The treatment group is used in
this example. First we compare two groups:
  >rat.data.L = c(0,0,0,1,1,1).
# B represents the number of permutations. The adj option
can be "fdr" for calculating the false discovery rate
(FDR) or "adjp" for controlling for the family wise error
rate (FWER):
  >deds.rat.data = deds.stat.linkC(rat.data, rat.data.L,
B = 1000, adj = "fdr").
# With the combined statistical scores, we can list the
top ranked genes by significant order. "genelist" is the
unique identifier:
  >topgenes(deds.rat.data, number = 20, genelist =
rat.data.gnames).
# Calculate the number of genes that have smaller p-values
than 0.01:
  >sum(deds.rat.data$p <0.01).
# Calculate the number of genes that have smaller p-values
than 0.05:
  >sum(deds.rat.data$p <0.05).
# Otherwise, Student's t-statistic can be calculated sepa-
rately in case you are only interested in the t-score:
  >t.score = comp.stat(rat.data, rat.data.L, test = "t").
```

25. The gene expression profiles of normal tissues are available at GNF (http://symatlas.gnf.org). In this example, expression profiles of 79 normal tissues are used to generate hierarchical trees *(41)*. The expression data are available for download at http://wombat.gnf.org/downloads/GNF1Hdata.zip. This file must be downloaded and unzipped locally.

```
# Change the working directory to where the GNF1Hdata.txt
is located. Load the expression data:
  >gnf.normal.tissue = read.delim("GNF1Hdata.txt",
row.names = 1).
# Load the cluster library:
  >library(cluster).
# Transpose rows and columns:
  >transposed.gnf = t(gnf.normal.tissue).
```

```
# Cluster the samples with the hierarchical method. First,
calculate the distance matrix with the dist() function.
The default distance metric is the Euclidian distance:
   >dist.matrix = dist(transposed.gnf, method = "euclidian").
# Calculate the average linkage hierarchical clustering:
   >hclust.average.gnf = hclust(dist.matrix, method =
"average").
   >plot(hclust.average.gnf, labels = FALSE, main = "Average linkage").
# Calculate the complete linkage hierarchical clustering:
   >hclust.complete.gnf = hclust(dist.matrix, method =
"complete").
   >plot(hclust.complete.gnf, labels = FALSE, main = "Complete linkage").
# Calculate the single linkage hierarchical clustering:
   >hclust.single.gnf = hclust(dist.matrix, method =
"single").
   >plot(hclust.single.gnf, labels = FALSE, main = "Single linkage").
```

26. The software is available at http://www/tm4.org/mev.html. It is written in Java language to support multiple operating systems.
27. The data are available for download at ftp://ftp.ncbi.nih.gov/pub/geo/data/geo/by_series/GSE803_family.soft.gz.
28. Since there are multiple probes per probe set in GeneChip, it is not always possible to design unique probes. Thus some probes are not specific but are shared by several probe sets. This might cause a certain amount of cross-hybridization. This uniqueness can be identified by the probe set ID. xxxxxx_at is used for a unique probe set. xxxxxx_x_at and xxxxxx_s_at refer to a nonunique probe design. However, the naming schemes vary among the different GeneChip types. A complete description is available in the Data Analysis Fundamentals Manual (http://www.affymetrix.com/Auth/support/downloads/manuals/data_analysis_fundamentals_manual.pdf)
29. The annotation data package has annotation information (http://www.bioconductor.org/data/metaData.html). For instance, the annotations of HG-U133A GeneChip can be retrieved as follows:

```
>library(hgu133a).
>hgu133a().
Quality control information for hgu133a.
Date built: Wed Sep 15 18:48:08 2004.
Number of probes: 22283.
Probe number mismatch: none.
Probe mismatch: none.
```

```
Mappings found for probe-based rda files:
    hgu133aACCNUM found 22283 of 22283.
    hgu133aCHRLOC found 19448 of 22283.
    hgu133aCHR found 20998 of 22283.
    hgu133aENZYME found 2360 of 22283.
    hgu133aGENENAME found 21042 of 22283.
    hgu133aGO found 16833 of 22283.
    hgu133aGRIF found 0 of 22283.
    hgu133aLOCUSID found 21505 of 22283.
    hgu133aMAP found 20918 of 22283.
    hgu133aOMIM found 14579 of 22283.
    hgu133aPATH found 4382 of 22283.
    hgu133aPMID found 20616 of 22283.
    hgu133aREFSEQ found 20468 of 22283.
    hgu133aSUMFUNC found 0 of 22283.
    hgu133aSYMBOL found 21042 of 22283.
    hgu133aUNIGENE found 20546 of 22283.
Mappings found for non-probe-based rda files:
    hgu133aCHRLENGTHS found 25.
    hgu133aENZYME2PROBE found 643.
    hgu133aGO2ALLPROBES found 5417.
    hgu133aGO2PROBE found 4032.
    hgu133aORGANISM found 1.
    hgu133aPATH2PROBE found 132.
    hgu133aPMID2PROBE found 62651.
```

30. GenMAPP is available for download at http://www.GenMapp.org/.
31. FatiGO is a web-based tool that is available at http://fatigo.bioinfo.cnio.es/.
32. The Database for Annotation, Visualization, and Integrated Discovery 2.0 (DAVID 2.0) is a web-based tool available at http://david.niaid.nih.gov/david/version2/index.htm. Expression Analysis Systematic Explorer (EASE) is a stand-alone application for use on Windows operating systems that includes a function that will download and parse the annotation from a public database.
33. GeneMerge is available for both online use and as a stand-alone package that can be installed locally. It is available at http://www.oeb.harvard.edu/hartl/lab/publications/GeneMerge/GeneMerge.html.
34. A comprehensive list of web-based and downloadable tools for microarray data analysis is available at http://www.geneontology.org/GO.tools.microarray.shtml.

Acknowledgments

The author would like to thank Dr. Peter J. Park for critical reading and support. This work was supported in part by NIH grant 5P50HL074734-02.

References

1. Janicki, S. M., Tsukamoto, T., Salghettie, S. E., et al. (2004) From silencing to gene expression: real-time analysis in single cells. *Cell* **116,** 683–698.
2. Storch, K. F., Lipan, O., Leykin, I., et al. (2002) Extensive and divergent circadian gene expression in liver and heart. *Nature* **417,** 78–83.
3. Kerr, M. K. and Churchill, G. A. (2001) Experimental design for gene expression microarrays. *Biostatistics* **2,** 183–201.
4. Wilson, C. L., Pepper, S. D., Hey, Y., and Miller, C. J. (2004) Amplification protocols introduce systematic but reproducible errors into gene expression studies. *Biotechniques* **36,** 498–506.
5. Zien, A., Fluck, J., Zimmer, R., and Lengauer, T. (2003) Microarrays: how many do you need? *J. Comput. Biol.* **10,** 653–667.
6. Affymetrix technical note. (2004) *GeneChip® Expression Platform: Comparison, Evolution, and Performance.*
7. Affymetrix technical manual. (2004) *GeneChip Expression Analysis, Data Analysis Fundamentals.* http://www.affymetrix.com/support/downloads/manuals/data_analysis_fundamentals_manual.pdf.
8. Geller, S. C., Gregg, J. P., Hagerman, P., and Rocke, D. M. (2003) Transformation and normalization of oligonucleotide microarray data. *Bioinformatics* **19,** 1817–1823.
9. Freudenberg, J., Boriss, H., and Hasenclever, D. (2004) Comparison of preprocessing procedures for oligo-nucleotide micro-arrays by parametric bootstrap simulation of spike-in experiments. *Methods Inf. Med.* **43,** 434–438.
10. Wu, Z. and Irizarry, R. A. (2004) Preprocessing of oligonucleotide array data. *Nat. Biotechnol.* **22,** 656–658; author reply, 658.
11. Bolstad, B. M., Irizarry, R. A., Astrand, M., and Speed, T. P. (2003) A comparison of normalization methods for high density oligonucleotide array data based on variance and bias. *Bioinformatics* **19,** 185–193.
12. Yang, Y. H., Dudoit, S., Luu, P., et al. (2002) Normalization for cDNA microarray data: a robust composite method addressing single and multiple slide systematic variation. *Nucleic Acids Res.* **30,** e15.
13. Choe, S. E., Boutros, M., Michelson, A. M., Church, G. M., and Halfon, M. S. (2005) Preferred analysis methods for Affymetrix GeneChips revealed by a wholly defined control dataset. *Genome Biol.* **6,** R16.
14. Johnson, R. A. and Wichern, D. W. (2002) *Applied Multivariate Statistical Analysis,* 5th ed. Prentice Hall, Englewood Cliffs, NJ.
15. Bo, T. and Jonassen, I. (2002) New feature subset selection procedures for classification of expression profiles. *Genome Biol.* **3,** 17.
16. Dudoit, S., Yang, Y. H., Speed, T. P., and Callous, M. J. (2002) Statistical methods for identifying differentially expressed genes in replicated cDNA microarray experiments. *Statistica Sinica* **12,** 111–139.
17. Storey, J. D. and Tibshirani, R. (2003) Statistical significance for genomewide studies. *Proc. Natl. Acad. Sci. USA* **100,** 9440–9445.

18. Yang, Y. H., Xiao, Y., and Segal, M. R. (2005) Identifying differentially expressed genes from microarray experiments via statistic synthesis. *Bioinformatics* **21**, 1084–1093.
19. Tusher, V. G., Tibshirani, R., and Chu, G. (2001) Significance analysis of microarrays applied to the ionizing radiation response. *Proc. Natl. Acad. Sci. USA* **98**, 5116–5121.
20. Baldi, P. and Long, A. D. (2001) A Bayesian framework for the analysis of microarray expression data: regularized t-test and statistical inferences of gene changes. *Bioinformatics* **17**, 509–519.
21. Kim, R. D. and Park, P. J. (2004) Improving identification of differentially expressed genes in microarray studies using information from public databases. *Genome Biol.* **5**, R70.
22. Eisen, M. B., Spellman, P. T., Brown, P. O., and Botstein, D. (1998) Cluster analysis and display of genome-wide expression patterns. *Proc. Natl. Acad. Sci. USA* **95**, 14863–14868.
23. Tibshirani, R., Walther, G., and Hastie, T. (2001) Estimating the number of clusters in a dataset via the gap statistic. *J. R. Statist. Soc. B.* **63**, 411–423.
24. Yeung, K. Y., Haynor, D. R., and Ruzzo, W. L. (2001) Validating clustering for gene expression data. *Bioinformatics* **17**, 309–318.
25. Pomeroy, S. L., Tamayo, P., Gassenbeek, M., et al. (2002) Prediction of central nervous system embryonal tumour outcome based on gene expression. *Nature* **415**, 436–442.
26. Yanai, I., Benjamin, H., Shmoish, M., et al. (2005) Genome-wide midrange transcription profiles reveal expression level relationships in human tissue specification. *Bioinformatics* **21**, 650–659.
27. Liu, G., Loraine, A. L., Shigeta, R., et al. (2003) NetAffx: Affymetrix probesets and annotations. *Nucleic Acids Res.* **31**, 82–86.
28. Mootha, V. K., Lindgren, C. M., Eriksson, K. F., et al. (2003) PGC-1alpha-responsive genes involved in oxidative phosphorylation are coordinately down-regulated in human diabetes. *Nat. Genet.* **34**, 267–273.
29. Segal, E., Friedman, N., Koller, D., and Regev, A. (2004) A module map showing conditional activity of expression modules in cancer. *Nat. Genet.* **36**, 1090–1098.
30. Affymetrix technical note. (2002) *GeneChip Eukaryotic Samll Sample Target Labeling Assay Version II.* http://www.affymetrix.com/support/technical/technotes/smallv2_technote.pdf.
31. Baugh, L. R., Hill, A. A., Brown, F. L., and Huator, C. P. (2001) Quantitative analysis of mRNA amplification by in vitro transcription. *Nucleic Acids Res.* **29**, E29.
32. Eberwine, J., et al. (2001) mRna expression analysis of tissue sections and single cells. *J. Neurosci.* **21**, 8310–8314.
33. Tietjen, I., Rihel, J. M., Cao, Y., Koentges, G., Zakhary, L., and Dulac, C. (2003) Single-cell transcriptional analysis of neuronal progenitors. *Neuron* **38**, 161–175.
34. Kong, S. W., Hwang, K. B., Kim, R. D., et al. (2005) CrossChip: a system supporting comparative analysis of different generations of Affymetrix arrays. *Bioinformatics* **21**, 2116–2117.

35. Park, P. J., Cao, Y. A., Lee, S. Y., et al. (2004) Current issues for DNA microarrays: platform comparison, double linear amplification, and universal RNA reference. *J. Biotechnol.* **112,** 225–245.
36. Yuen, T., Wurmbach, E., Pfeffer, R. L., Ebersole, B. J., and Sealfon, S. C. (2002) Accuracy and calibration of commercial oligonucleotide and custom cDNA microarrays. *Nucleic Acids Res.* **30,** e48.
37. Kong, S. W., Bodyak, N., Yue, P., et al. (2005) Genetic expression profiles during physiological and pathological cardiac hypertrophy and heart failure in rats. *Physiol. Genomics* **21,** 34–42.
38. Li, C. and Wong, W. H. (2001) Model-based analysis of oligonucleotide arrays: expression index computation and outlier detection. *Proc. Natl. Acad. Sci. USA* **98,** 31–36.
39. Huber, W., von Heydebreck, A., Sultmann, H., Poustka, A., and Vingron, M. (2002) Variance stabilization applied to microarray data calibration and to the quantification of differential expression. *Bioinformatics* **18 Suppl 1,** S96–104.
40. Cope, L. M., Irizarry, R. A., Jaffee, H. A., Wu, Z., and Speed, T. P. (2004) A benchmark for Affymetrix GeneChip expression measures. *Bioinformatics* **20,** 323–331.
41. Su, A. I., Cooke, M. P., Ching, K. A., et al. (2002) Large-scale analysis of the human and mouse transcriptomes. *Proc. Natl. Acad. Sci. USA* **99,** 4465–4470.
42. Draghici, S. (2003) Data analysis tools for DNA microarrays. Chapman & Hall/CRC, Boca Raton, FL, p. 393.
43. Reimers, M. and Weinstein, J. N. (2005) Quality assessment of microarrays: visualization of spatial artifacts and quantitation of regional biases. *BMC Bioinformatics* **6,** 166.

II

CARDIAC GENE REGULATION
GENE-SPECIFIC mRNA MEASUREMENT IN THE MYOCARDIUM

6

Measurement of Cardiac Gene Expression by Reverse Transcription Polymerase Chain Reaction (RT-PCR)

Nicola King

Summary

Two methods applicable to the measurement of cardiac gene expression by reverse transcription polymerase chain reaction (RT-PCR) are presented. Each method describes a specific technique and includes information on how to optimize the experiments and how to trouble-shoot any problems. The first method illustrates how conventional RT-PCR was used to detect expression of a specific gene in rat heart. The second method explains how to investigate the amount of gene expressed under different conditions utilizing the new technique of quantitative real-time RT-PCR. More specifically, this second method was used to investigate the effect of a relevant cardiac insult, ischemia reperfusion, on gene expression, although the technique could be readily adapted to other cardiac insults or to compare different hearts or treatments. The gene investigated in these examples was *SNAT3*, which is responsible for the cotransport of sodium predominantly with glutamine in exchange for hydrogen ions. The results suggested that a 20-fold increase in *SNAT3* expression occurs during ischemia, which reduced to baseline levels on reperfusion.

Key Words: RT-PCR; quantitative real-time RT-PCR; LightCycler™; cardiac gene expression; ischemia reperfusion; amino acid transport; *SNAT3*.

1. Introduction

Probably the two most common and important reasons for applying reverse transcription polymerase chain reaction (RT-PCR) to investigations of cardiac gene expression are detection and quantification. The aim in detection is to address the question of whether mRNA from a specific gene is expressed in heart. The attraction of answering this using RT-PCR arises from the greater sensitivity of this technique compared with other methods such as Northern

From: *Methods in Molecular Biology, vol. 366: Cardiac Gene Expression: Methods and Protocols*
Edited by: J. Zhang and G. Rokosh © Humana Press Inc., Totowa, NJ

blotting or RNase protection assay *(1)*. Following on from this, the goal in quantification is to identify how much of the gene is present and whether this alters under various therapeutic or stressful conditions or between different hearts. In this case the advantage of RT-PCR is that for a time during the reaction there is linear amplification in which the amount of product is directly proportional to the amount of starting material *(1–7)*. This type of measurement has become more accessible in recent years following the development of novel thermocyclers with built-in fluorimeters that allow the investigator to track the reaction as it happens *(2–7)*.

The methods described in this chapter chart the application of conventional RT-PCR to detect gene expression in the rat heart and real-time quantitative RT-PCR to measure possible changes to that gene's expression during ischemia reperfusion. The examples presented investigate *SNAT3*, also known as *SN1* or, in humans, as *SLC38A3 (8,9)*. This gene encodes an electroneutral amino acid transporter, which mediates the cotransport of sodium predominantly with glutamine in exchange for hydrogen ions *(8,9)*. In other words, it is one of the genes responsible for the classically described system N amino acid transport *(8)*. The methods described are, however, equally appropriate to any gene that excites the cardiac researcher's interest.

2. Materials

2.1. Equipment for Conventional RT-PCR

1. Pipets capable of dispensing volumes <1 µL.
2. Sterile filter pipet tips.
3. Benchtop centrifuge (preferably refrigerated).
4. Thermocycler (e.g., MJ Research PTC-200 Peltier Thermal Cycler).
5. Electrophoresis equipment.
6. UV transilluminator.
7. Whirly mixer.
8. Heating block.
9. Mortar and pestle.
10. Spectrophotometer capable of measuring at 260 and 280 nm.
11. 20-Gage needles.
12. 1-mL Syringes.
13. Sterile 1.5- and 0.5-mL Eppendorf tubes.
14. Male Wistar rats (250–300 g).
15. –70°C Freezer.

2.2. Solutions, Kits, and Reagents for Conventional RT-PCR

1. Liquid nitrogen.
2. RNeasy® Fibrous Tissue Mini Kit (Qiagen, cat. no. 74704).
3. TriReagent (Sigma, cat. no. T9424).

4. RETROscript™ 1st Strand Synthesis Kit for RT-PCR (Ambion, cat. no. 1710).
5. PCR Master Mix (Promega, cat. no. M7502).
6. Molecular weight ladder (Roche XIV).
7. 1% Agarose gel made in 1X TAE buffer containing 0.5 µg/mL ethidium bromide. 50X TAE buffer is prepared by combining 50 mL 0.5 M EDTA (pH 8.0) solution with 2 M Tris(hydroxymethyl)aminomethane (TRIS) and 28 mL glacial acetic acid in a total volume of 500 mL.

2.3. Additional Materials for Real-Time Quantitative RT-PCR

1. LightCycler™ with associated computer and software (e.g., Roche).
2. LightCycler Centrifuge adaptors supplied in an aluminum cooling block (Roche, cat. no. 1 909 312).
3. LightCycler Capillaries (Roche, cat. no. 1 909 339).
4. Langendorff perfusion system.

2.4. Additional Solutions, Kits, and Reagents for Real-Time Quantitative RT-PCR

1. QuantiTect SYBR Green I (Q-SYBR) PCR Master Mix (Qiagen, cat. no. 204143).
2. QIAquick Gel Extraction kit (Qiagen, cat. no. 28704).
3. Krebs solution containing: 118 mM NaCl, 25 mM NaHCO$_3$, 4.8 mM KCl, 1.2 mM KH$_2$PO$_4$, 1.2 mM MgSO$_4$, 11 mM glucose, and 1.2 mM CaCl$_2$.

3. Methods

3.1. Conventional RT-PCR: Detection of SNAT3 RNA in Rat Heart

1. Rats were killed by cervical dislocation.
   a. Total RNA was extracted from rat heart using the RNeasy Fibrous Tissue Mini Kit according to the manufacturer's instructions.
   b. Total RNA from rat brain was extracted using TriReagent also according to the manufacturer's instructions (*see* **Notes 1** and **2**).
   c. The amount of starting material was 10 to 20 mg, and the tissue was initially crushed under liquid nitrogen using a mortar and pestle followed by homogenization with a 20-gage needle and syringe.
   d. The end point of this procedure was a 30 µL suspension of RNA in RNase-free water.
2. The absorbance of 2 µL of the RNA suspension diluted into 100 µL Rnase-free water was measured at 260 and 280 nm (*see* **Note 3**). The optical density ratio OD$_{260}$/OD$_{280}$ provided an indication of the purity of isolation, with a ratio over 1.7 considered acceptable.
3. The concentration of RNA was calculated from the equation:

$$\frac{40 \times OD_{260} \times 100}{2} = \text{ng RNA/µL}$$

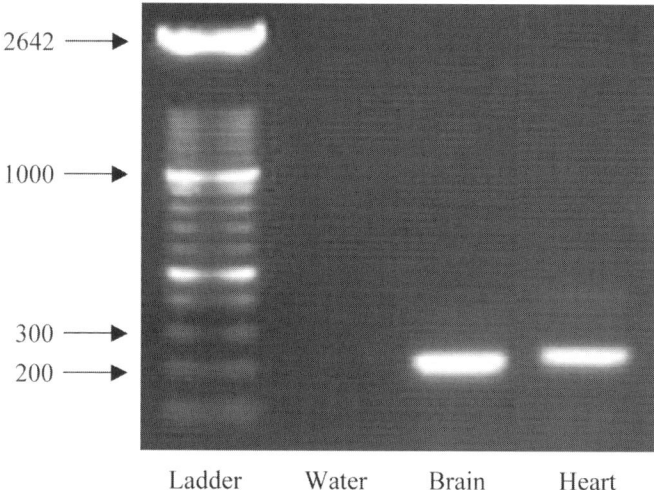

Fig. 1. Expression of *SNAT3* in the rat heart. Photograph of products taken after conventional RT-PCR using primers directed against *SNAT3* and gel electrophoresis. The single band present for samples of rat heart and brain (used as a positive control) suggested products of approximately the correct length as indicated by the DNA ladder. The numbers on the left indicate sizes of bands on the ladder in base pairs.

4. The RNA was reverse-transcribed into complementary DNA using the RETROscript 1st Strand Synthesis Kit for RT-PCR. This was carried out according to the manufacturer's instructions wherein Method 1a was followed, oligo(dT) was used as the primer, and the starting amount of RNA was 1 to 2 μg (*see* **Note 4**). The samples were stored at –20°C until use.
5. PCR was accomplished using the PCR Master Mix, which contained nuclease-free water, *Taq* DNA polymerase in a proprietary buffer, dNTPs, and $MgCl_2$.
6. A cocktail was prepared containing sufficient volume for four samples (heart, brain, water as a negative control, and a spare [*see* **Note 5**]) comprising: 25 μL PCR Master Mix per sample, 22 μL nuclease-free water per sample (provided in the Promega kit), 1 μL forward primer per sample, and 1 μL reverse primer per sample (*see* **Note 6**).
7. Then 49 μL of the cocktail was pipeted into three individual PCR tubes, 1 μL of the cDNA sample or water was added to each tube, and the samples were centrifuged briefly to settle their contents.
8. These tubes were placed in a thermocycler preset with the following program (*see* **Note 7**): 94°C for 3 min; 35 cycles of 94°C for 30 s, 58°C for 45 s, and 72°C for 45 s; and finally 72°C for 7 min.
9. At the end of the PCR reaction, 10 μL gel loading dye (provided in the RETROscript kit) was added to each tube.

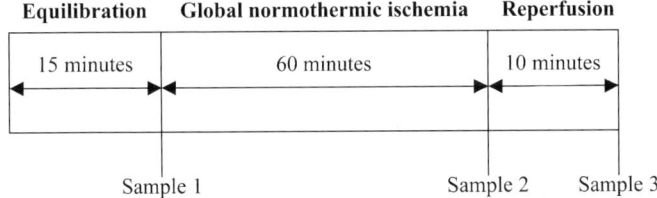

Fig. 2. Illustration of the protocol used to expose isolated and perfused rat hearts to ischemia and reperfusion. Samples 1 to 3 were snap-frozen in liquid nitrogen and stored at −70°C until use.

10. The entire tube contents were electrophoresed at 100 V for approx 1 h with an ethidium bromide-stained 1% agarose gel prepared in 1X TAE solution with an appropriate molecular weight ladder.
11. The gel was visualized on a UV transilluminator and photographed (e.g., AlphaImager™ 2200 and software version 5.5) (*see* **Note 8** and **Fig. 1**).

3.2. Quantitative Real-Time RT-PCR:
Effect of Ischemia Reperfusion on SNAT3 Expression in Rat Heart

3.2.1. Sample Preparation

1. Rats were killed by cervical dislocation and their hearts excised into ice-cold Krebs solution. Hearts were rapidly cannulated via the aorta and perfused in the Langendorff mode *(10)*. The time-course followed, together with the points selected for sampling, are shown in **Fig. 2** (*see* **Note 9**).
2. Total RNA was extracted using the RNeasy Fibrous Tissues Mini Kit as described in **Subheading 3.1., step 2** above.
3. The amount of RNA in each sample was quantified by spectrophotometry as outlined in **Subheading 3.1., step 3** above.
4. The samples were reverse-transcribed using the RETROscript 1st Strand Synthesis Kit for RT-PCR as described in **Subheading 3.1., step 4** above. The only and a very important exception was that the amount of RNA added to the reverse transcription reaction was exactly 1.3 µg (*see* **Note 10**).

3.2.2. Primer Quality Control and Standard Preparation

1. A cocktail was prepared on ice (*see* **Note 11**) for three samples (heart, water as a negative control, and a spare). This contained: 5 µL 2X Q-SYBR PCR Master Mix per sample, 0.5 µL forward primer per sample, 0.5 µL reverse primer per sample, and 2 µL RNase-free water (supplied in the Qiagen kit) per sample.
2. Then 8 µL of this cocktail was pipeted into the bowl (*see* **Note 12**) of two pre-cooled LightCycler capillary tubes.
3. Then 2 µL heart cDNA (this can be the same cDNA as prepared in **Subheading 3.1., step 4** above or freshly prepared cDNA) or water was added to each capillary tube, and the capillary was capped.

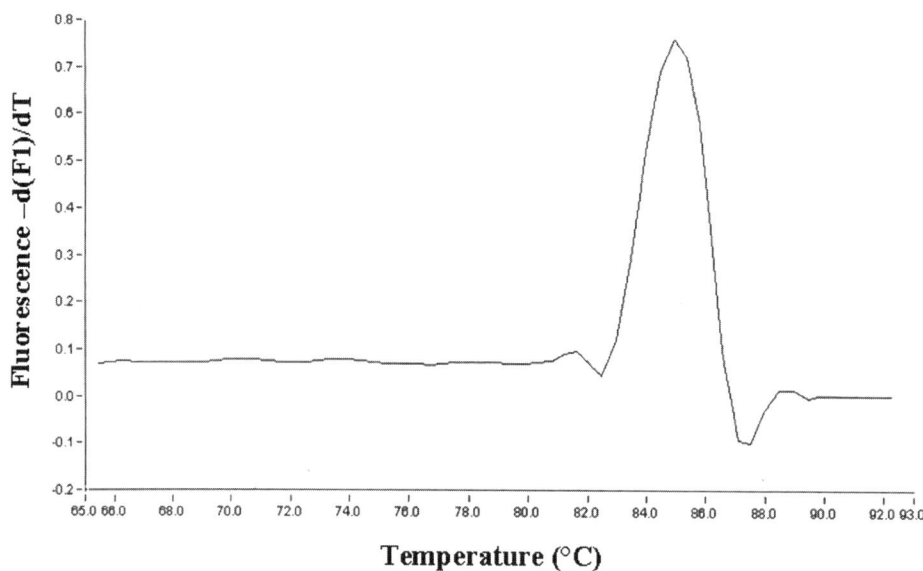

Fig. 3. Melting curve analysis following real-time quantitative RT-PCR of *SNAT3* from rat heart. An integration of fluorescence and temperature plotted against temperature as provided by the LightCycler software. Peaks occur when SYBR Green I dye was released from the double-stranded DNA products upon denaturing during the melting cycle. This sample showed one peak indicative of a single product (*see* **Note 12**).

4. The capillary tubes were centrifuged briefly (10–20 s) at low speed (200*g*).
5. The tubes were placed into the LightCycler and run with the following programs:
   a. Amplification cycle: 95°C for 15 min and the 50 cycles of 95°C for 15 s, 58°C for 20 s, and 72°C for 15 s. A single fluorescence acquisition at 72°C was used in each cycle.
   b. Melting program: 95°C for 0 s, 63°C for 30 s; **step 3b**, heat to 95°C at 0.1°C/s and hold for 0 s; 40°C for 30 s. Continuous fluorescence acquisition was used during **step 3b**.
6. The results of the melting program were analyzed (**Fig. 3**).
7. The LightCycler capillary tubes were removed from the LightCycler, their caps were removed, and the tubes were placed upside down in 1.5-mL Eppendorf tubes. These were centrifuged briefly (10–20 s at 200*g*).
8. Then 2 µL gel loading dye (from RETROscript) was added to each tube followed by electrophoresis on an ethidium bromide-stained 1% agarose gel.
9. A scalpel was used to excise the single-band product at approx 200 bp from the gel.
10. The DNA contained in that gel slice was extracted and purified using the QIAquick Gel Extraction Kit in accordance with the manufacturer's instructions.

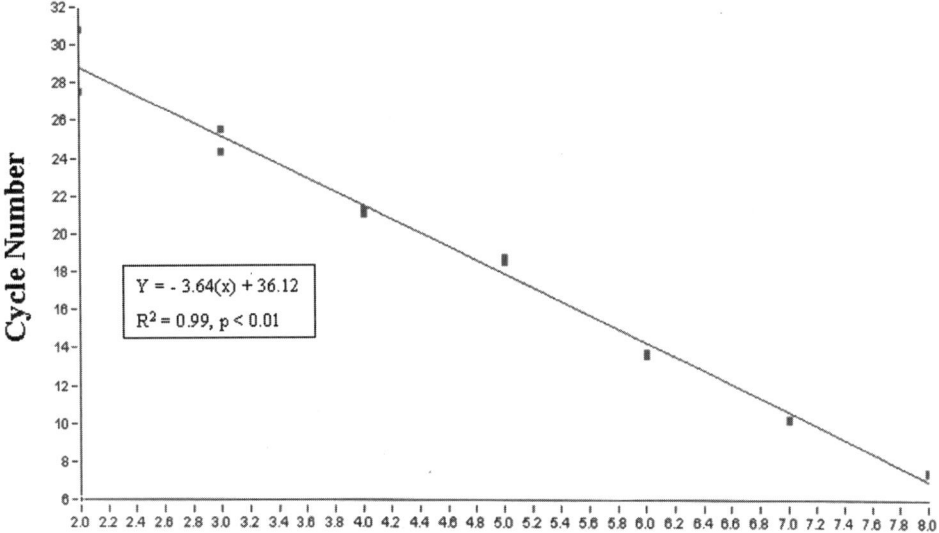

Fig. 4. Amplification of *SNAT3* in standards and test samples during quantitative real-time RT-PCR. During the reaction the level of fluorescence in each trace is proportional to the amount of product. Each trace's pattern was examined with the LightCycler software to determine the best single level of fluorescence (called the crossing point), which occurred during the linear growth phase in every sample. The different number of cycles required for each standard to achieve this level (as shown in **Fig. 5**) will form the basis for calculating the number of *SNAT3* copies in each test sample. This is a black and white reproduction. The LightCycler software assigns each sample a different color, which is then used throughout the experiment and during analysis.

11. The absorbance of 2 µL of the purified DNA diluted into 100 µL nuclease-free water was measured at 260 and 280 nm. From this the number of copies of *SNAT3* DNA in the sample was determined according to the following:

$$\frac{OD_{260} \times 50 \times 100}{2} = \text{ng DNA/µL}$$

Avogadro's number = 6.02×10^{23}.
∴ 1 mole = 6.02×10^{23} molecules = molecular weight of PCR product (in grams).
The molecular weight of the *SNAT3* PCR product based on the average weight of a base pair and the length of the PCR product was $660 \times 211 = 139{,}260$ g.

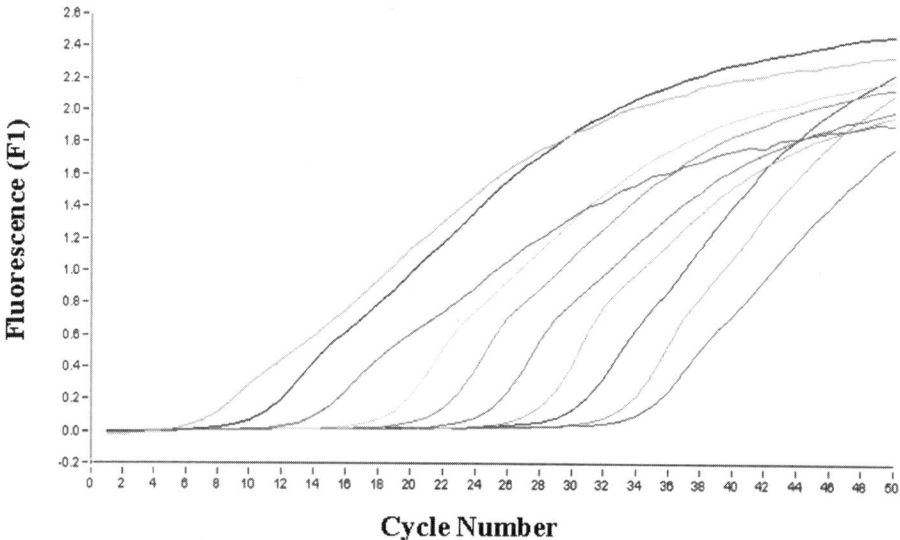

Fig. 5. Standard curve for *SNAT3*. There was a linear relationship between the number of cycles required to reach the crossing point and the number of *SNAT3* copies in each standard. The equation describing this relationship was used to calculate the number of *SNAT3* copies in each of the test samples.

$\therefore$ 139,260 g = 6.02 × 10^{23} molecules/copies.
$\therefore$ 1 g = 6.02 × 10^{23}/139,260 = 4.32 × 10^{18} copies.
$\therefore$ 1 ng = 4.33 × 10^{9} copies.

12. Based on these calculations, a top standard containing 1×10^8 copies was prepared. This top standard was used in serial 10-fold dilutions to obtain a series of standards from 1×10^8 to 1×10^2 (*see* **Note 14**).

3.2.3. Quantitative Real-Time PCR of Samples and Standards

1. This reaction involved 11 samples: the three samples prepared in **Subheading 3.2.2.**; a negative control; and 7 standards (1×10^8, 1×10^7, 1×10^6, 1×10^5, 1×10^4, 1×10^3, and 1×10^2 copies).
2. Therefore a cocktail was prepared for 12 samples comprising: 0.5 µL forward primer per sample, 0.5 µL reverse primer per sample, 5 µL Q-SYBR mix per sample, and 2 µL DNase free water (provided in the Q-SYBR Kit) per sample.
3. Then an 8 µL cocktail was pipeted into the bowls of 11 LightCycler capillary tubes.
4. Next, 2 µL of the relevant sample was added to each LightCycler capillary tube, and the tubes were centrifuged briefly at 200*g* for 10 to 20 s.
5. The tubes were placed into the LightCycler and run with the identical amplification and melting programs as described in **Subheading 3.2.2., step 5** above.

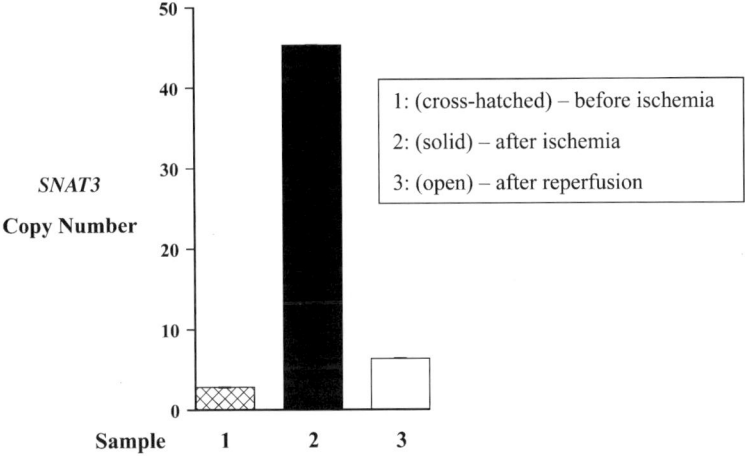

Fig. 6. Effect of ischemia reperfusion on *SNAT3* expression in rat heart. These results suggest that *SNAT3* expression is induced during ischemia.

3.2.4. Analysis

1. At the end of the PCR, the melting curves were analyzed for peak number.
2. Graphs of the time-course of the reaction and the standard curve were obtained using the LightCycler software (*see* **Figs. 4** and **5**).
3. The data in these graphs were used in the LightCycler software to calculate the number of copies of *SNAT3* DNA in the samples taken before and after ischemia and at the end of reperfusion (**Fig. 6**).

4. Notes

1. Several of the procedures described here used kits. In addition to supplying most or all of the necessary reagents, these kits were always accompanied by comprehensive instructions, trouble-shooting guides, and links to the companies' web sites for more information and technical advice. This additional support was very helpful.
2. The RNeasy Fibrous Tissue Mini Kit is specifically designed for tissues containing abundant contractile proteins, connective tissue, and collagen such as heart but not brain. Attempts to isolate total RNA from brain using this kit resulted in poor yields and low purity; therefore RNA from rat brains was isolated using TriReagent (Sigma).
3. When measuring the absorbance, it was necessary to mix the RNA or DNA sample thoroughly before placing the cuvet into the spectrophotometer in order to obtain a stable reading. This was particularly important with regard to the real-time quantitative RT-PCR, in which an equal amount of RNA was reverse-transcribed for each sample.

4. During reverse transcription using the RETROscript kit, the initial heat was 82°C. (A range of 70–85°C was suggested by the kit instructions with the higher temperatures required for RNAs with abundant secondary structure or a high G-C content.) The 1-h incubation was carried out at 42°C. (A range of 40–44°C was suggested by the kit instructions with the possibility of increasing to 50°C if the results were poor.)
5. For each experiment, the cocktail contained sufficient volume for all the samples to be tested plus one. This precaution shielded the investigator from small inaccuracies in sample handling.
6. Primer sequences were designed with the aid of the web site, http://FRODO.wi.mit.edu. This was achieved by searching with the complete coding sequence of rat *SNAT3* (accession number NM 057139), as obtained from http://www.ncbi.nlm.nih.gov/entrez/query.fcgi?db=Nucleotide. The only constraint, which was adopted during the search, was the length of the desired product. This was set according to the requirements of real-time quantitative RT-PCR, which ideally needed a product size of 50 to 250 bp. The primers were received lyophilized, from which a concentrated stock was prepared by dissolving the primers at 1 µg/µL in nuclease-free water. This concentrated stock was diluted 1 in 10 before use in the PCR. All stocks were stored at –20°.
7. The annealing temperature of 58°C was set at 5°C below the primer T_m. The primer T_m was included in the information supplied from the manufacturer, Genosys (Sigma).
8. Purification of the gel bands followed by processing for DNA sequencing could be used to confirm the products' identities.
9. During normal perfusion, this solution was oxygenated with 95% O_2/5% CO_2 (pH 7.4), delivered by pump to the heart at 11 mL/min and maintained at 37°C *(10,11)*. Global normothermic ischemia is achieved by switching off the pump with the heart bathed in solution. Draining the bathing solution and restarting the pump initiated reperfusion *(10,11)*.
10. Setting the level of RNA was considered to be a simple and reproducible option for standardizing the amount of sample entered into the real-time quantitative RT-PCR. Alternatively, the standardization could be achieved by measuring both the gene of interest and a housekeeping gene. The results would then be expressed relative to the amount of the housekeeper.
11. The cocktail was prepared on ice, because the Taq DNA polymerase contained in the Q-SYBR kit was "hot start."
12. The LightCycler capillaries consisted of a narrow tube inserted into a plastic bowl. The samples were dispensed into the bowl of the capillary and later centrifuged in order to settle the contents into the base of the narrow tube. These capillaries were fragile and fiddly to manipulate, requiring great care in their handling particularly during loading and unloading of the LightCycler carousel. They were, however, supplied with special centrifuge adaptors and a dedicated cooling block.
13. The appearance of more than one peak in the melting curve analysis was undesirable. Any extra peaks would suggest the formation of additional or contaminating products including, for example, primer dimers. In some cases these

extraneous products can be excluded from the measurement by adjusting the temperature of the fluorescence acquisition. This could, however, compromise the efficiency of the desired products' amplification since all products would be competing for the Taq DNA polymerase. We would solve the problem of extra peaks by repeating the experiment with different primers.
14. This procedure provided sufficient volume for several reactions. To obtain a durable and plentiful supply of standards, the DNA could be ligated into a vector and cloned into competent cells.

Acknowledgments

This work was supported by the British Heart Foundation, of which N. King is an Intermediate Research Fellow. The author would also like to thank Mrs. Hua Lin for her hard work and technical assistance, as well as Drs. John McGivan and M. Saadeh Suleiman for critical reading of the manuscript.

References

1. Bustin, S. A. (2000) Absolute quantification of mRNA using real-time reverse-transcription polymerase chain reaction assays. *J. Mol. Endocrinol.* **25,** 169–193.
2. Morrison, T. B., Weiss, J. J., and Wittwer, C. T. (1998) Quantitation of low-copy transcripts by continuous SYBR Green I monitoring during amplification. *Biotechniques* **24,** 954–958.
3. Wittwer, C. T., Ririe, K. M., Andrew, R. V., David, D. A., Gundry, R. A., and Balis, U. J. (1997) The LightCycler: a microvolume multisample fluorimeter with rapid temperature control. *Biotechniques* **22,** 176–181.
4. Wilhelm, J. and Pingoud, A. (2003) Real-time polymerase chain reaction. *Chembiochem* **4,** 1120–1128.
5. Mackay, I. M., Arden, K. E., and Nitsche, A. (2002) Real-time PCR in virology. *Nucleic Acids Res.* **30,** 1292–1305.
6. Klein, D. (2002) Quantification using real-time PCR technology: applications and limitations. *Trends Mol. Med.* **8,** 257–260.
7. Mocellin, S., Rossi, C. R., Pilati, P., Nitti, D., and Marincola, F. M. (2003) Quantitative real-time PCR: a powerful ally in cancer research. *Trends Mol. Med.* **9,** 189–195.
8. MacKenzie, B. and Erickson, J. D. (2004) Sodium-coupled neutral amino acid (System N/A) transporters of the SLC38 gene family. *Pflugers Arch.* **447,** 784–795.
9. Boulland, J. L., Rafiki, A., Levy, L. M., Storm-Mathisen, J., and Chaudhry, F. A. (2003) Highly differential expression of SN1, a bi-directional glutamine transporter, in astroglia and endothelium in the developing rat brain. *Glia* **41,** 260–275.
10. King, N., Lin, H., McGivan, J. D., and Suleiman, M.-S. (2004) Aspartate transporter activity in hypertrophic rat heart and ischaemia-reperfusion injury. *J. Physiol.* **556,** 849–858.
11. Javadov, S. A., Lim, K. H. H., Kerr, P. M., Suleiman, M.-S., Angelini, G. D., and Halestrap, A. P. (2000) Protection of hearts from reperfusion injury by propofol is associated with inhibition of the mitochondrial permeability transition. *Cardiovasc. Res.* **45,** 360–369.

7

Quantitative (Real-Time) RT-PCR in Cardiovascular Research

Kevin John Ashton and John Patrick Headrick

Summary

Quantitative (real-time) PCR (qPCR) represents a highly sensitive, sequence-specific, and reproducible technique for the gel-free detection and quantitation of nucleic acids. Owing to its large dynamic range and throughput, this approach has become the chosen method for rapid quantification of mRNA levels in biological samples. The sensitivity of this method permits the reliable detection of low concentrations of initial template, while delivering a linear range of up to 10 orders of magnitude in copy number. This chapter details the basic methodology behind key components of a qPCR experiment, including sample preparation, fluorescent chemistries, primer/probe design, and data analysis applicable to cardiovascular research.

Key Words: Real-time quantitative polymerase chain reaction; reverse transcription; SYBR green I; 5'-nuclease probes; TaqMan; molecular beacons; standard curve; comparative delta C_T; primer design.

1. Introduction

Quantitative (real-time) polymerase chain reaction (PCR) relies on the ability to measure the accumulation of PCR product during each amplification cycle. This is achieved by measuring the increase in a fluorescent signal, which is proportional to the amount of double-stranded DNA present. These fluorescent approaches include nonspecific intercalating dyes (e.g., SYBR Green I) or sequence-specific fluorescent oligonucleotide probes (e.g., 5' nuclease assay, molecular beacons). The advent of this technology allows the accurate quantitation of gene expression and single nucleotide polymorphism (SNP) analysis *(1–3)*. This chapter concentrates on the application of two-step quantitative (real-time) reverse transcriptase-PCR (qRT-PCR) for the measurement of gene expression in cardiac tissues (although appropriate for other tissues also).

From: *Methods in Molecular Biology, vol. 366: Cardiac Gene Expression: Methods and Protocols*
Edited by: J. Zhang and G. Rokosh © Humana Press Inc., Totowa, NJ

2. Materials
2.1. Equipment

1. Real-time PCR instrument (e.g., Bio-Rad iCycler/MyiQ, Applied Biosystems 7700/7000/7900HT, Stratagene Mx4000/Mx3000P, and others).
2. PCR primer design software (optional; e.g., PrimerExpress, Beacon Designer, Primer3, and others).
3. Mechanical rotor homogenizer.
4. Pipets: 20, 200, and 1000 µL.
5. UV spectrophotometer or fluorescent plate reader.
6. Water bath or heating block.

2.2. Reagents

1. Reaction tubes (2, 1.5, and 0.6 mL; DNA and RNA-free).
2. Pipet tips: aerosol-resistant filter-tipped for 20, 200, and 1000 µL.
3. Cardiac tissue or cells (snap-frozen in liquid nitrogen).
4. RNase decontaminant (e.g., RNaseZAP, RNase AWAY).
5. Phenol/GITC reagent (e.g., TRIzol, TRI reagent).
6. Chloroform.
7. Isopropanol (isopropyl alcohol).
8. Ethanol (ethyl alcohol).
9. DEPC-treated water (see **Note 1**).
10. RNasin ribonuclease inhibitor (40 U/µL).
11. 10X DNase buffer; rDNase I enzyme (2 U/µL), DNase Inactivation Reagent (Ambion).
12. 5X RT Buffer; recombinant MMLV-RT reverse transcriptase enzyme (e.g., SuperScript II/III, StrataScript, and others).
13. 100 mM Dithiothreitol (DTT).
14. TE Buffer (10 mM Tris-HCl, pH 7.5; 1 mM EDTA).
15. PicoGreen® and/or RiboGreen® quantitation reagents (Molecular Probes).
16. 2X Probe PCR Mastermix: 20 mM Tris-HCl, pH 8.4, 100 mM KCl (2X PCR buffer), 10 mM MgCl$_2$, 0.4 mM dNTPs (each), 0.5 U hot start *Taq* DNA polymerase, 2X passive reference dye (see **Note 2**).
17. 2X SYBR Green I PCR Mastermix: 20 mM Tris-HCl, pH 8.4, 100 mM KCl, 6 mM MgCl$_2$, 0.4 mM dNTPs (each), 0.5X SYBR Green I, 0.5 U hot start *Taq* DNA polymerase, 2X passive reference dye (see **Note 2**).
18. Primers: specific primers complementary to the 5' and 3' ends of the target sequence to be amplified must be suitable for real-time PCR application (see **Subheading 3.4.**), resuspended in TE buffer.
19. Fluorescently labeled oligonucleotide probe for use in 5'-nuclease or molecular beacon assays resuspended in TE buffer (see **Subheading 3.4.** and **Note 3**).
20. Nuclease-free water.
21. Optical 96-well PCR reaction plates or tubes.
22. Optical caps or sealing film.

3. Methods

The methods described below outline (1) RNA isolation, (2) cDNA synthesis, (3) choice of fluorescent chemistry, (4) primer/probe design, (5) preparation of PCR standards, (6) quantitative RT-PCR, (7) data analysis, (8) standard curve analysis, and (9) the comparative C_T method.

3.1. RNA Isolation

The following protocol is a modification of methods developed by Chomczynski and Sacchi utilizing a monophasic phenol/guanidine isothiocyanate (GITC) reagent (e.g., TRIzol, TRI reagent) *(4)*. We have used this procedure to isolate total RNA from whole mouse hearts (average weight 75–125 mg; RNA yield typically 90–120 µg). Phenol/GITC reagent should be used at a volume of 1 mL per 50 to 100 mg of tissue or 10^7 cells.

1. Decontaminate homogenizer, workbench, tube racks, and pipets with RNase decontaminant before use; wear gloves at all times.
2. Add 625 µL of phenol/GITC reagent to a 2-mL tube containing liquid nitrogen snap-frozen tissue/cells.
3. Homogenize heart completely for 4 to 5 min with a mechanical rotor homogenizer until no visible clumps remain. Add an additional 625 µL of phenol/GITC reagent, to yield a final volume of approx 1.25 mL.
4. Use a 3-mL syringe with an 18-gage needle to "syringe" homogenize tissue/cells further and shear genomic DNA.
5. Incubate at room temperature for 10 min to permit disruption of cells.
6. Centrifuge at 12,000g for 5 min at 4°C to pellet insoluble material.
7. Transfer supernatant to a fresh 2-mL tube.
8. Incubate at room temperature for 5 min to permit complete dissociation of nucleoprotein complexes.
9. Add 250 µL (0.2X vol) of chloroform, and shake vigorously by hand for 15 to 20 s. Incubate at room temperature for 5 min.
10. Centrifuge at 12,000g for 15 min at 4°C to separate phases (upper clear aqueous phase, interphase, lower red phenol/chloroform phase).
11. Collect as much as possible of the upper clear aqueous phase (containing RNA) into a fresh 2-mL tube, without disturbing the other phases (*see* **Note 4**).
12. Add 625 µL (0.5X vol) of isopropanol to precipitate RNA. Mix briefly and incubate on ice for 10 min (*see* **Note 5**).
13. Centrifuge at 12,000g for 10 min at 4°C to pellet RNA.
14. Remove the supernatant without disturbing the gel-like RNA pellet.
15. Add 1.25 mL (1X vol) of 70% ethanol (prepared in DEPC-treated water) to wash pellet, mixing briefly.
16. Centrifuge at 7500g for 5 min at 4°C to pellet RNA.
17. Remove the supernatant and repeat ethanol wash step and centrifugation.
18. Remove as much supernatant as possible without disturbing RNA pellet, and air-dry pellet for 5 to 10 min, without drying completely (*see* **Note 6**).

19. Resuspend the RNA pellet in 40 µL DEPC-treated water. To prevent degradation, add 1 µL (40 U) RNasin ribonuclease inhibitor (40 U/µL).
20. Incubate at 55°C for 5 min, with intermittent vortexing to dissolve RNA fully.
21. Store at –80°C or proceed immediately to DNase treatment (see **Note 7**).
22. Add to RNA sample in the following order, 1 µL DEPC-treated water, 5 µL 10X DNase buffer, 3 µL (6 U) rDNase I enzyme (2 U/µL). Mix gently and spin briefly.
23. Incubate at 37°C for 30 min to degrade traces of genomic DNA.
24. Resuspend DNase Inactivation Reagent by vortexing, and add 10 µL to DNase-treated RNA.
25. Incubate at room temperature for 2 min with occasional mixing.
26. Centrifuge at 10,000g for 1.5 min and transfer supernatant (containing RNA) to a new tube.
27. Quantitate purified RNA (e.g., UV spectrophotometer or RiboGreen®). An RNA concentration of 0.5 µg/µL or above is ideal. RNA integrity should be assessed by electrophoresis or an Agilent Bioanalyzer.

3.2. cDNA Synthesis

The following method describes the synthesis of cDNA for use in two-step qRT-PCR. The reverse transcription is performed as a separate step from the PCR amplification (see **Notes 8** and **10**).

1. Decontaminate workbench, tube racks and pipets with RNase decontaminant before use; wear gloves at all times.
2. All solutions are stored at –20°C. Before use, warm them to room temperature (except MMLV-RT reverse transcriptase and RNasin ribonuclease inhibitor) and set on ice. Vortex and spin all solutions briefly before use.
3. Total RNA is thawed quickly and stored on ice at all times. Freeze immediately and store at –80°C after use. Limit repeated freeze/thawing of RNA, as this can result in significant degradation of RNA integrity.
4. Prepare a master mix for all RT reactions by adding the necessary volumes of the required solutions to one tube: 1.0 µL dNTPs (10 mM each), 0.5 µL random primers (250 ng) (see **Note 9**), and 2.5 µL DEPC-treated water to give a 4.0 µL final volume per reaction.
5. Add 4.0 µL of this master mix to 8.0 µL total RNA (up to 1 µg) to give a 12.0 µL final reaction volume.
6. Incubate this mixture at 65°C for 5 min to denature RNA secondary structures. Place tube on ice for 5 min.
7. Prepare a master mix for all RT reactions by adding the necessary volumes of the required solutions to one tube: 4.0 µL 5X RT Buffer, 2.0 µL DTT (0.1 mM), 1.0 µL RNasin ribonuclease inhibitor (40 U/µL).
8. Add a 7.0-µL aliquot of master mix to the chilled RT mix, and incubate at 25°C for 2 min to allow primer annealing.
9. Add 1.0 µL (200 U) MMLV-RT reverse transcriptase (200 U/µL).
10. Mix gently by pipeting up and down; the final reaction volume is 20 µL.

11. Incubate this mixture at 25°C for 10 min, followed by 42°C for 2 h. The incubation temperature of 42°C may differ according to the RT enzyme used.
12. Inactivate RT enzyme by incubating at 85°C for 5 min, and chill on ice for 5 min.
13. Dilute 1:10 with the addition of 190 μL nuclease-free water; store at –20°C.

3.3. Fluorescent Chemistries

The first step before proceeding with real-time PCR is choosing a fluorescent chemistry to be used. Although many options are available, they can be basically divided into two groups: (1) nonspecific intercalating dyes and (2) sequence-specific oligonucleotide probes.

3.3.1. Intercalating Dyes

SYBR Green I is the most common intercalating dye used. This dye fluoresces 50 to 100 times more in the presence of double-stranded DNA, which is generated during the extension step of each PCR cycle (*see* **Fig. 1A**). Thus the level of fluorescence is proportional to the concentration of amplicon produced. The advantage of this approach is that it is relatively inexpensive compared with fluorescent probes, extremely flexible, and requires very little optimization. However, multiplex reactions are not possible, and nonspecific products including secondary amplicons and primer dimers can cause problems. However, inclusion of dissociation/melt curves in the protocol can permit detection of nonspecific amplification (although they increase assay times compared with probes). SYBR Green I is also the least sensitive of approaches.

3.3.2. Fluorescent Probes

In addition to the two primers, a third oligo is added—in general, probes are short oligonucleotide sequences complementary to a fragment (flanked by the primers) of the target gene, with a covalently attached fluorescent dye. All probes rely on the close proximity of a quencher molecule to the fluorescent dye to inhibit fluorescence under specific conditions. In comparison with intercalating dyes, probes are more sensitive and have a lower background—multiplex reactions are possible (two to five simultaneous assays). However, primer design is more complex, and reactions can require greater optimization, in addition to extra cost. Many probe chemistries are available, including:

1. *5' Nuclease (TaqMan/hydrolysis) probes:* the probe is complementary to a region within the target sequence, the 5' end is labeled with a fluorescent reporter dye, and the 3' end is attached to the quencher molecule. (The 3' end is also blocked to prevent probe extension.) Fluorescence is emitted when the probe is hydrolyzed by *Taq* DNA polymerase, separating the fluorescent dye from the quencher (*see* **Fig. 1B**). Thus, the concentration of probe hydrolyzed (leading to fluorescence emission) is directly proportional to PCR amplicon concentration. This is the

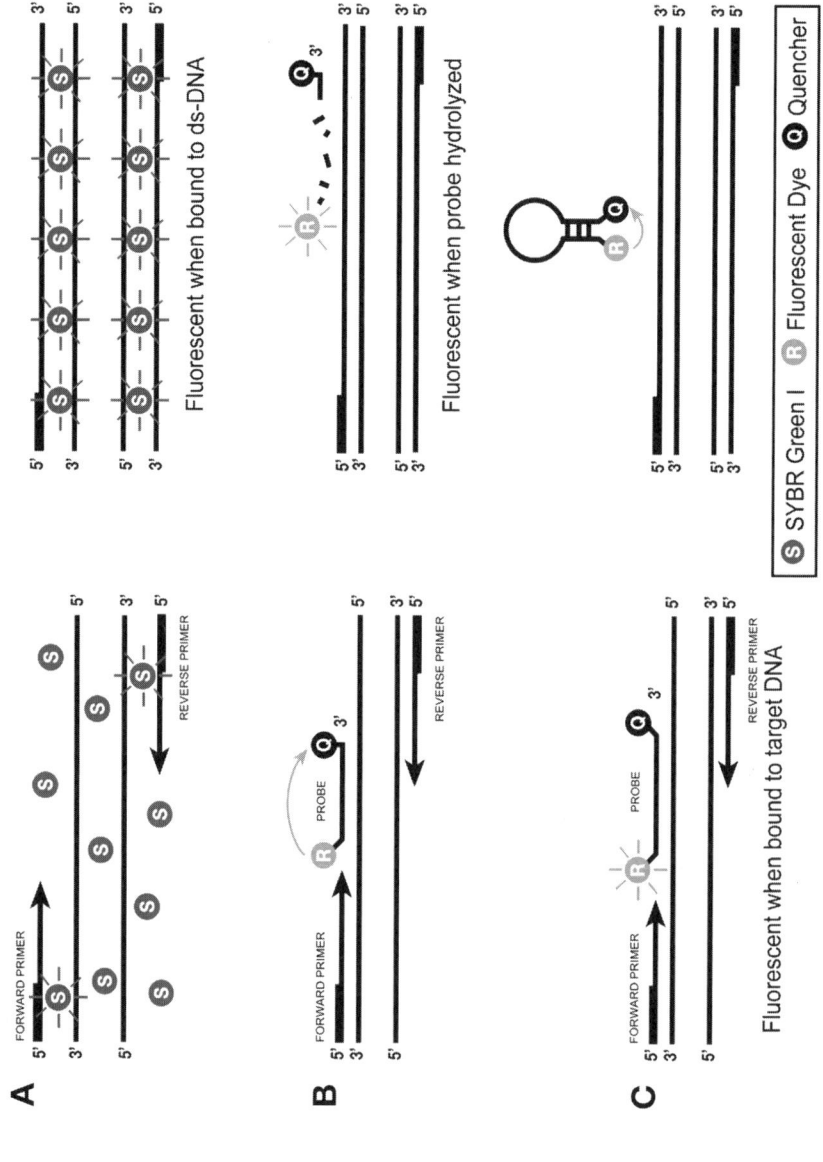

Fig. 1. Schematic of fluorescent chemistries used in qPCR. (**A**) SYBR Green I intercalating dye. (**B**) 5' Nuclease/TaqMan probes. (**C**) Molecular beacons.

easiest probe to design and optimize, but other probe chemistries may provide greater sensitivity with lower background.
  b. *Molecular beacons:* the probe is similar in design to a 5' nuclease probe with the addition of four to six bases at each end, which self-anneal to form a hairpin. When free in solution as a hairpin, the reporter dye is in close proximity to the quencher, thus absorbing the fluorescence emission. When the molecular beacon is hybridized to a target sequence during the annealing step, the reporter dye is separated from the quencher. Thus, the concentration of probe hybridized (leading to fluorescence emission) is directly proportional to PCR amplicon concentration (*see* **Fig. 1C**).
  c. Many other probe chemistries exist including Hybridization/FRET, LUX, Scorpion, and others. Refer to **refs.** *1* to *3* for an explanation of how these approaches work and their applications.

3.4. Primer/Probe Design

Commercial software packages are available for primer and probe design including Primer Express (Applied Biosystems) and Beacon Designer (Premier BioSoft). Other general primer design programs can be used to design primers for use in SYBR Green I assays including: Primer3 *(5)* and Primer Quest *(6)*. In addition, off-the-shelf prevalidated primers/probe assays are also commercially available for thousands of human, rat, and mouse genes (www.allgenes.com). Alternatively, many companies provide custom assay design services.

The following are general guidelines used to design primers or primers/probes for use in SYBR Green I (probe design not required) and 5' nuclease assays and molecular beacons. For the design of probes for other chemistries, follow the manufacturer's guidelines.

3.4.1. Primer Design

1. The amplicon should be between 75 and 150 bp in size.
2. The optimal primer length is 20 bp (range 17–25 bp) and the GC content within 30 to 80%.
3. The T_m of primers should be within 58 to 60°C (1°C difference between them). For molecular beacons, the T_m of primers are designed within 54 to 56°C.
4. The last five bases at the 3' terminal of each primer should have no more than two Gs or Cs.
5. Avoid secondary structures (e.g., primer dimers and hairpins).
6. Avoid runs of more than three identical nucleotides, especially for G.
7. When possible, choose a primer that spans at least one exon-exon junction to prevent potential amplification of contaminating genomic DNA.

3.4.2. 5' Nuclease Probe Design

1. The T_m should be within 68 to 70°C (8–10°C greater than the primer T_m).
2. The optimal probe length is 20 bp with MGB (range 17–25 bp) *(7)*. The GC content should be within 30 to 80%.

3. Avoid runs of more than three identical nucleotides, especially for G.
4. Place the probe as close as possible to either primer without overlapping.
5. The probe is selected from the strand with more Cs than Gs and no G on the 5' terminal.
6. Avoid secondary structures and complementarity with primers.

3.4.3. Molecular Beacon Design

1. The T_m should be 7 to 10°C greater than the primer T_m (prior to adding stem sequences).
2. The GC content should be within 30 to 80% for the probe and 70 to 80% for the stem.
3. The probe is selected from the strand with more Cs than Gs.
4. Avoid runs of more than three identical nucleotides, especially for G.
5. Place the probe near the center of the amplicon.
6. Avoid secondary structures and complementarity with primers.
7. Once the probe sequence is selected, two complementary stem sequences (5–7 bp) are added to each side of the probe sequence.
8. Use a DNA folding program to estimate the T_m of the stem (approx 7–10°C greater than the primer T_m) *(8)*.
9. When selecting a stem sequence, place a C on the 5' terminal of the probe.
10. Refer to **ref. 9** for further details on molecular beacon design and synthesis.

Primers and probes designed within these guidelines will work under the universal thermal cycling conditions described in **Subheading 3.6.** It is also recommended that a BLAST search (www.ncbi.nlm.nih.gov/BLAST) for primer and probe sequences be performed (prior to synthesis) to ensure that oligos do not hybridize to unintended targets.

3.5. Preparation of PCR Standards

This step is only necessary if quantitation is to be performed via the standard curve method (*see* **Subheading 3.8.**). Disregard if you are quantitating via the comparative C_T method (*see* **Subheading 3.9.**). Standard curves can be generated from a known number of molecules of purified PCR product (*see* **Note 11**).

1. Perform a PCR for each primer pair/probe to be used.
2. Electrophorese amplicons on a 2% agarose/TAE gel.
3. With a fresh scalpel blade, excise the specific amplicon band only and place the gel slice in a fresh 2-mL tube.
4. Extract DNA from the gel slice (e.g., spin column).
5. Quantitate the amplicon concentration via appropriate methods (e.g., PicoGreen® or UV spectrometry).
6. Calculate the copy number according to the following generic formula:

$$\text{Amount (copies/}\mu\text{L)} = \frac{6.023 \times 10^{23} \text{ (copies/mol)} \times C \text{ (g/}\mu\text{L)}}{\text{MW (g/mol)}}$$

where C = concentration of amplicon in g/µL as determined by quantitation and MW = molecular weight of the PCR product (size in base pairs × 658 g).

7. Make PCR standard to a concentration of 1×10^9 copies/5 µL.
8. Perform serial dilutions from 1×10^8 to 1×10 copies per 5 µL.
9. Add sodium azide as a preservative, final concentration 0.025% w/v.
10. Store in aliquots at –20°C for up to 6 mo.

3.6. Quantitative RT-PCR

Although the strategy of qRT-PCR is relatively new, in the short period since its introduction many chemistries and approaches have been developed. The following methods describe the measurement of expression of two transcripts, the target gene and an invariant control (*see* **Note 12**) as a singleplex (*see* **Note 13**) in three control samples, three treated samples, and a negative/no-template control (*see* **Note 14**). PCR standards (10^6–10^2 for target gene and 10^7–10^3 for invariant control) are included for standard curve analysis (*see* **Subheading 3.8.**) Standards can be omitted if you are quantitating via the comparative C_T method (*see* **Subheading 3.9.**) Both nonspecific intercalating dyes (i.e., SYBR Green I) and sequence-specific oligonucleotide probes (i.e., 5' nuclease and molecular beacon assays) are described. Although the reaction setups are extremely similar, there are important differences, as outlined below.

Owing to the sensitivity of qPCR, the use of master mixes is highly recommended (*see* **Notes 15** and **16**). Ideally, PCR should be performed in a dedicated room or at least a clean workspace. However, the use of reagents and pipets also dedicated for PCR is essential. Aerosol-resistant filter-containing pipet tips will prevent aerosols from pipets from contaminating reagents and reactions. In addition, gloves should always be worn when handling PCR reagents and setting up reactions.

1. All solutions are stored at –20°C. Before use, warm them to room temperature and set on ice. Vortex and spin all solutions briefly before use.
2. The following master mix is prepared for the analysis of three control samples, three treated samples, and a no-template control. Assays are performed in triplicate for our target gene and invariant control. Scale volumes appropriately for extra samples.

	1X Reaction (µL)	84X Reaction (µL)
2X Real-time PCR mix (*see* **Notes 2** and **17**)	10.0	840.0
Nuclease-free water	2.0	168.0
Final volume	**12.0**	**1008.0**

3. Remove two 480-µL aliquots of master mix (performing 36 reactions per transcript, enough for 40 reactions) into separate 1.5-mL tubes.

Table 1
Primer and Probe Concentrations for qPCR

Fluorescent approach	Stock primer conc. (μM)	Final primer conc. in 20 μL (nM)	Final probe conc. (μM)	Stock probe conc. in 20 μL (nM)
SYBR Green I	1	50	N/A	N/A
5'-Nuclease assay	18	900	4	200
Molecular beacon	8	400	4	200

4. To the first aliquot add 40 µL of each target gene primer (forward and reverse) and probe (not applicable for SYBR Green I; substitute with nuclease-free water). See **Table 1** for stock and final reaction concentrations (see **Notes 18** and **19**). Vortex and spin briefly. The final volume is 15 µL per reaction.
5. Repeat **step 4** for the second aliquot with primers/probe for invariant control.
6. In 12 separate 0.6-mL tubes, prepare 12X aliquots of 48 µL of target gene master mix; repeat for invariant control.
7. To each tube add 16 µL of cDNA (five standards and six unknown samples [three control and three treated]); to the final tube add 16 µL nuclease-free water.
8. Aliquot 3X 20 µL of each reaction into a 96-well reaction plate according to **Fig. 2**.
9. Seal the plate well with optical caps or film (see **Note 20**).
10. Spin plate at 1000 rpm for 2 min to collect samples.
11. Insert plate into real-time PCR instrument and lock lid.
12. In the real-time PCR software, set up the plate template and well definitions (see **Fig. 2**) and the thermal cycling protocol for 20 µL final reaction volume as below. (Ensure that the appropriate fluorescent dye is defined for each well and that fluorescence detection is set during the appropriate step.)

3.6.1. Two-Step PCR Protocol (SYBR Green I and 5' Nuclease Assay)

Taq DNA polymerase activation (see **Note 21**)	95°C	3–10 min
Followed by 40 cycles of:		
Denaturation	95°C	15 s
Anneal/extension (fluorescent detection step)	60°C	60 s

3.6.2. Three-Step PCR Protocol (Molecular Beacons Only)

Taq DNA polymerase activation (see **Note 21**)	95°C	3–10 min
Followed by 40 cycles of:		
Denaturation	95°C	30 s
Anneal (fluorescent detection step)	55°C	60 s
Extension	72°C	30 s

	1	2	3	4	5	6	7	8	9	10	11	12
A	Target Gene STD1	Target Gene STD2	Target Gene STD3	Target Gene STD4	Target Gene STD5		Target Gene CTRL1	Target Gene CTRL2	Target Gene CTRL3	Target Gene TRT1	Target Gene TRT2	Target Gene TRT3
B	Target Gene STD1	Target Gene STD2	Target Gene STD3	Target Gene STD4	Target Gene STD5		Target Gene CTRL1	Target Gene CTRL2	Target Gene CTRL3	Target Gene TRT1	Target Gene TRT2	Target Gene TRT3
C	Target Gene STD1	Target Gene STD2	Target Gene STD3	Target Gene STD4	Target Gene STD5		Target Gene CTRL1	Target Gene CTRL2	Target Gene CTRL3	Target Gene TRT1	Target Gene TRT2	Target Gene TRT3
D	Invariant Control STD1	Invariant Control STD2	Invariant Control STD3	Invariant Control STD4	Invariant Control STD5		Invariant Control CTRL1	Invariant Control CTRL2	Invariant Control CTRL3	Invariant Control TRT1	Invariant Control TRT2	Invariant Control TRT3
E	Invariant Control STD1	Invariant Control STD2	Invariant Control STD3	Invariant Control STD4	Invariant Control STD5		Invariant Control CTRL1	Invariant Control CTRL2	Invariant Control CTRL3	Invariant Control TRT1	Invariant Control TRT2	Invariant Control TRT3
F	Invariant Control STD1	Invariant Control STD2	Invariant Control STD3	Invariant Control STD4	Invariant Control STD5		Invariant Control CTRL1	Invariant Control CTRL2	Invariant Control CTRL3	Invariant Control TRT1	Invariant Control TRT2	Invariant Control TRT3
G												
H		Target Gene NTC	Target Gene NTC	Target Gene NTC				Invariant Control NTC	Invariant Control NTC	Invariant Control NTC		

Fig. 2. PCR plate template demonstrating layout and well definitions.

131

3.6.3. Dissociation/Melt Curve (Only Required When Using SYBR Green I)

Denaturation	95°C	60 s
Equilibriation	55°C	60 s
Ramp (fluorescent detection step)	55–95°C	ramp 0.5°C/10 s

3.7. Data Analysis

3.7.1. Baseline Correction and Threshold Determination

Most analysis software will automatically perform background and baseline subtraction of the raw fluorescent data. In addition, basic data analysis relies on setting a baseline and threshold for each assay. This is often performed automatically, but manual adjustment can frequently improve the quality and visualization of the data. (Refer to software manuals for exact details.)

1. The baseline is often set to a default range of cycles 3 to 15. Manual adjustment is often necessary if detectable amplification occurs prior to 15 cycles (e.g., when using 18S rRNA as invariant control).
2. Manually set the baseline to cover as many cycles as possible, setting to a maximum of one to two cycles prior to the first noticeable increase in fluorescent signal above background. Setting of the baseline is best performed with the linear view of the *y*-axis (*see* **Fig. 3A** and **Note 22**).
3. The threshold bar is often set to a default of 10 times the standard deviation (SD) of the mean background fluorescence (defined by the baseline). To set the threshold bar manually, select the logarithmic view of the *y*-axis. Place the threshold bar so that it crosses the linear component (corresponding to the exponential amplification) of all the amplification plots, typically the halfway point (*see* **Fig 3B**). The threshold bar should never be set outside the region of exponential amplification.
4. The C_T (threshold cycle) value is generated for each amplification curve, corresponding to the cycle at which each sample crosses the threshold. This value is the basis of all further calculations. Export well definitions and C_T values into a spreadsheet for further analysis.
5. NB: Non-template controls absent of cDNA should demonstrate no amplification (*see* **Note 23**).

3.7.2. Dissociation Curve Analysis

Dissociation/melt curve analysis is necessary when one is performing quantitation via intercalating dyes such as SYBR Green I. Because these dyes are nonspecific, any secondary PCR products including primer dimers can contribute to the fluorescent signal and thus affect accurate quantitation. In this approach the PCR amplicons are slowly heated from 60 to 95°C while you measure the fluorescence. As a double-stranded PCR amplicon reaches its spe-

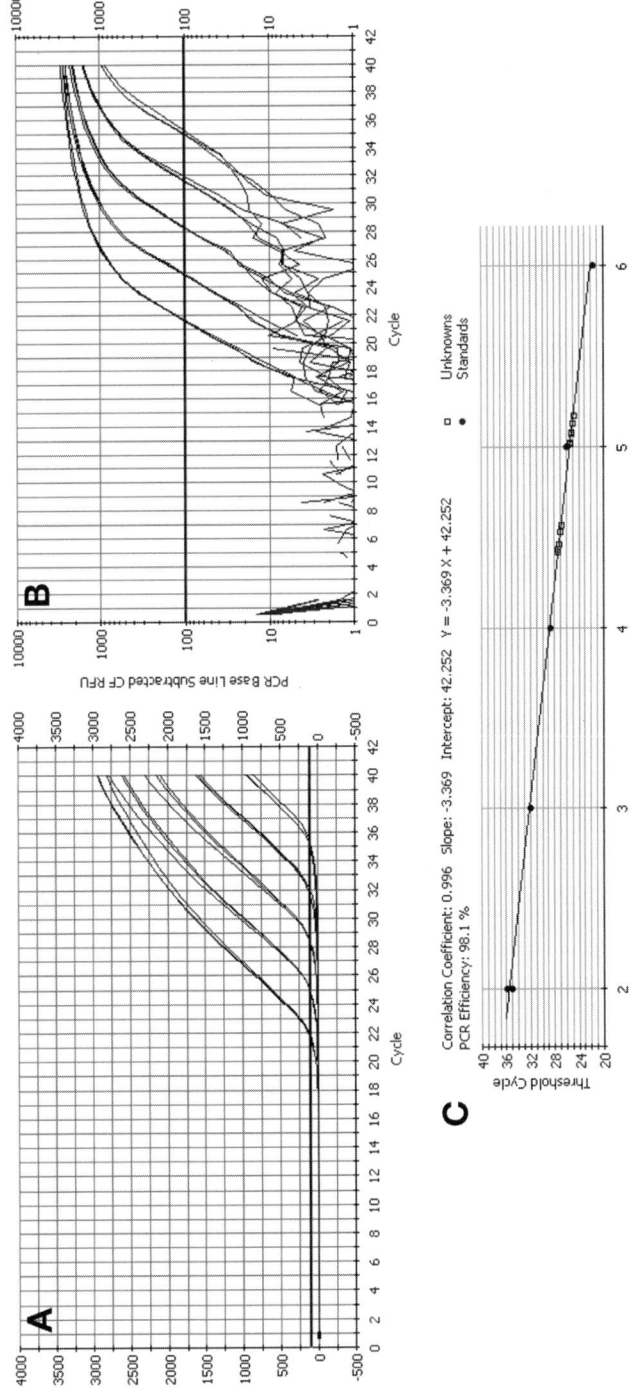

Fig. 3. Construction of a standard curve for qPCR analysis. (**A**) Amplification plots demonstrating ideal baseline setting. (**B**) Semilog view indicating threshold determination during exponential phase of each plot. (**C**) Typical standard curve showing near perfect fit (0.996) and PCR efficiency (98.1%).

cific denaturation temperature (the exact melt temperature or T_m is dependent on GC content, sequence, and length), it will separate into single strands. At this temperature the fluorescent signal will decline rapidly. (SYBR Green I is fluorescent only when attached to double-stranded DNA; *see* **Fig. 4A**.) A negative first derivative of the melt curve data (rate of change of fluorescence over temperature or –dF/dT vs temperature) presents the data as peaks. If a single amplicon is produced, then a single peak will be observed at its specific T_m (*see* **Fig. 4B**). More than one specific product will demonstrate more than one peak (*see* **Fig. 4C**).

Often non-template controls will demonstrate an increase in fluorescent signal after 35 cycles when one is using SYBR Green I. This is owing to the production of primer dimers and not the specific amplicon. Primer dimers typically melt at a lower T_m than the specific amplicon because of their shorter length (*see* **Fig. 4D**). Non-template controls should be accepted if they demonstrate no amplification or primer dimer only (i.e., absence of specific amplicon). Quantitative data from each sample should only be accepted when a single peak is observed at the expected T_m, i.e., no secondary peaks including primer dimers (*see* **Note 23**). It is also advisable to run agarose gels periodically to demonstrate that the specific amplicon is of the desired fragment size.

3.8. Standard Curve Analysis

This method requires running standard curves for both the target gene and invariant control. Standard curves can be constructed from a dilution series (use at least five different concentrations) of a cDNA sample that expresses both transcripts, PCR standards (*see* **Subheading 3.5.**) or cRNA (*see* **Note 11**). The expression level for both transcripts for all samples should fall within the limits of your standard curve. The C_T values of the unknown samples are derived from the standard curve to calculate absolute values.

1. Prepare quantitation PCR master mixes as outlined in **Subheading 3.6.**
2. In the software, use well definitions (i.e., STANDARD, UNKNOWN, and so on) to assign samples.
3. Run quantitative PCR amplification.
4. Perform basic data analysis as appropriate (*see* **Subheading 3.7.1.**).
5. By the use of well definitions, most analysis software will construct standard curves and calculate starting quantities (SQ) of unknown samples automatically. Alternatively, export C_T values into a spreadsheet to create standard curves and calculate unknown SQ (e.g., TREND function).
6. **Figure 3C** demonstrates a typical standard curve. The correlation coefficient should be between 0.98 and 1.00 to demonstrate a good fit of the data.

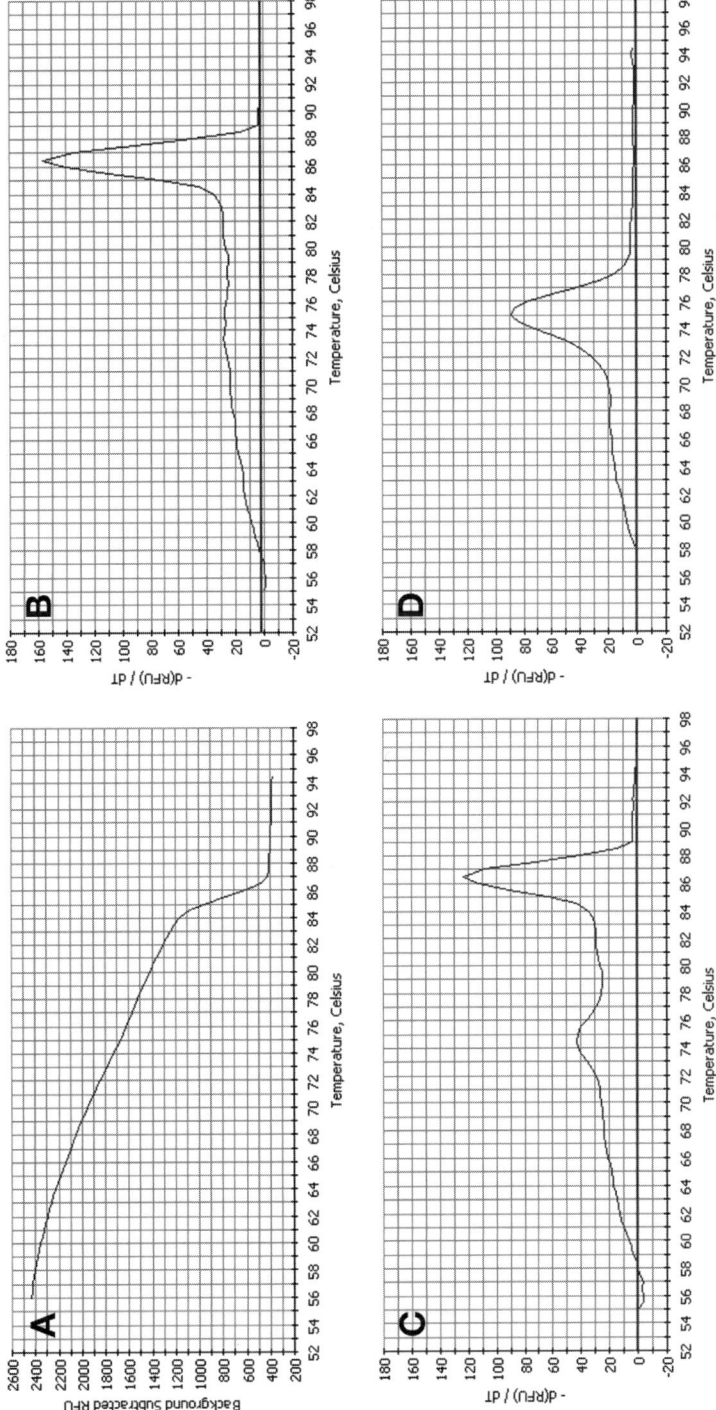

Fig. 4. Dissociation/melt curve plots. (**A**) Fluorescence vs temperature. (**B**) First derivative plot demonstrating a single sharp peak/amplicon, T_m 85.5°C. (**C**) Amplification of amplicon and secondary product/primer dimer, T_m 85.5°C and 75°C, respectively. (**D**) No-template control demonstrating amplification of primer dimer only, T_m 75°C.

Table 2
Determination of Differential Expression via Standard Curve Analysis[a]

	Target gene			Invariant control				Differential expression		
Sample	Mean C_T	SD	CV (%)	Mean SQ	Mean C_T	SD	CV (%)	Mean SQ	Target/ 1E+07 invariant	Mean	Ratio
Control											
1	27.26	0.04	0.1	2.81E+04	13.19	0.03	0.2	6.52E+07	4310	4065	1.00
2	27.37	0.04	0.1	2.60E+04	13.03	0.08	0.6	7.33E+07	3548		
3	27.01	0.09	0.3	3.33E+04	12.96	0.06	0.5	7.69E+07	4336		
Treatment											
1	25.07	0.09	0.4	1.24E+05	13.23	0.05	0.4	6.37E+07	19,439	19,266	4.74
2	25.34	0.03	0.1	1.03E+05	13.19	0.04	0.3	6.55E+07	15,786		
3	25.06	0.11	0.4	1.25E+05	13.42	0.07	0.5	5.55E+07	22,573		

[a]Refer to **Table 3** for individual C_T values of each sample triplicate.

7. If not automatically calculated, the PCR efficiency can be derived from the slope ($[10^{(-1/\text{slope})}] - 1$). Thus a slope of -3.32 demonstrates almost 100% efficiency (*see* **Note 24**).
8. Absolute copy numbers for the target gene are then normalized to the invariant control expression (*see* **Table 2**).
9. Alternatively, the copy number value can be normalized to the number of cells, tissue weight, RNA, or cDNA concentration used.
10. In the example shown in **Table 2**, after normalization with the invariant control, the expression of the target gene was 4.74-fold upregulated compared with the control group.

3.9. Comparative C_T Method

Although the determination of exact transcript copy numbers is attractive to most investigators, the difference in expression of a gene between a control/unstimulated/normal and treatment/stimulated/diseased sample(s) is often sufficient to answer specific hypotheses. The expression data are presented as a ratio and not an absolute amount. The comparative C_T method circumvents the need for standard curve construction; thus this approach is more straightforward and has a higher throughput than the method described in **Subheading 3.8**. The following strategy was originally described by Livak and Schmittgen (*9*).

1. Perform basic data analysis as appropriate (*see* **Subheading 3.7.1.**).
2. Import the well definitions and C_T values into a spreadsheet; arrange the data in triplicates and sample groups for each gene investigated (*see* **Table 3**).
3. Calculate the mean C_T, SD, and coefficient of variance (CV) for each triplicate for both the target gene and the invariant control. We routinely remove a single outlier from each triplicate if the CV is greater than 1.0%.
4. Calculate the mean C_T and SD for both the control and treatment groups for both transcripts.
5. Subtract the mean C_T of the target gene from the mean C_T of the invariant control to normalize samples; this is referred to as the delta C_T or ΔC_T. The ΔC_T SD can be calculated according to **Note 25**.
6. Calculate the $\Delta\Delta C_T$ by subtracting the treatment group ΔC_T from the control group ΔC_T.
7. To determine the ratio change, perform the following expression: $2^{-\Delta\Delta C_T}$ (*9*) (*see* **Note 24**).
8. The significance of this differential expression can be determined from the individual mean C_T values of each triplicate using appropriate statistical tests (e.g., *t*-test, ANOVA, and so on).
9. In the example shown in **Table 3**, after normalization with the invariant control, the expression of the target gene was 4.83-fold upregulated compared with the control group.

Table 3
Determination of Differential Expression via Comparative C_T Method

		Target gene				Invariant control					Differential expression		
Sample	C_T	Mean C_T	SD	CV (%)	C_T	Mean C_T	SD	CV (%)	ΔC_T	Mean	$\Delta\Delta C_T$	$2^{-\Delta\Delta C_T}$
Control												
1	27.24	27.26	0.04	0.1	13.16	13.19	0.03	0.2	14.06	14.15	0.00	**1.00**
1	27.23				13.20							
1	27.30				13.22							
2	27.33	27.37	0.04	0.1	13.05	13.03	0.08	0.6	14.34			
2	27.38				12.94							
2	27.40				13.10							
3	27.00	27.01	0.09	0.3	12.89	12.96	0.06	0.5	14.04			
3	26.92				12.99							
3	27.10				13.01							
Treatment												
1	25.17	25.07	0.09	0.4	13.24	13.23	0.05	0.4	11.85	11.88	−2.27	**4.83**
1	24.99				13.27							
1	25.06				13.17							
2	25.35	25.34	0.03	0.1	13.15	13.19	0.04	0.3	12.15			
2	25.36				13.19							
2	25.30				13.22							
3	25.15	25.06	0.11	0.4	13.48	13.42	0.07	0.5	11.64			
3	24.94				13.44							
3	25.08				13.34							

4. Notes

1. Treatment of distilled water with 0.1% diethylpyrocarbonate (DEPC) removes traces of RNases. Add DEPC (1 mL/L of water), shake well, and leave overnight. Autoclave treated water for 20 min to destroy DEPC. **Caution:** take necessary safety precautions when handling DEPC stock, as it is highly toxic. Alternatively, DEPC-treated water can be obtained commercially.
2. Ready-to-use master mixes containing a hot start *Taq* DNA polymerase, nucleotides, magnesium chloride, and reaction buffers are also commercially available. In addition, they may also include a passive reference dye (e.g., ROX or fluorescein). Master mixes are also available that contain SYBR Green I. It is possible to formulate your own master mixes using standard PCR reagents as previously described. Although this formulation will work well for most real-time PCR assays it may be necessary to adjust the $MgCl_2$ concentration if the desired results are not obtained (3–7 mM; generally 3 mM for SYBR Green I and 5 mM $MgCl_2$ for probes). The choice of passive reference dye depends on the real-time PCR instrument used; for Applied Biosystems and Stratagene instruments, ROX is used (final concentration 300 and 30 nM, respectively), whereas Bio-Rad platforms utilize fluorescein (final concentration 12.5 nM).
3. Synthesized 5'-nuclease probes and molecular beacons should be 5'-labeled with an appropriate fluorophore (most commonly FAM or VIC) and 3'-labeled with an appropriate nonfluorescent quencher (NFQ). The inclusion of a minor groove-binding (MGB) moiety in 5'-nuclease probes increases the stability and specificity of the probe, with the added advantage of shorter probe design *(7)*.
4. To limit the transfer of contaminating genomic DNA in the RNA, it is best to leave some of the aqueous phase behind. The interphase and lower organic phase can be stored at –20°C for subsequent isolation of protein and DNA.
5. For low quantities of tissue (<10 mg) or cells (<10^4), add 10 μg RNase-free glycogen (volume of 800 μL phenol/GITC reagent) to the aqueous layer prior to precipitation with isopropyl alcohol. Glycogen acts as a carrier, assisting in precipitation and visualization of the RNA pellet.
6. A completely dry RNA pellet is difficult to resuspend, demonstrating an A_{260}/A_{280} ratio of <1.6.
7. No RNA isolation procedure yields RNA devoid of contaminating genomic DNA. Thus traces of DNA must be removed via DNase I treatment, prior to sensitive assays such as qRT-PCR. The removal of DNase I in the reaction is also important; heat inactivation in the presence of a chelating agent (e.g., EDTA) can be performed. However, the chelation agent may have inhibitory effects on the RT-PCR if also not removed. We utilize a novel rDNAse I and inactivation reagent to remove traces of DNA (<5%) from up to 200 μg of total RNA. This approach does not require heat denaturation, chelating agents, or any subsequent processing, which can be detrimental to the RNA sample. Alternatively, any RNA-grade DNase I enzyme can be utilized with subsequent purification via RNA spin columns. This second approach can result in the loss of approx 20% of the RNA sample after cleanup.

8. Standard RT-PCR can be performed using one-step and two-step strategies. In one-step RT-PCR, the RT and PCR steps are combined in a single reaction tube, with each step utilizing a separate enzyme (reverse transcriptase and *Taq* DNA polymerase, respectively). However, since the RT reaction is primed using the same primers as the PCR assay at lower temperature than the PCR amplification, mispriming can occur resulting in nonspecific products. In two-step RT-PCR, each reaction is performed separately; thus different primers can be used for each step (circumventing mispriming). In addition, a sufficient amount of cDNA can be synthesized in the RT step to allow multiple PCR assays of different target genes from the same cDNA sample.

9. Alternatively, oligo(dT) can be used to prime the RT reaction when both the target gene and invariant control are polyadenylated. Longer cDNA fragments are synthesized, although this is not necessary for quantitative PCR (amplicons generally 50–150 bp in size). The use of random primers generates a higher cDNA yield and can be used to quantitate nonpolyadenylated transcripts (e.g., 18S rRNA). The 5'-end of an mRNA population is also better represented compared with using oligo(dT). A mixture of both oligo(dT) and random primers can be advantageous in some approaches. The use of gene-specific primers is the most specific for an individual transcript, but the synthesized cDNA is limited to a single transcript. This approach is used when one is performing one-step RT-PCR.

10. As it is virtually impossible to eliminate DNA from RNA preparations, it may be desirable to run a –RT control (mock RT containing all reaction components except RT enzyme). Any amplification from this control indicates the presence of contaminating genomic DNA. This is not necessary if you are using exon-exon spanning primers.

11. Standards can also be synthesized from cRNA; this has the added advantage of measuring the efficiency of the RT reaction. The synthesis of cRNA standards is more tedious and difficult to prepare. In addition, cRNA standards are extremely prone to degradation.

12. The choice of invariant control (also referred to as a housekeeping gene or endogenous control) is of extreme importance. It is vital that the expression of the invariant control (as the name suggests) remain constant over the conditions to which the target gene is tested. Expression values for the target gene investigated are normalized to the invariant control to correct for minor variations in reaction kinetics, experimental conditions, and cDNA template quantity. Thus the expression of this transcript must be virtually identical across all investigated samples. β-Actin, cyclophilin, glyceraldehyde-3-phosphate dehydrogenase (GAPDH), and $β_2$-microglobulin are commonly used invariant controls. Numerous studies have, however, demonstrated the differential expression of one or more of these so-called housekeeping genes under different stimuli (reviewed in **refs. *1–3***). We have also observed changes in β-actin and GAPDH with ischemia-reperfusion, for example, in heart samples *(11)*. Ribosomal transcripts including 28S and 18S rRNA are more stable to changes in gene expression. However,

their use can be problematic owing to the high expression of the transcript. It is recommended that an evaluation of several invariant controls be performed prior to target gene investigation for any given system/treatment/ stimuli/disease state. Finally, the target and invariant control must have nearly identical PCR efficiencies; otherwise efficiency correction is necessary (*see* **Note 24**).
13. Multiplex PCR is the amplification of more than one primer/probe set in the same tube. This is most commonly used in probe-based assays in which each specific probe is labeled with a different fluorescent reporter dye (e.g., FAM, VIC, NED, Cy5). Although this approach sounds attractive, considerable design and optimization is required to ensure the absence of cross-reaction amplification, identical PCR efficiencies, and noncompetition of different amplicons for accurate quantitation. Unless the target genes are routinely investigated, the optimization of this approach can often be more time consuming and expensive than simply running the genes individually. For strategies on performing multiplex qRT-PCR, refer to ABI User Bulletin 5 *(12)*.
14. A cross-plate calibrator is required if the number of samples in an assay take up more than one reaction plate. This calibrator consists of a positive control cDNA that is run on every reaction plate. All samples are calibrated to the positive control to compensate for interassay variability. If you are utilizing standard curves, this is not necessary.
15. Owing to the sensitivity of qRT-PCR, any variability in reaction conditions should be kept to a minimum. To minimize sample-to-sample and well-to-well variation and improve reproducibility, the use of master mixes is encouraged. The inclusion of a passive reference dye (e.g., ROX or fluorescein depending on real-time PCR instrument) is added to the master mix to reduce well-to-well variations.
16. Although quantitative (real-time) PCR circumvents the routine use of gels, it is recommended that gels be run when one is using a new primer/probe set to ensure the correct amplicon size and primer specificity (absence of secondary products and primer dimers).
17. Some commercially available master mixes contain uracil-N-glycosylase (UNG) and dUTP. Using this approach, dUTP is incorporated into the amplicon, allowing an optional pretreatment of subsequent reactions (2-min incubation at 50°C; 10-min inactivation at 95°C). UNG hydrolyzes carried-over PCR products (containing dUTP) if they are suspected of contaminating new reactions. UNG cannot be used with one-step RT-PCR. Quantitative PCR negates the further processing of PCR products (e.g., gel electrophoresis); thus reaction plates should rarely need to be opened after amplification (and never in the PCR setup area), usually negating the need for UNG.
18. The concentrations of primers and probe used in **Table 1** are guidelines only and may differ for each primer set. These concentrations, especially for probe-based assays, can often be reduced without significantly affecting the C_T or amplification efficiency. For SYBR Green I assays, increasing the primer concentration too much increases the potential for primer dimer formation; the final concentrations should be kept within 50 to 300 n*M* each.

19. Several approaches can be evaluated to eliminate the production of secondary products/primer dimers. Increase the anneal/elongation temperature, decrease the primer concentration, adjust the $MgCl_2$ concentration (3–7 mM), and/or reduce the cycle number from 40. (This may have an impact on detection of lowly expressed transcripts.) If these approaches fail, it may be necessary to redesign the primers to another region of the gene.
20. When using optical film, it is important to seal the plates well; seal from the middle of the plate to the outer edges, taking care not to stretch the film. Improper sealing can result in the evaporation of sample, particularly in the outer wells. Also take care not to touch the top or bottom of the reaction plate; handle by the edges only.
21. The time required to activate hot-start *Taq* DNA polymerase differs according to the manufacturer.
22. It may be necessary to omit early cycles (the first five to eight) from the baseline determination. Fluorescence can fluctuate during these cycles owing to disequilibrium of the reaction components. The baseline should be set so that it is flat and does not demonstrate a slow rise before detectable amplification is observed. This may be difficult if amplification curves span a large dynamic range as in standard curve construction.
23. A detectable C_T value in a no-template control denotes the detection of an amplification product, an indicator of contamination. If you are using SYBR Green I, check the dissociation curves to determine whether the amplicon is the actual target sequence or primer dimer.
24. The comparative C_T method of quantitation assumes the exact doubling of every target sequence (as designated by the $2^{-\Delta\Delta C_T}$ expression) during the exponential phase of amplification (thus 100% efficiency). When PCR efficiency deviates from 100%, accurate quantitation is compromised. Providing the PCR efficiency is between 90 and 110% (difference of <10% between the target gene and invariant control), then efficiency correction is not necessary. The efficiency of each assay can be calculated by constructing standard curves or measuring the slope of each amplification plot. Several methods of efficiency correction exist to improve quantitative real-time PCR data *(13,14)*. The use of standard curves to calculate expression in unknown samples automatically takes into account the efficiency of each assay.
25. $\Delta C_T \text{ SD} = \sqrt{\text{SD}_{\text{target gene}}^2 + \text{SD}_{\text{invariant control}}^2}$

References

1. Bustin, S. A. (2000) Absolute quantification of mRNA using real-time reverse transcription polymerase chain reaction assays. *J. Mol. Endocrinol.* **25,** 169–193.
2. Bustin, S. A. (2002) Quantification of mRNA using real-time reverse transcription PCR (RT-PCR): trends and problems. *J. Mol. Endocrinol.* **29,** 23–39.
3. http://www.gene-quantification.info/
4. Chomczynski, P. and Sacchi, N. (1987) Single-step method of RNA isolation by acid guanidinium thiocyanate-phenol-chloroform extraction. *Anal. Biochem.* **162,** 156–159.

5. Rozen, S. and Skaletsky, H. J. (2000) Primer3 on the WWW for general users and for biologist programmers, in *Bioinformatics Methods and Protocols: Methods in Molecular Biology* (Krawetz, S. and Misener, S., eds.), Humana Press, Totowa, NJ, pp. 365–386, http://frodo.wi.mit.edu/primer3/primer3_code.html.
6. http://biotools.idtdna.com/primerquest/.
7. Kutyavin, I. V., Afonina, I. A., Mills, A., et al. (2000) 3'-Minor groove binder-DNA probes increase sequence specificity at PCR extension temperatures. *Nucleic Acids Res.* **28,** 655–661.
8. Zuker, M. (2003) Mfold web server for nucleic acid folding and hybridization prediction. *Nucleic Acids Res.* **31,** 3406–3415.
9. http://www.molecular-beacons.org/.
10. Livak, K. J. and Schmittgen, T. D. (2001) Analysis of relative gene expression data using real-time quantitative PCR and the 2(-Delta Delta C(T)) method. *Methods* **25,** 402–408.
11. Ashton, K. J., Holmgren, K., Peart, J., et al. (2003) Effects of A_1 adenosine receptor overexpression on normoxic and post-ischaemic gene expression. *Cardiovasc. Res.* **57,** 715–726.
12. ABI User Bulletin 5: Multiplex PCR with TaqMan® VIC Probes. (2001) http://docs.appliedbiosystems.com/pebiodocs/04306236.pdf.
13. Peirson, S. N., Butler, J. N., and Foster, R. G. (2003) Experimental validation of novel and conventional approaches to quantitative real-time PCR data analysis. *Nucleic Acids Res.* **31,** e73.
14. Pfaffl, M. W., Horgan, G. W., and Dempfle, L. (2002) Relative expression software tool (REST) for group-wise comparison and statistical analysis of relative expression results in real-time PCR. *Nucleic Acids Res.* **30,** e36.

8

RNase Protection Assay for Quantifying Gene Expression Levels

Yongxia Qu and Mohamed Boutjdir

Summary

Quantifying the level of mRNA is central to the study of mammalian gene expression. Conventional approaches such as Northern blotting are often prone to low sensitivity and reproducibility. The RNase protection assay (RPA) provides a sensitive alternative for the detection and quantification of mRNA. The RPA is based on the hybridization in solution of a labeled single-stranded antisense RNA probe with a target mRNA. After hybridization, single-strand specific RNases are then used to digest away unhybridized RNA. The hybrid can be resolved by a denaturing gel. Subsequent detection will reveal the appropriate-sized gel band corresponding to the target mRNA. The major advantage of RPA is the high sensitivity and the simultaneous detection and quantification of multiple mRNA targets in a single RNA sample. The primary limitation of RPA is the lack of information on transcript size.

Key Words: mRNA; gene expression; heart; internal controls; RPA.

1. Introduction

Human heart diseases such as hypertrophy and heart failure are often associated with changes in expression of steady-state mRNA levels. These changes correspond to alterations in protein levels and myocardial function and may have clinical implications regarding etiology, clinical state, or prognosis. A variety of techniques exist to measure gene expression, including Northern blot analysis, reverse transcription polymerase chain reaction (RT-PCR), and RNase protection assay (RPA). RPA is an extremely sensitive method for the detection, quantitation, and characterization of specific mRNAs as low as 4000 copies in a complex mixture of total cellular RNA or mRNA *(1–3)*. The basis of RPA is solution hybridization of a labeled antisense cRNA probe (radiolabeled or nonisotopic labeled) to an RNA sample. After hybridization,

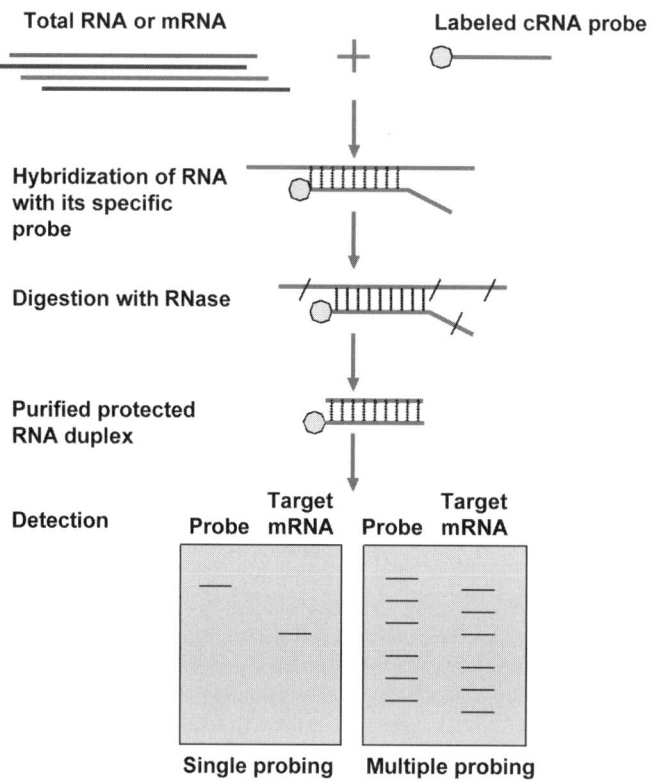

Fig. 1. Diagram of RNase protection assay.

the mixture is treated with single-strand specific RNase to degrade all remaining single-stranded RNA. Labeled probe that is hybridized to complementary RNA from the sample will be protected from RNase digestion and can be separated on a polyacrylamide gel and visualized either by autoradiography (radioactively labeled probes) or by a secondary detection procedure (nonradioactively labeled probes). The undigested cRNA probe will contain a stretch of plasmid sequence and is therefore larger than the mRNA, which it protects from the digestion by RNase. It will migrate slower than the protected fragments and is used for their identification (**Fig. 1**). When the probe is present in molar excess over the target in the hybridization reaction, the intensity of the protected fragment will be directly proportional to the amount of target RNA in the sample mixture *(1–4)*.

Quantitation of mRNA level by RPA requires the use of adequate internal controls. RPA increases substantially the degree of precision with mRNA lev-

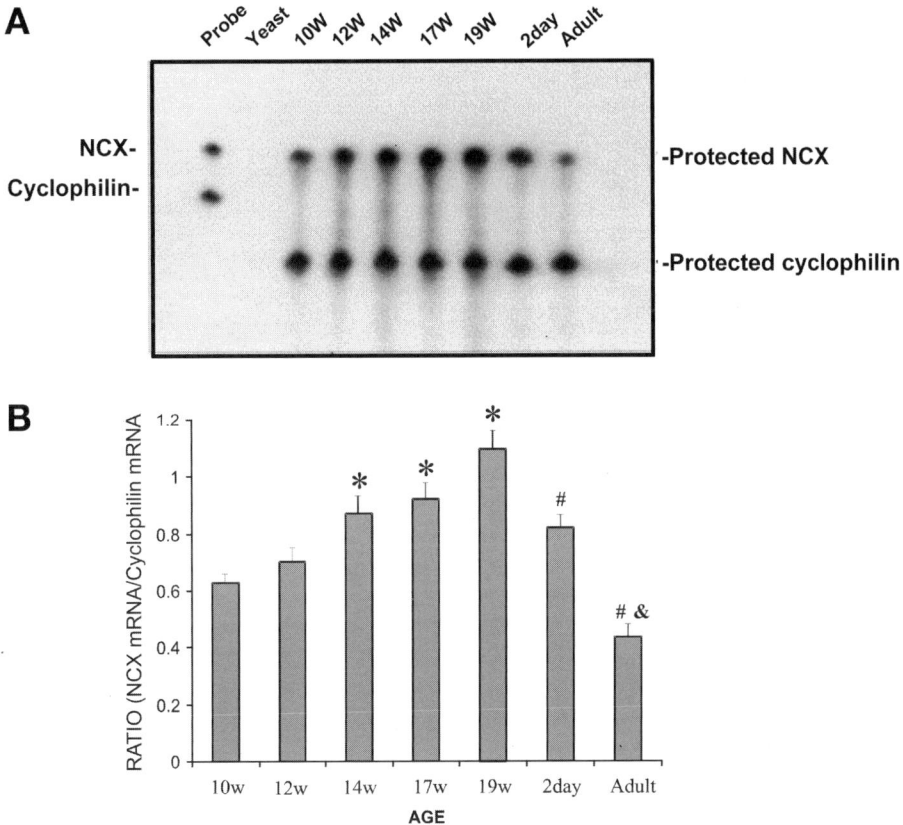

Fig. 2. Gene expression of Na/Ca exchanger (NCX) during human heart development by the RNase protection assay using cyclophilin mRNA as internal control. RPA was carried out using 10 µg total RNAs with [^{32}P]UTP-labeled NCX cRNA probe. (**A**) The first lane shows the full-length NCX and cyclophilin probes. The other bands are the protected fragments corresponding to NCX and cyclophilin mRNA. (**B**) Densitometric analysis of RPA. Density of protected NCX bands at different developmental stages is normalized to that of cyclophilin. (From **ref. 8**.)

els of a specific gene when compared from sample to sample using the internal control to normalize for differences in sample concentration and loading. The most commonly used internal controls include the constitutively expressed housekeeping genes such as glyceraldehyde-3-phosphate dehydrogenase (GAPDH), β-actin, cyclophilin, and 18S and 28S ribosomal RNA (5–7). **Figure 2** shows an example of Na/Ca exchanger (NCX) gene expression during development using RPA and cyclophilin mRNA as an internal control (8).

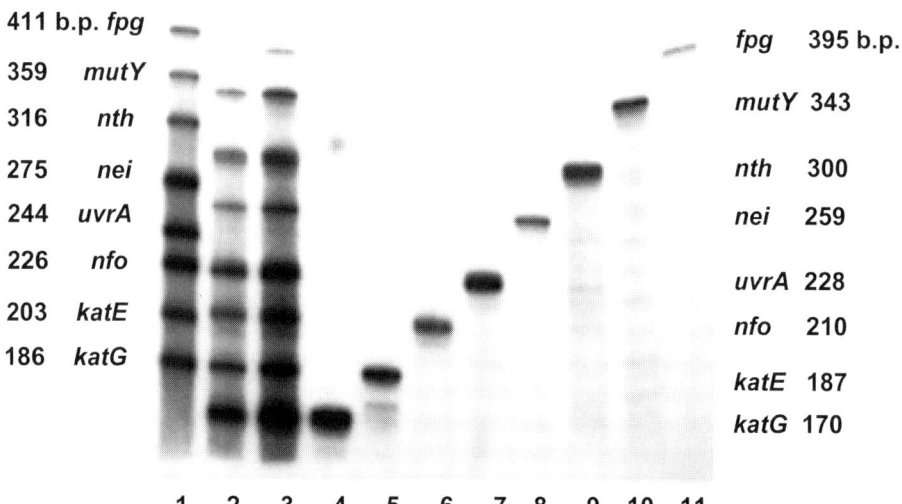

Fig. 3. Full-length antisense RNA probes and protected products from a multiprobe RNase protection assay. Lane 1, full-length probes (to the left of lane 1 are the names of the genes corresponding to each probe and size of each probe); lanes 2 and 3, protected products using 10 µg (lane 2) and 20 µg (lane 3) of *E. coli* RNA with 25,000 cpm of each full-length probe; lanes 4 to 11, protected products of each individual probe with 20 µg of RNA. To the right are the names of the genes corresponding to each probe and size corresponding to each protected product. (Reprinted with permission from **ref. 9**.)

The major advantage of RPA over the other techniques is that the expression of up to 10 or 12 mRNAs can be studied simultaneously in a given sample, as long as the probes are designed to produce different-sized, protected RNA fragments that can be separated on a denaturing polyacrylamide gel *(1,9–12)*. It is especially applicable for measuring mRNAs of multiple genes simultaneously in small amounts of biopsies. **Figure 3** illustrates the simultaneous analysis of eight genes in a given RNA sample by Gifford CM et al. *(9)*.

The advantages and disadvantages of RPA compared with other molecular methods used to measure mRNA *(13,14)* are summarized in **Table 1**.

2. Materials

A standard RPA involves the following steps: (1) RNA isolation, (2) probe generation and purification, (3) hybridization, (4) RNase digestion, and (5) separation and detection of protected fragments. In all steps of the protocol, standard precautions should be used to avoid RNase contamination and exposure of personnel to radioactivity. The material needed in RPA is addressed below for each step. There are numerous commercially available kits for RPA.

Table 1
Characteristics of Different Techniques for Gene Expression

Characteristic	RPA	Northern blot	RT-PCR
Relative/absolute quantification	Yes	Yes	Yes
Simultaneously detect multiple mRNAs	Yes	No	Limited
Sensitivity (copy of mRNA detectable)	4000–5000	10,000	50–100
Complex/labor intensive	Yes	Yes	No
Maximal sample size (RNA)	100 µg	30 µg	Not critical
Probability of inaccurate measurement owing to gene polymorphisms	High	Low	Low
Mapping studies	Yes	No	No
Tolerable partial RNA degradation	Yes	No	Yes
Size determination of mRNA transcripts	No	Yes	No
Resolve comigrating mRNA	Yes	No	Yes
Detection of alternatively spliced transcripts	No	Yes	No

In this chapter, we describe the material and methods using the RPA III™ Ribonuclease Protection Assay Kit from Ambion (Austin, TX) as an example.

2.1. RNA Isolation

1. Commercial RNA isolation kits: for total RNA isolation, there are numerous commercial kits such as the RNAzol™ B kit (TEL-TEST) *(8)*, RNAwiz™ (Ambion, cat. no. 9736) *(15,16)*, RNAgents® Total RNA Isolation System (Promega, cat. no. Z5110), and others. All these kits allow for isolation of high-quality total RNA to be used in RPA in less than 1 h to a few hours.
2. Chloroform.
3. Isopropanol.
4. ACS grade ethanol.
5. Agarose.
6. Ethidium bromide. [**Caution:** Carcinogenic; handle according to Material Safety Data Sheets (MSDS).]
7. Some RNA isolation procedures include phenol and chloroform. **Caution:** Organic extractions should be done in a fume hood using suitable protective clothing and eye protection.

2.2. Probe Generation and Purification

2.2.1. In Vitro Transcription

1. In vitro transcription kits: many commercial in vitro transcription kits are available. Generally, in vitro transcription kits include one of the T7/T3/SP6 RNA polymerases, NTPs, dithiothreitol (DTT)-containing transcription buffer, gel

loading buffer, control template, nuclease-free water, elution buffer, and so on. Examples include the Maxiscript in vitro transcription kit from Ambion and the AmpliScribe™ T7-Flash™ Transcription Kit from Epicentre.
2. Linearized cDNA or PCR template downstream of T7, or T3 or SP6 promotor (*see* **Note 2**).
3. [^{32}P]UTP: 10 mCi/mL, 800 Ci/mmol (Amersham).

2.2.2. Gel Purification

1. Denaturing gel preparation: acrylamide/bis-acrylamide 19:1, TEMED, urea, ammonium persulfate. Acrylamide and bis-acrylamide are slowly deaminated to acrylic acid; the reaction is catalyzed by light and alkali.
   a. A 15 mL solution is sufficient for a 13 cm × 15 cm × 0.75 mm gel. The components are: 7.2 g urea, 1.5 mL 10X TBE, 2.5 mL 30% acrylamide (acrylamide/bis-acrylamide 19:1); add dH$_2$O to 15 mL.
   b. Stir at room temperature, and then add 20 µL 10% ammonium persulfate and 6 µL TEMED; mix and pour the gel.
   c. Check the pH of the solution (neutral) and store it in the dark at 4°C.
   d. Use sequencing grade reagents.
   e. Discard the solution when a precipitate forms.
   f. Allow all unused acrylamide to polymerize before disposal.
2. Gel running buffer preparation: Tris-HCl, boric acid, EDTA.
3. Yeast RNA.
4. Gel loading and gel elution buffer preparation: usually comes with the RPA kits.
5. Tris-HCl-saturated phenol.

2.3. Probe/Template Hybridization

1. Sample RNA: total RNA from **Subheading 2.1.**
2. ^{32}P-labeled cRNA probe (from **Subheading 2.2.**).
3. 5*M* NH$_4$OAc.
4. Ethanol.
5. Yeast RNA.
6. Hybridization buffer: comes with RPA kit.

2.4. RNases Digestion

All the reagents listed below come with the Ambion RPA III™ kit.
1. Digestion buffer.
2. Yeast RNA.
3. RNase A/T1 mix: 250 U/mL RNase A and 10,000 U/mL RNase T1.
4. RNase inactivation/precipitation solution.

2.5. Separation and Detection of Protected Fragments

1. Denaturing gel, gel loading buffer, and gel running buffer: same as in **Subheading 2.2.2.**

3. Methods
3.1. Total RNA Isolation

In cardiovascular research, most of the total RNA will be prepared from heart tissue or cardiomyocytes in culture subjected to different experimental conditions. RNA isolation procedures have improved greatly over the past two decades in terms of simplicity and speed. The RNA isolation kits listed in **Subheading 2.1.**, **item 1** allow for isolation of high-quality total RNA to be used in RPA in less than 1 h to a few hours. The RPA tolerates partially degraded RNA; however, it is always worthwhile to check the RNA quality by the 260/280 ratio (>1.8) and by ethidium bromide-stained native 1% agarose gel (*see* **Note 1**). Smearing of the ribosomal RNA bands indicates degradation.

3.2. cRNA Probe Preparation

Probe preparation will be described using ^{32}P as an example (*see* **Note 3**). Reagents used are from Ambion *(17–19)* unless otherwise indicated.

3.2.1. In Vitro Transcription

1. Assembly of in vitro transcription reaction (20 µL): 1 µg DNA template (*see* **Note 2**), 2 µL 10X transcription buffer (at room temperature), 1 µL 10 mM ATP, 1 µL 10 mM CTP, 1 µL 10 mM GTP, 5 µL [^{32}P]UTP (10 mCi/mL, 800 Ci/mmol), 1 µL RNAse inhibitor, 2 µL T7 or T3 or SP6 enzyme mix (depends on template), and nuclease-free water to 20 µL
2. Mix and incubate at 37°C for 60 min.
3. Heat-denature the transcription reaction at 95°C for 2 min (to ensure the transcript doesn't protect its DNA template). Chill on ice, add 1 µL DNase I, and incubate at 37°C for 20 min.
4. Add 1 µL of 0.5 M EDTA to block the potential heat-induced RNA degradation.

3.2.2. Probe Purification

To avoid high background in RPA, we recommend gel purification of the probes to obtain primarily full-length probe (*see* **Note 3**).

1. Prepare 5% acrylamide/8 M urea denaturing polyacrylamide as described in **Subheading 2.2.2.**
2. Loading and running the gel: add 20 µL gel loading buffer to the probe from in vitro transcription, and heat at 95°C for 10 min. Load the gel and run at 300 V in 1X TBE buffer.
3. Autoradiography.

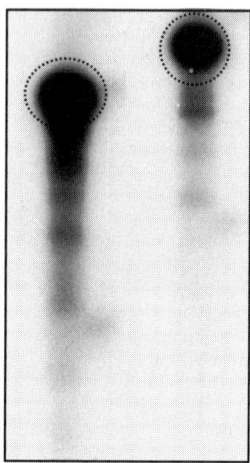

Fig. 4. Autoradiography of two different size cRNA probes from in vitro transcription in the presence of [^{32}P]UTP. Bands in the circle indicate the full-length probes.

     a. Remove one glass from the gel, wrap the gel entirely in the plastic wrap, and expose to film to get a clear band.
     b. Be careful to mark the orientation of the film.
     c. Align the film carefully with the gel and excise the band.
     d. The full-length transcript is usually the most slowly migrating, most intense band on the gel (circled bands of **Fig. 4**).
4. Recovery of probe: incubate the gel slice with 350 µL elution buffer at 37°C overnight. Take the gel slice out and centrifuge briefly.
5. Concentrate the probe (optional).
     a. Add 300 µL phenol, mix, and microcentrifuge for 4 min.
     b. Take the supernatant, add 10 µg yeast RNA and 750 µL cold 100% ethanol, mix, and keep at –20°C for 1 h.
     c. Centrifuge at 4°C for 15 min at maximal speed in desktop centrifuge.
     d. Carefully remove all residue of supernatant, and resuspend the pellet in 50 µL gel elution buffer.
     e. Take 2 µL out for quantification by scintillation counting (cpm/µL).

3.3. Probe/Template Hybridization

1. Mix sample RNA and labeled probe: Sample RNA can be up to 100 µg. The most commonly used amount is 10 or 20 µg total RNA. Use about 2 to 8×10^4 cpm probe per 10 µg total sample RNA (*see* **Note 4**).
2. Setup controls.
     a. Prepare two control tubes containing the same amount of labeled probes plus yeast RNA equivalent to the highest amount of sample RNA.

b. One tube will serve as a no target control to check the function of RNase, the probe self-hybridization, and the presence of residue template DNA in the probe preparation. Ideally, after digestion with RNase, there should be no signal in this lane of the gel.
c. The other tube will serve as a no RNase control to check the probe integrity. This lane should show a single band at the expected probe size.
d. Some investigators also include a positive control using synthetic sense cRNA.

3. Coprecipitate the probes and sample RNAs
   a. Add NH_4OAc to a final concentration of 0.5 M. Add 2.5 vol of EtOH, mix, and incubate at –20°C for at least 15 min.
   b. Centrifuge at maximal speed in desktop centrifuger at 4°C for 15 min.
   c. Remove all traces of EtOH and dry the pellet for 5 min.
4. Resuspend the pellet: resuspend the pellets in 10 µL hybridization buffer. Vortex to make sure the pellet resolves completely and microcentrifuge for a few seconds to collect the liquid at the bottom of the tube.
5. Denature RNA: heat-denature RNA at 95°C for 4 min. Vortex and microcentrifuge briefly. Incubate overnight at 42°C (*see* **Note 4f** and **g**).

3.4. RNase Digestion (see Note 5)

After hybridization, the specific probe will hybridize to the target mRNA to create a double-stranded RNA that is resistant to RNase digestion. A mixture of RNase A and RNase T1 is then used to remove all the unhybridized segments. The remaining protected fragments will then be precipitated, washed, and resuspended to be separated and detected.

1. To each sample RNA tube and one yeast RNA control tube (no target control), add: 150 µL RNase digestion buffer and 1.5 µL RNase A/T1 mix.
2. For the second yeast RNA tube, add only RNase digestion buffer with no RNase (no RNase control).
3. Incubate at 37°C for 30 min.
4. Add 225 µL RNase inactivation/precipitation solution, and incubate at –20°C for at least 15 min. We recommend adding 20 to 50 µg of yeast RNA to increase the size and visibility of the final pellets.
5. Centrifuge at maximal speed in desktop centrifuge at 4°C for 15 min.
6. Remove the supernatant completely and wash the pellet with 70% EtOH. (Note: the RNase inactivation/precipitation solution from Ambion will replace the conventional proteinase K and phenol-chloroform steps to allow the entire procedure be performed in a single tube.)
7. Resuspend the pellet in 4 to 10 µL of gel loading buffer.

3.5. Separation and Detection of Protected Fragments (see Note 6)

1. Heat the sample from **Subheading 3.4.** at 95°C for 3 min.
2. Load the sample immediately in 5% denaturing polyacrylamide gel. For the no RNase control, load only 10% to avoid obscuring the signal from adjacent lanes.

3. Run the gel at 200 V until the blue dye reaches the bottom of the gel.
4. Transfer and dry the gel on chromatography paper.
5. Expose the dried gel directly to film. We routinely develop the film after overnight exposure to evaluate the signal density to decide the exposure duration.
6. Perform densitometric analysis of the protected bands.

4. Notes

A standard RPA started with sample RNA isolation can be completed in 3 d. The exposure time for autoradiography depends on the abundance of the target mRNA.

1. Sample RNA preparation for RPA: Total RNA preparation should be free of protein and DNA. For rare mRNA, mRNA could be used instead of total RNA.
2. Template preparation for in vitro transcription.
   a. The optimal probe size for RPAs is 200 to 500 nucleotides. The shorter the probe, the less sensitive the assay will be. Probe longer than 1000 bases will be difficult to separate on polyacrylamide gels.
   b. To prepare an antisense cRNA probe, the plasmid templates for in vitro transcription need to be linearized by restriction enzyme digestion at or near the 5' end of the insert. Phenol extraction and ethanol precipitation is recommended after plasmid linearization.
   c. Plasmid DNA template must be completely linearized. Circular plasmid templates will generate extremely long, heterogeneous RNA transcripts.
   d. DNA template for in vitro transcription can be generated using PCR to add phage promoters to specific DNA sequences without the need to subclone into a phage transcription vector *(20,21)*.
3. Probe preparation.
   a. In aqueous solution, RNA-RNA hybrids are more stable than RNA-DNA hybrids *(22)*. Thus, antisense cRNA probe is routinely used for RPA.
   b. ^{32}P-labeled probes coupled with autoradiographic detection provide the highest degree of sensitivity and resolution currently available in hybridization assays *(2,23)*.
   c. The transcription reagents (transcription buffer, nucleotides) should be at room temperature prior to adding template to avoid precipitating DNA at low temperature.
   d. The ^{32}P-labeled probe does not last longer than 1 or 2 d because of radiolysis. Use the probes as soon as possible after gel purification.
   e. Use of [^{32}P]UTP that has decayed one half-life may lead to decreased probe labeling and increased lane background.
   f. To avoid high background in RPA, we highly recommend gel-purifying the probes to get primarily full-length probe.
   g. Add yeast RNA as a carrier to facilitate probe precipitation.

h. There is always a tradeoff when using hot and cold UTP. Increasing the unlabeled UTP will increase the full-length probe synthesis but will decrease the sensitivity of the probe.
   i. It is critical to eliminate DNA template by DNase I digestion after the RNA transcription reaction. Leftovers of template DNAs will show full-length sized bands in all lanes of autoradiography, because they will hybridize with the cRNA probes.
   j. The probe should be longer than the protected fragment to allow separation of the full-length probe and the protected fragments.
   k. Nonradioactive labeled probes: RPAs using ^{32}P-labeled probes are so far the most sensitive. For investigators concerned with safety issues associated with radioactivity, nonradioactive probe labeling and RPA kits are also commercially available. Nonradioactive labeled probes are stable for at least 1 yr, and the detection time is shortened.
4. Hybridization.
   a. Sample RNA can be with either total RNA or mRNA.
   b. RPA tolerates partially degraded RNA. However, it is always worthwhile to check the total RNA quality by ethidium bromide-stained native 1% agarose gel.
   c. The amount of sample RNA required will depend on the abundance of the mRNA being detected and on the specific activity of the probe. Assays can be carried out using total RNA up to 100 µg. However, the most commonly used quantity is 5 to 20 µg of total RNA.
   d. For quantitative purposes, it is important that the labeled probe be present in molar excess over the target mRNA. In most cases, 2 to 8×10^4 cpm of high-specific-activity probe per 10 mg total RNA is sufficient to be in molar excess for fairly abundant mRNAs.
   e. For accurate and reproducible quantification of target mRNAs, all the RNA must be completely dissolved in hybridization buffer.
   f. Hybridization temperature: this temperature can be varied between 42°C and 68°C for certain RNAs for optimization.
   g. For multiple probe detection, increasing hybridization stringency by raising the hybridization temperature to 56°C or 68°C can eliminate the cross-hybridization without affecting probe-target interactions.
5. Digestion
   a. Choice of RNases: RNase A/T1 is the most common choice for RPAs *(24)*.
   b. RNase conditions: the amount of RNase to be used has to be determined empirically. It is usually easier to change the RNase concentration before altering the hybridization temperature.
6. Detection
   a. Let the gel warm up to room temperature from –80°C and wipe off any condensation before film development.
   b. The no target/+RNase lane should have no signal, and the no target/no RNase lane should show a band corresponding to full-length probe (*see* **Figs. 2** and **3**).

c. If severe degradation of probe in the yeast RNA/–RNase lane occurs, then ribonuclease contamination of the tubes and pipet tips should be suspected.
d. If the no target/+RNase control lanes show protected bands, then the function of RNase, the probe self-hybridization, and leftover template DNA in the probe preparation should be suspected.
e. Protected fragments are smeared or consist of multiple bands: sample RNA degradation, excess probe, degradation of the probes, and so forth should be suspected. Check the RNA integrity, increase the hybridization temperature, gel-purify the probe, decrease the probe concentration, increase the RNase digestion stringency, and use mRNA instead of total RNA.
f. No protected bands at all, but the probe is intact: it is possible that the gene of interest is not expressed, or sample RNA is degraded, or pellet loss has occurred or the final pellet did not resuspend in the loading buffer well enough, or the probe sequence differs from the target mRNA sequence in the expected protected fragment region. Alternatively, the target mRNA is rare and the experimental condition is not sensitive enough. In this case, increase the total sample RNA, increase the probe sensitivity by using only radiolabeled but none of the unlabeled nucleotide in the transcription reaction, or prepare a longer probe.
g. Full-length probe is seen in all lanes: check the RNase function, reduce the probe used, and treat RNA probe preparation with DNase I.
h. For abundant mRNA, a few hours of exposure will give strong signals. Less abundant mRNA will need 5 d to up to 2 wk of exposure time.
i. Internal control: the signal for the internal control in RPA is used to normalize each sample for variations in the amount of starting RNA. It is therefore necessary to identify the appropriate control RNA for the particular set of experimental RNA samples to be studied. The prime requirement for any control transcript is maintenance of a consistent expression level under all experimental conditions. The most commonly used internal controls include the constitutively expressed housekeeping genes such as GAPDH, β-actin, cyclophilin, and 18S and 28S ribosomal RNA. However, these popular internal controls are not always constant. Both β-actin, cyclophilin and GAPDH mRNA levels may fluctuate under a variety of conditions. For example, β-actin mRNA levels were found to increase in rat myocardium following abdominal aortic banding *(25)*, and GAPDH mRNA levels increase to hypoxic conditions *(26)*. rRNAs have their own drawback of being ribosomal RNA but not mRNAs. Thus, the widely used internal controls are not always ideal normalizers. Although none of the controls appears to be ideal, there are several choices for the best candidate. First, the least variant of the internal controls are rRNAs. In addition, since rRNA makes up 80% of a total RNA sample, the concentration of total RNA is rather based on the rRNAs. Second, more than one internal control can be used. Third, levels of the mRNA of interest can be simply normalized to be RNA loaded (μg) per lane and repeated multiple times.

j. The most commonly used internal control templates listed in **Note 6i** are commercially available. Since the internal controls usually express at moderate or high levels in the cells, low-specific-activity probes are needed for the internal controls. This can be achieved by adding the cold UTP in the in vitro transcription reaction. Altering the specific activity of probes reduces their signal intensity to the level that rare and abundant transcripts can be detected with a single exposure.

References

1. Melton, D. A., Krieg, P. A., Rebagliati, M. R., Maniatis, T., Zinn, K., and Green, M. R. (1984) Efficient in vitro synthesis of biologically active RNA and RNA hybridization probes from plasmids containing a bacteriophage SP6 promoter. *Nucleic Acids Res.* **12,** 7035–7056.
2. Sambrook, J., Fritsch, E. F., and Maniatis, T. (eds.) (1989) *Molecular Cloning: A Laboratory Manual*, 2nd ed., Cold Spring Harbor Laboratory Press, Cold Spring Harbor, NY.
3. Gilman, M. (1989) Ribonuclease protection assay. In: *Current Protocols in Molecular Biology*, vol. 2 (Ausubel, F. M., Roger Brent, R., Kingston, R. E., et al., eds.), John Wiley & Sons, New York, pp. 4.7.1–4.7.8.
4. Einspanier, R. and Plath, A. (1998) Detection mRNA by use of the ribonuclease protection assay (RPA), in *Molecular Biomethods Handbook* (Rapley, R. and Walker, J. M., eds.), Humana Press, Totowa, NJ, pp. 51–58.
5. Suzuki, T., Higgins, P. J., and Crawford, D. R. (2000) Control selection for RNA quantitation. *Biotechniques* **29,** 332–337.
6. Zhong, H. and Simons, J. W. (1999) Direct comparison of GAPDH, beta-actin, cyclophilin, and 28S rRNA as internal standards for quantifying RNA levels under hypoxia. *Biochem. Biophys. Res. Commun.* **259,** 523–526.
7. Spanakis, E. (1993) Problems related to the interpretation of autoradiographic data on gene expression using common constitutive transcripts as controls. *Nucleic Acids Res.* **21,** 3809–3819.
8. Qu, Y., Ghatpande, A., el-Sherif, N., and Boutjdir, M. (2000) Gene expression of Na/Ca exchanger during development in human heart. *Cardiovasc. Res.* **45,** 866–873.
9. Gifford, C. M., Blaisdell, J. O., and Wallace, S. S. (2000) Multiprobe RNase protection assay analysis of mRNA levels for the *Escherichia coli* oxidative DNA glycosylase genes under conditions of oxidative stress. *J. Bacteriol.* **182,** 5416–5424.
10. Melton, D. A., Krieg, P. A., Rebagliati, M. R., Maniatis, T., Zinn, K., and Green, M. R. (1984) Efficient in vitro synthesis of biologically active RNA and RNA hybridization probes from plasmids containing a bacteriophage SP6 promoter. *Nucleic Acids Res.* **12,** 7035–7056.
11. Ngai, J., Dowling, M. M., Buck, L., Axel, R., and Chess, A. (1993) The family of genes encoding odorant receptors in the channel catfish. *Cell* **72,** 657–666.
12. Hobbs, M. V., Weigle, W. O., Noonan, D. J., et al. (1993) Patterns of cytokine gene expression by CD4+ T cells from young and old mice. *J. Immunol.* **150,** 3602–3614.

13. Ambion, Inc. Strategies for detecting mRNA Northern blotting, nuclease protection assays, in situ hybridization, and RT-PCR. Technote 6(3), http://ambion.com/techlib/tn/63/631.html.
14. Rottman, J. B. (2002). The ribonuclease protection assay: a powerful tool for the veterinary pathologist. *Vet. Pathol.* **39,** 2–9.
15. Brenner, R., Yu, J. Y., Srinivasan, K., et al. (2000) Complementation of physiological and behavioral defects by a slowpoke Ca-activated K channel transgene. *J. Neurochem.* **75,** 1310–1319.
16. Simonic, T., Duga, S., Strumbo, B., Asselta, R., Ceciliani, F., and Ronchi, S. (2000) cDNA cloning of turtle prion protein. *FEBS. Lett.* **469,** 33–38.
17. Yan, N., Moscovitz, R., Udenfriend, S., and Tate, S. S. (1992) Distribution of mRNA of a Na-independent neutral amino acid transporter cloned from rat kidney and its expression in mammalian tissues and *Xenopus laevis* oocytes. *Proc. Natl. Acad. Sci. USA* **89,** 9982–9985.
18. Xu, J., Matsuzaki, K., McKeehan, K., Wang, F., Kan, M., and McKeehan, W. L. (1994) Genomic structure and cloned cDNAs predict that four variants in the kinase domain of serine/threonine kinase receptors arise by alternative splicing and poly(A) addition. *Proc. Natl. Acad. Sci. USA* **91,** 7957–7961.
19. Takenaka, M., Bagnasco, S. M., Preston, A. S., et al. (1995) The canin betaine gamma-amino-n-acid transporter gene: divers mRNA isoforms are regulated by hypertonicity and are expressed in a tissue-specific manner. *Proc. Natl. Acad. Sci. USA* **92,** 1072–1076.
20. Mullis, K. B. and Faloona, F. (1987) Specific synthesis of DNA in vitro via a polymerase catalyzed chain reaction. *Methods Enzymol.* **155,** 335–350.
21. Stoflet, E. S., Koeberl, D. D., Sarkar, G., and Sommer, S. S. (1988) Genomic amplification with transcript sequencing. *Science* **239,** 491–494.
22. Alphey, L. and Parry, H. D. (1995) Making nucleic acid probes, in *DNA Cloning I: Core Techniques* (Glover, D. M. and Hames, B. D., eds.), IRL, Oxford, UK, pp. 121–141.
23. Keller, G. H. and Manak, M. M. (eds.) (1989) *DNA Probes.* Stockton, New York.
24. Winter, E., Yamamoto, F., Almognesa, C., and Perucho, M. (1985) A method to detect and characterize point mutations in transcribed genes: amplification and overexpression of the mutant c-Ki-ras allele in human tumor cells. *Proc. Nat. Acad. Sci.* USA **82,** 7575–7579.
25. Chapman, D., Weber, K. T., and Eghbali, M. (1990) Regulation of fibrillar collagen types I and III and basement membrane type IV collagen gene expression in pressure overloaded rat myocardium. *Circ. Res.* **67,** 787–794.
26. Graven, K. K. and Farber, H. W. (1998) Endothelial cell hypoxic stress proteins. *J. Lab. Clin. Med.* **132,** 456–463.

9

In Situ Hybridization

A Technique to Study Localization of Cardiac Gene Expression

Thierry P. Calmels and David Mazurais

Summary

In situ hybridization allows the detection of specific gene transcripts in tissues, cells or, chromosomes. In the cardiovascular field, this powerful and rapid methodology provides precious insights into the complex gene organization and expression within an heterogeneous cell population. This technique is particularly useful to elucidate the genes and pathways involved in cardiac cells processes (differentiation, proliferation, apoptosis) or in the development of cardiovascular pathologies. *In situ* hybridization allows the precise localization of gene transcripts to the different heart regions and to individual cell types such as working cardiomyocytes, cells from conductive tissues and blood vessels displaying specific functions. This chapter describes the different technical procedures that are of crucial importance to carry on sensitive and specific *in situ* hybridization experiments in heart samples. The detection of transcripts within paraformaldehyde-fixed, paraffin-embedded cardiac tissue samples is illustrated here with the detection of cardiac sphingosine-1-phosphate receptor expression.

Key Words: *In situ* hybridization; heart tissues; cardiac gene expression; gene transcript localization; probe synthesis; paraformaldehyde tissue fixation; paraffin-embedded tissue; S1P receptors.

1. Introduction

In situ hybridization, as the name suggests, allows specific nucleic acid sequences (either mRNA in the cytoplasm or DNA in the nucleus) to be detected in morphologically preserved chromosomes, cells, or tissue sections. The basic principle of this technique, originally developed by Pardue and Gall *(1)* and John and colleagues *(2)*, relies on the fact that DNA and RNA will undergo hydrogen bonding to complementary sequences of DNA or RNA.

Hybridization on membrane supports requires the isolation of DNA or RNA, separation on a gel, blotting onto nitrocellulose or nylon, and probing with a complementary sequence. In contrary, *in situ* hybridization is a powerful technique offering a unique approach for studying the macroscopic distribution and cellular localization of DNA and RNA sequences directly in a heterogeneous cell population. In combination with immunocytochemistry, *in situ* hybridization relates microscopic topological information to gene activity at the mRNA level. However, contrary to the immunological approach, which necessitates a tedious development of specific antibodies, *in situ* hybridization represents a rapid analysis using specific probes generated from fragments of known DNA sequence. In the past three decades, many improvements have been made on technologies and methodologies related to *in situ* hybridization that has been carried out in different tissues. Such methodological improvements have notably led to the detection of low-abundance RNA/DNA in the range of 10 to 20 copies of mRNA or DNA per cell *(3)*. Giving precious spatial information, this technique is particularly useful in providing a better understanding of the cardiac genes and the pathways involved in cardiac cell differentiation, embryogenesis, or the development of cardiovascular pathologies. Based on the massive increase in available human sequences, the development of new approaches such as microarray technologies provides novel insights by quantitative large-scale gene expression analyses for the diagnosis and treatment of cardiovascular diseases and for the study of cardiac myocyte differentiation *(4,5)*. Thus *in situ* hybridization represents a powerful tool to investigate spatial gene expression. It can help investigators to understand better how transcriptome regulation in various cell types from different regions in the heart may mediate various pathologies. *In situ* hybridization indeed allows the precise localization of gene transcripts to heart-specific regions (e.g., the ventricle, atrium, auricle, septum) and/or to specific cell types (working cardiomyocytes, conductive tissues, blood vessels).

Because of the originality and characteristics of an *in situ* approach, specific difficulties encountered in signal detection must be overcome. Although the global expression of a specific gene can be easily detected by Northern blot and/or reverse transcription polymerase chain reaction (RT-PCR), the amount of a specific transcript may be found at very low concentration within each individual cell such as cardiomyocyte, smooth muscle, or endothelial cell. Likewise, transcripts may also be protected within cellular structures or masked by associated proteins.

Therefore, it is important to consider the most sensitive protocol providing good cell permeability and accessibility of the labeled probe to the targeted nucleotide sequence without damaging the structural integrity of the analyzed tissues. In our attempts to study the expression of several genes in human cardiovascular tissues *(6,7)*, we have developed a sensitive and selective procedure allowing the detection of low transcript levels in fixed tissue. The expression of human cardiac sphingosine-1-phosphate (S1P) receptors (S1P1 and S1P3) is described to illustrate the procedure given here.

2. Materials

Normal heart sections (Hybrid Ready Tissues) originating from human donors were obtained from Novagen (Madison, WI). Rat heart sections were prepared in the laboratory.

2.1. Special Equipment

1. Hypercassettes™, β-max film, super Fix MRP (fixater), Roentoroll (developer), and film hyperprocessor (Amersham Bioscience, Cleveland, OH).
2. Microscope: after development, sections were analyzed on a Nikon (Eclipse E 800, Sony camera DXC-950P) microscope.
3. Computer-based image analysis system: Biocom VisioL@b 2000, version V4,00 SR-5 (Biocom, Les Ulis, France).

2.2. Chemicals and Reagents

1. Proteinase K, nucleotides, RNA polymerases, and yeast tRNA (Boehringer Mannheim, Indianapolis, IN).
2. Sephadex column (Bio-Spin/Micro Bio-Spin 30; Bio-Rad, Hercules, CA).
3. [^{35}S]Uridine triphosphate (UTP; Dupont NEN, Boston, MA).
4. Ilford K5 nuclear track emulsion (Polysciences, Warrington, PA), developer D-19 (Kodak).
5. RNAse inhibitor (RNAsin), restriction enzymes, RNase-free DNase (RQ-1 DNase), dithiothreitol (DTT) (Promega, Madison, WI).
6. Acetic anhydride, diethylpyrocarbonate (DEPC), dextran sulfate, RNase A, Denhardt's 50X solution, ethanol, xylene, chlorhydric acid, triethanolamine (Sigma, St. Louis, MO).
7. pBluescript II SK+ cloning vector (Stratagene, La Jolla, CA).
8. Paraplast: Paraplast Tissue Embedding Medium (Sherwood, St. Louis, MO).
9. SuperFrost/Plus slides (Fisher, Hampton, NH).
10. Mounting medium (Pertex, Microm, France).
11. Sterile filters (0.22-μm filter system, Corning).

2.3. Buffers

Until hybridization, it is very important to ensure that all procedures are carried out in RNAse-free conditions *(8)*. For this reason, it is highly recommended to wear gloves and to decontaminate laboratory instruments with detergent and DEPC water before sterilization.

Phosphate-buffered saline (PBS), NaCl, and triethanolamine solutions must be prepared with DEPC-treated distilled water, filtered through sterile filters

(0.22-μm filter system, Corning), and sterilized. Ethanol dilutions must also be prepared using DEPC-treated distilled water.

1. 0.1 M PBS (pH 7.4): weigh 0.262 g monohydrate NaH_2PO_4, 1.146 g anhydrous Na_2HPO_4, and 0.9 g of NaCl. Dissolve in distilled water, qsp 100 mL. This buffer is stable at 4°C for up to 2 mo after autoclaving.
2. 0.9% NaCl: weigh 900 mg NaCl per 100 mL distilled water. This is stable at 4°C for up to 2 mo after autoclaving.
3. 0.1 M Triethanolamine (pH 8.0). This solution must be freshly prepared and used in the same working day. Add 18.57 g triethanolamine to 900 mL DEPC water. Dissolve and adjust to pH 8.0 using NaOH. Adjust to 1 L with DEPC water and sterilize in an autoclave.
4. 4% Paraformaldehyde (PAF) in PBS. **Caution:** This solution must be prepared carefully. Paraformaldehyde is toxic.
   a. Weigh 4 g of paraformaldehyde into a conical flask.
   b. Add 60 mL of sterile PBS and cover with parafilm.
   c. In a fume hood, warm it up to 60°C and mix.
   d. Add a few drops of 1 M NaOH and continue to agitate until it dissolves completely.
   e. Then switch off the heat and adjust the pH to 7.2 to 7.4.
   f. Make up the volume to 100 mL with PBS.
   g. It is important to use freshly prepared formaldehyde since the pH from a stock bottle changes over time and may lead to inefficient fixation as well as degradation of tissue and RNA.
5. 40 μg/mL Proteinase K stock solution in 10 mM EDTA, 100 mM Tris-HCl, pH 7.5.
6. Washing solution 1: 5X SSC (SSC 1X: 15 mM trisodium citrate, 150 mM NaCl, pH 7.0), 10 mM DTT.
7. Washing solution 2: 2X SSC, 50% deionized formamide, 10 mM DTT.
8. NTE buffer: 0.5 mM NaCl, 10 mM Tris-HCl, 5 mM EDTA, pH 8.0.
9. Kodak D-19 filtered solution: 16% in ddH_2O.
10. Stop solution: 1% acetic acid and 1% glycerol in ddH_2O.

3. Methods

The methods described here outline (1) slide preparation, (2) treatments prior hybridization, (3) cRNA probe synthesis, (4) hybridization, (5) washing, and (6) detection of the hybridization signal. The timing of **steps 1** to **7** below is estimated for 20 to 30 slides and may fluctuate depending on the number of the slides processed.

1. Tissue fixation, overnight, and slide preparation: 2 to 3 d.
2. Treatments prior to hybridization (dewaxing, postfixation, permeabilization, acetylation, and dehydration): 3 h.
3. Probe preparation: 3 to 5 h.

4. Hybridization: overnight.
5. Washing steps: 3 h.
6. Emulsion dipping: 5 h.
7. Development of emulsion-dipped slides: 1 h.

3.1. Slide Preparation

3.1.1. Human Heart Sections

Normal human heart sections (Novagen, Madison, WI) originate from tissues fixed in 4% paraformaldehyde, embedded in paraffin, and sectioned at 5 μm thickness (*see* **Note 1**).

3.1.2. Rat Heart Sections

Some experiments are also performed using rat heart samples prepared in the laboratory (*see* **Note 2**).

1. Rats are sacrificed with as little stress as possible to avoid dramatic fluctuations of mRNA levels. All procedures carried out on animals were conducted in agreement with the institutional guidelines for the use of animals and their care (European Communities directive 86/609/EEC).
2. Hearts are briefly but properly rinsed in sterile PBS to remove blood (a putative source of artifactual hybridization signal).
3. Hearts are rapidly fixed in freshly prepared paraformaldehyde (4% PAF in PBS) to avoid RNA degradation. Heart fixation occurs overnight at 4°C to ensure RNA stability.
    Note: During the following ethanol, xylene, or paraplast incubations, the solution volume/tissue ratio should be 10:1 or greater. Take care when changing solutions to avoid damaging the heart.
4. The heart is dehydrated by serial ethanol incubations:

    70% Ethanol: 30 min at room temperature (RT)
    80% Ethanol: 30 min at RT
    95% Ethanol: 30 min at RT (twice)
    100% Ethanol: 45 min at RT (twice

    A fixed heart is stable for several months at −20°C in 100% ethanol.
5. Clearing: to prepare the specimen for paraplast infiltration, ethanol is removed by two successive incubations of the heart in fresh xylene.

    100% Xylene: 30 min at RT
    100% Xylene: 45 min at RT
6. Infiltration: immerse the heart in infiltrating molten paraplast for 30 min at 58°C. The paraplast melting temperature should not exceed 62°C. Replace the paraplast bath by fresh infiltrating paraplast and let it incubate for 1 h at 58°C (twice).
7. Embedding: the heart is finally embedded in fresh paraplast using an adequate mould at RT. To determine the future orientation of the sections, take care to note

the position of the heart in the paraplast block. At this step, the embedded heart is stable for several months at 4°C.
8. Sectioning: the embedded rat heart is mounted onto microtome blocks and sectioned at 5 to 7 µm thickness. We have found this to be the optimal thickness for rat heart to achieve both good signal and resolution. A clean container filled with sterile distilled water at 40°C is used to spread sections on SuperFrost/Plus slides. Dried sections can be stored for a few days at 4°C under dry atmosphere.

3.2. Treatments Prior to Hybridization

3.2.1. Dewaxing and Rehydration

Human and rat heart sections are dewaxed and rehydrated using the following steps:

1. Remove the slides from 4°C storage and allow them to warm up to RT.
2. Dewax sections by three 5-min washes using fresh xylene.
3. Rehydration in graded ethanol solutions:
   100% Ethanol: 5 min at RT (twice).
   95% Ethanol: 5 min at RT.
   85% Ethanol: 5 min at RT.
   70% Ethanol: 5 min at RT (twice).
   50% Ethanol: 5 min at RT.
   30% Ethanol: 5 min at RT.

   Note that it is important to prevent sections from drying out during these successive steps to prevent a significant increase in signal background.
4. Rinse sections in 0.9% NaCl for 3 min at RT.

3.2.2. Postfixation

Heart sections are then postfixed to preserve the morphology of the tissue:

1. Wash slides with PBS for 5 min at RT.
2. Immerse sections in 4% PAF for 10 min.
3. Rinse sections for 5 min twice in PBS.

3.2.3. Permeabilization

To optimize probe penetration in the tissue sections and to increase the accessibility of the target RNA by unmasking nucleic acids from associated proteins, we perform chlorhydric acid and proteinase K treatments:

1. Immerse slides in 0.2 M HCl for 10 min at RT to denature proteins.
2. Rinse slides for 2 min twice in PBS.
3. Immerse sections in proteinase K digestion buffer containing 40 µg/mL proteinase K for 20 min at RT.
4. Rinse sections for 5 min twice in PBS.

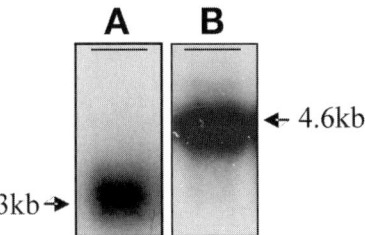

Fig. 1. Northern blot analysis of S1P1 (**A**) and S1P3 (**B**) transcripts in human heart shows that the two insert sequences used as template for S1P1 and S1P3 riboprobe synthesis are specific to their respective transcript targets (*see* **ref. 7** for detailed protocol).

5. Fix the sections again in 4% PAF for 5 min to stop proteinase K activity and to ensure firm attachment of the sections onto the slides.
6. Rinse sections for 5 min twice in PBS.

3.2.4. Acetylation

The aim of this step is to acetylate positive groups of proteins in the tissue and on the slide in order to reduce background owing to nonspecific electrostatic binding of the negatively charged probe (*9*).

1. Incubate sections for 15 min in 0.1 M triethanolamine (pH 8.0) supplemented with 0.25% acetic anhydride. Acetic anhydride has to be added to the triethanolamine just before section incubation, as the half-life of acetic anhydride in solution is very short.
2. Rinse slides in distilled water for 2 min.

3.2.5. Dehydration

1. Dehydrate the sections through an ethanol series:
   - 50% Ethanol: 2 min at RT.
   - 70% Ethanol: 5 min at RT (twice).
   - 85% Ethanol: 2 min at RT.
   - 90% Ethanol: 2 min at RT.
   - 100% Ethanol: 2 min at RT (twice).
2. Air-dry sections in a clean area before hybridization.

3.3. cRNA Probe Preparation

To localize transcripts that are rarely expressed in cardiovascular tissues, a sensitive approach based on radioactive RNA probe synthesis is more suitable (*see* **Note 3**). S1P1 and S1P3 inserts of adequate size (*see* **Note 3**) were cloned in pBluescript II SK+ plasmid and their respective specificity tested by Northern blot analysis (**Fig. 1**; *see* **Note 7**).

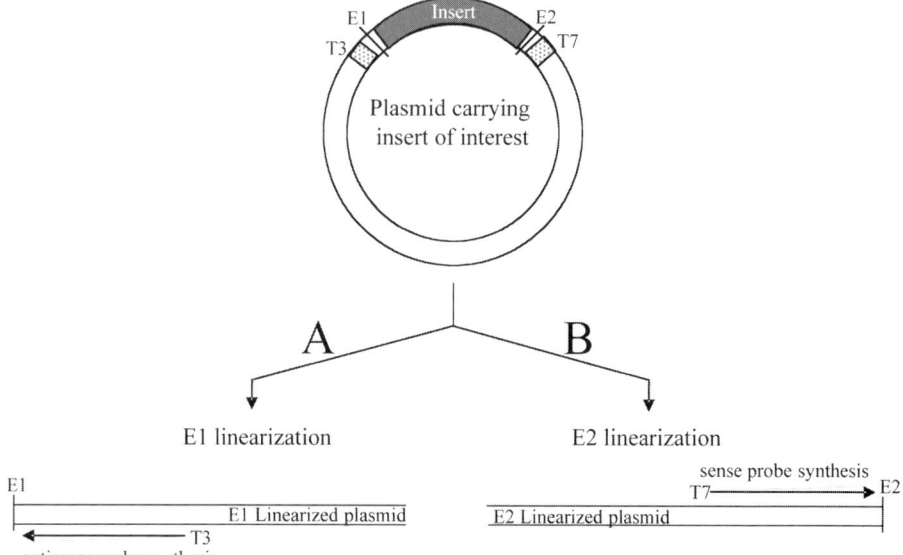

Fig. 2. Principle of sense and antisense cRNA probes synthesis. **(A)** Plasmid is linearized with restriction enzyme 1 (E1), and the antisense probe is synthesized using T3 polymerase. **(B)** Plasmid is linearized with restriction enzyme 2 (E2), and the sense probe is synthesized using T7 polymerase.

3.3.1. Plasmid Digestion

1. First, 20 µg of each plasmid are linearized following optimal digestion conditions furnished by restriction enzyme manufacturers. Two restriction enzymes located at both extremities of the insert must be selected for each plasmid to give rise to the antisense and sense probes, respectively (**Fig. 2**). Synthesis of sense probe is used as negative control to estimate background signal (*see* **Note 7**).
2. To control the efficiency of plasmid linearization, an aliquot of the digestion product is analyzed on 0.6 to 1% agarose gel in parallel with uncut plasmid. The migration distance for linearized plasmid appears generally longer than that for native supercoiled plasmid (**Fig. 3**).
3. After digestion, the plasmid is purified by ethanol precipitation, resuspended in DEPC water, and quantified by optical densitometry.

3.3.2. Probe Synthesis

The next steps involve probe labeling using $\alpha[^{35}S]$UTP (*see* **Note 4**).

1. Probe synthesis must be performed in RNase-free, siliconized microfuge tubes: 1 µg of each linearized plasmid is incubated for 2 h at 37°C (heating block) in a solution containing 1X transcription buffer, 30 mM DTT, 0.4 mM rATP, 0.4 mM rGTP,

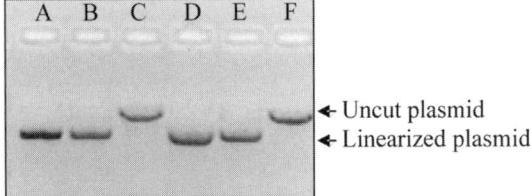

Fig. 3. Analysis on agarose gel of the two plasmids digests containing S1P1 (lanes A–C) and S1P3 (lanes D–F) inserts, respectively. Each plasmid is linearized with adapted restriction enzymes to generate template for sense and antisense riboprobe synthesis (*see* **Fig. 2**). The digestion products (lanes A, D: enzyme 1 and B, E: enzyme 2) as well as uncut plasmids (lanes C and F) were run on agarose gel. The efficiency of plasmid digestion is shown by the difference of the migration distance between digested and uncut supercoiled plasmids.

0.4 mM rCTP, 5 µCi/µL α[^{35}S]UTP, 1.6 U/µL RNAse inhibitor, and 0.4 U/µL of appropriate T7 or T3 RNA polymerase. This protocol allows the random incorporation of labeled UTP throughout the length of the designed probe (**Fig. 2**).
2. Centrifuge the tube (30 s) to recapture the condensation. Then digest the DNA templates with 10 U of RQ-1 DNAse (Promega) for 15 min at 37°C. This step gets rid of the template DNA, which will otherwise compete with the labeled probe to the target mRNA in the hybridization reaction.
3. Probe isolation is achieved on a equilibrated sephadex column (Bio-Spin/Micro Bio-Spin 30) according to the manufacturer's instructions to separate the labeled probe from the unincorporated [^{35}S]UTP. At this step, the incorporation percentage of radioactive label and the specific activity of each radiolabeled riboprobe are calculated to determine whether the labeling reaction has been successful. It is expected to reach up to 40 to 80% of total [^{35}S]UTP incorporation into the probe.
4. Add 10 µg yeast tRNA to the sample. Precipitate each probe by adding 2 vol of cold 100% ethanol and place at −20°C for 2 h.
5. Capture the cRNA by centrifugation (12,000*g*) at 4°C for 30 min. Remove the supernatant and resuspend the pellet containing each cRNA probe in 100 mM DTT to reach 10^6 cpm/µL of probe. Store the labeled probe at −20°C until hybridization.

3.4. Hybridization

After air-drying, tissue sections are directly hybridized under cover slips (*see* **Note 5**):

1. Prepare the hybridization buffer containing 50% formamide, 0.3 M NaCl, 20 mM Tris-HCl, pH 8.5, 5 mM EDTA, 10% dextran sulfate, 1X Denhardt's solution, 0.5 µg/µL yeast tRNA, and 10 mM DTT. Filter (0.22 µm) and preheat at 55°C. The volume of hybridization buffer for the entire experiment is evaluated by considering 50 to 100 µL of total hybridization volume by slide.

2. Calculate the volume of cRNA needed to reach 2×10^4 cpm/μL for each probe in the hybridization mix. This concentration has been optimized to obtain the best signal-to-background ratio with the aim of detecting rare transcripts in cardiac tissues. Heat aliquots for 5 min at 65°C in a heating block before adding them in an adequate volume of preheated hybridization buffer. Vortex and centrifuge briefly.
3. Transfer about 50 to 100 μL of hybridization solution containing 2×10^4 cpm/μL [^{35}S]UTP-labeled riboprobe onto slides and place the cover slip on delicately. Note that it is important to avoid formation of air bubbles between the section and the cover slip; otherwise some regions will not be accessible to the target probe. Hybridize the sections overnight at 55°C in an oven under humid atmosphere (wet paper).

3.5. Posthybridization Washing

All the washing steps are performed in slide racks designated for posthybridization (*see* **Note 5**):

1. Turn on three water baths (37°C, 55°C, and 65°C) and prewarm the solutions for washing (washing solution 1 at 55°C, washing solution 2 at 65°C, and RNase digestion in NTE buffer at 37°C).
2. Remove the cover slips during the washing step in solution 1 at 55°C (15 min).
3. Wash the slides for 30 min with washing solution 2 at 65°C.

   It is noteworthy that some RNA probes may intrinsically generate nonspecific hybridization signal leading to high background. However such background can be removed by RNase treatments that do not digest probe-target (RNA-RNA) hybrids.
4. Incubate slides for 10 min in NTE buffer at 37°C. Treat sections with 20 μg/mL RNAse A in NTE for 30 min at 37°C.
5. To stop the reaction, rinse slides for 15 min in NTE at room temperature.
6. Wash the sections through three separate washes: in washing solution 2 for 30 min at 65°C, in 2X SSC for 10 min at RT, and finally in 0.1X SSC for 10 min at RT.
7. Dehydrate the slides in ethanol series containing 0.3 *M* ammonium acetate:

   30% Ethanol: 1 min at RT.
   50% Ethanol: 1 min at RT.
   70% Ethanol: 5 min at RT (twice).
   85% Ethanol: 2 min at RT.
   90% Ethanol: 2 min at RT.
   100% Ethanol: 5 min at RT (twice).
8. Then air-dry the slides and tape them onto cassettes for film exposure.

3.6. Detection

After hybridization, the distribution of hybridized radioactive probe is detected by first exposing slides to X-ray film and then by dipping the slides in

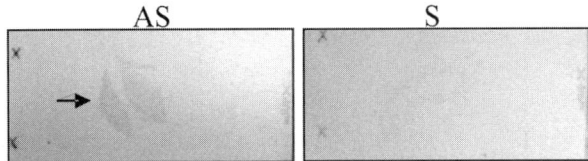

Fig. 4. Autoradiogram obtained after exposition of S1P1 hybridized heart sections to film for 4 d. AS and S represent results obtained with antisense and sense probes, respectively. Contrary to sense probe, widespread hybridization signal can be detected throughout the tissue with the antisense probe (black arrow). The low intensity of the signal predicts a long time needed for emulsion exposition.

photographic emulsion. X-ray film affords a quick means of visualization (1–4 d) and predicts the time necessary for the emulsion exposition (**Fig. 4**). However, it does not allow precise tissue localization, unlike dipped slides, which offer high resolution by microscopic analysis (*see* **Note 6**).

3.6.1. X-Ray Film Exposure

1. Appose the dried tissue sections to β-max film and store in cassettes at room temperature for 4 d. Time of exposition is dependent on the abundancy of transcript studied.
2. Develop the films in developer for 2 min at RT, followed by a brief wash in H_2O, and fix the film with fixater for 5 min. Rinse the film in H_2O and air-dry.

3.6.2. Emulsion Dipping

1. Preparation: preheat water bath to 42°C and proceed in the darkroom with safe red light. Dilute emulsion with ddH_2O (1 vol/1 vol). For 20 to 30 slides, pour 10 mL of ddH_2O into a 50-mL tube and fill carefully to 20 mL with Ilford K5 nuclear track emulsion by taking it piece by piece. Incubate in a water bath (42°C) for 20 min.
2. Carefully rotate tube 10 times and pour 10 mL emulsion mix into adapted glass vial.
3. Dip slides slowly in emulsion mix, and then allow excess emulsion to drip into vial.
4. Let slides dry out in a vertical position in the darkroom for 3 to 4 h. (Switch off or remove the water bath to prevent water evaporation.)
5. Place dried slides in black boxes with desiccant and store at 4°C for 2 to 10 wk.
6. Avoid any scratches on the emulsion during development. Develop the slides in freshly prepared 16% Kodak D-19 in ddH_2O at RT for 2 min.
7. Stop the reaction by incubating slides in stop solution (1% acetic acid and 1% glycerol in ddH_2O) for 1 min.
8. Fix the slides for 2 min in 30% sodium thiosulfate (prepared in ddH_2O).
9. Rinse the slides in water for 10 min (twice).

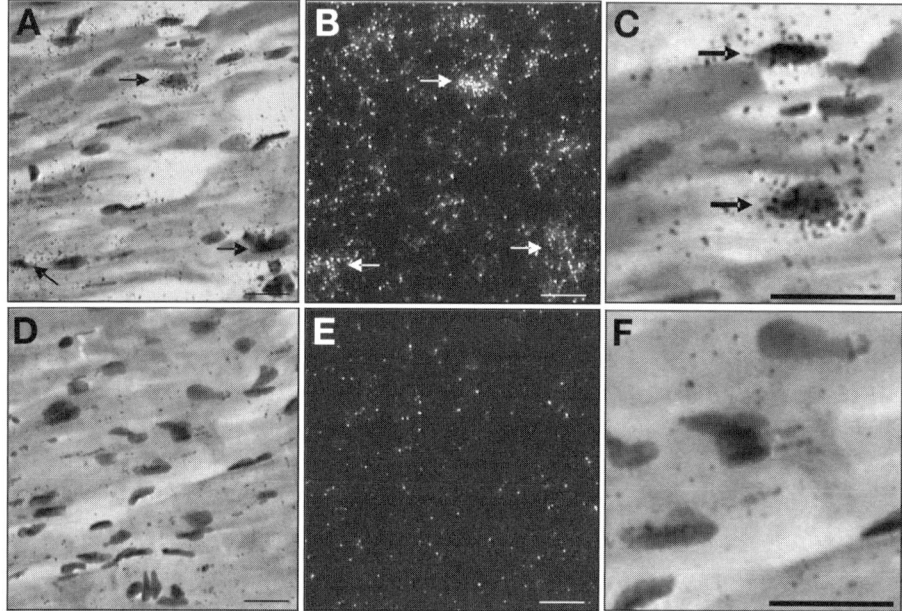

Fig. 5. Light microscopy analysis of S1P1 expression in cardiomyocytes of the human left ventricle. Brightfield (A zoomed in C, D zoomed in F) and darkfield (B,E) illuminations showing the specific signal with the antisense probe (**A–C**). Note the large density of S1P1 messengers around the cardiomyocyte nuclei (arrows). Control sections incubated with the sense probe do not reveal any significant signal (**D–F**). Scale bar = 20 μm.

10. Counterstaining: dip the sections in 0.02% toluidine blue solution. Rinse counterstained sections in water 5 min (twice) and let the slide air-dry. Counterstaining the sections greatly facilitates the anatomical identification of gene-expressing regions (**Figs. 5** and **6**).
11. Incubate sections in xylene for 10 min (twice) and mount the slides in Pertex (Microm, France).

3.7. Results

3.7.1. Specificity of the Probes

The specificity of the hybridization signal obtained with the probe of interest (S1P1 and S1P3 antisense probes) is evaluated first by comparing labeling obtained with corresponding sense probes in adjacent sections. Microscopic analysis of sections shows that no significant signal is observed with either S1P1 or S1P3 sense probes. This finding, combined with preliminary Northern blot studies indicating that used S1P1 and S1P3 probes recognized

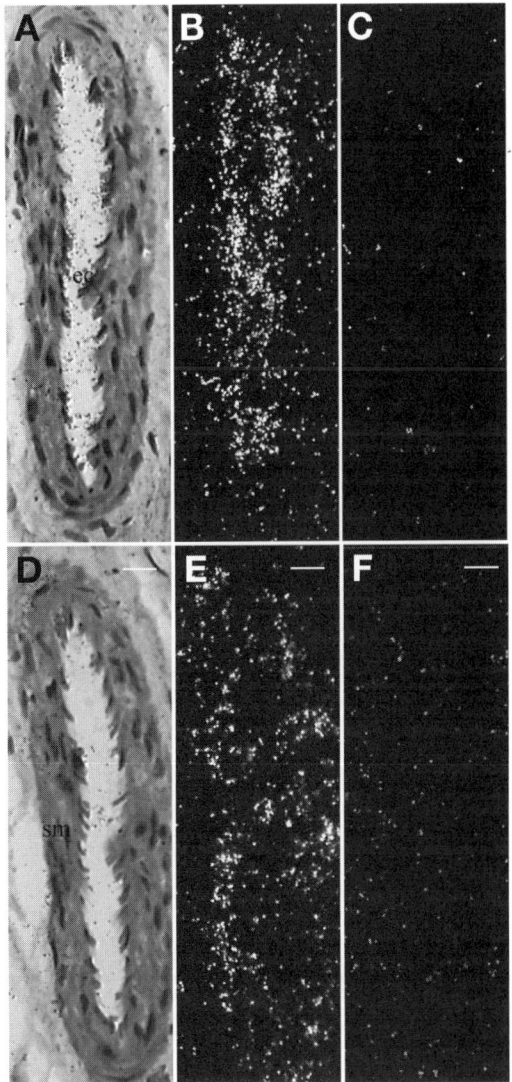

Fig. 6. Light microscopy analysis of S1P1 (**A–C**) and SIP3 (**D–F**) expression in coronary vessel walls of human left ventricle. (**A–C**) Brightfield (**A**) and darkfield (**B**) photomicrographs illustrate localization of S1P1 messengers in the endothelial cell layer of vessel (ec) using antisense probe. Darkfield analysis of the sense probe hybridization (**C**) does not reveal any significant signal. (**D–F**) Brightfield (**D**) and darkfield (**E**) analysis of EDG3 messenger localization revealing hybridization signals in the smooth muscle layers of vessel (sm). No significant hybridization signal is observed with sense probe (**F**). The differential expression of S1P1 and S1P3 detected in vessel layers indicate that no cross-hybridization has occurred between S1P1 and S1P3 probes. Scale bar = 20 μm.

respective transcript targets of the correct molecular size (**Fig. 1**), suggests that the probes are specific. Moreover, as S1P1 and S1P3 belong to the same gene family and present a good level of sequence homology (52%), it was important to avoid potential mRNA cross-hybridization.

S1P1 hybridization signal is mainly detected in cardiomyocytes as well as in the endothelial cell layer of vessels; S1P3 is found to be expressed specifically in smooth muscle cell layers of vessels (**Figs. 5** and **6**). These results, showing differential expression of S1P1 and S1P3 in cardiovascular tissues, strongly indicate that no cross-hybridization occurred between EDG family transcripts during the procedure.

3.7.2. Sensitivity

The penetration of probes into the tissue is crucial to achieve sensitive detection. However, it is also important that the permeabilization treatment preserve the morphology of the tissue (*see* **Note 2**). As probe penetration and tissue fragility could vary among different cell types in the tissue, it is important to develop a procedure that takes this heterogeneity into account. To minimize this difficulty, the tissue is treated with an adapted proteinase K concentration, which enhances probe penetration. The sample is reasonably fixed by 4% PAF to ensure sample morphology. Therefore, the procedure allows one to detect hybridization signals in cardiomyocytes as well as in vascular endothelial and smooth muscle cells (**Figs. 5** and **6**).

3.7.3. Troubleshooting

3.7.3.1. ABSENCE OF SIGNAL

One of the first reasons for the absence of specific signal is target RNA or cRNA probe degradation. The time required for tissue sampling and fixation as well as for the RNase-free prehybridization has to be optimal to prevent target transcript digestion. Similarly, the probes are prone to RNase degradation as well, and RNase-free conditions during their synthesis and purification steps are therefore necessary. The absence of mRNA target expression in the sections analyzed may represent another possible explanation for the lack of hybridization signal. To confirm this latter hypothesis, investigators should test the probe on a control tissue known to express the transcripts of interest (*see* **Note 7**).

3.7.3.2. HIGH NONSPECIFIC BACKGROUND

If a strong background is observed on the tissue in parallel with specific labeling but is absent around the sections, this indicates a problem with probe hybridization. To reduce such background, the investigator should increase the stringency of the hybridization and posthybridization steps, by increasing the temperature or modifying the salt concentration (*see* **Note 5**). In the absence of specific signal, the riboprobe might have been degraded, and it is necessary to resynthesize the probe. Finally, the presence of background on the entire slide (around the section) can be explained by the emulsion having been accidentally damaged (light exposition, scratches).

4. Notes

There are almost as many *in situ* hybridization protocols with discrete discrepancies as there are experiments that have been performed (*see* **Table 1**). It is thus very important to understand and control the different stages in the process. This section presents an overview of alternative *in situ* hybridization procedures and describes advantages and disadvantages of the various methodologies that can be used on different heart tissue samples.

1. *Sample origin:* during the past few years, most of the published studies describing better investigation of cellular or molecular factors in cardiac function were performed on animal models (*see* **Table 1**). Animals represent the most commonly used physiological and pharmacological models. Unlike easily available human tumor biopsy samples, nonpathological cardiac tissue is evidently hard to come by. Within the medical environment, it might be possible to obtain pathological hearts from patients undergoing cardiac transplantation or bypass surgery. In this case, sampling procedures have to be followed carefully to preserve the reproducibility of results.

    In the last few years, several manufacturers have developed ready-to-use human tissue sections that can be used to study spatial gene expression in cardiovascular tissues. Interestingly, some suppliers very recently developed tissue microarrays that provide researchers the opportunity to screen a large number (50–300 human sections arrayed on a single slide) of samples for gene expression. Normal or diseased samples are generally fixed, embedded in paraplast, sectioned, and mounted onto slides under experimental conditions suitable for *in situ* hybridization. Using those sections, researchers will have to take the fixation protocols used into consideration for the following slide treatments.

    Whatever the nature of the samples, it is important to keep in mind that messengers are synthesized and degraded at a high rate and that their number can dramatically drop in vivo until tissue fixation. Therefore, the quicker the tissue is fixed after sampling, the lower the artifactual fluctuations.

2. Tissue fixation and treatment: A critical step for any *in situ* hybridization experiment is to achieve proper tissue fixation. On the one hand, the fixation method should provide accessibility of the probe to the target sequence since DNA and RNA target sequences are surrounded by proteins (extensive crosslinking of these proteins could mask the target nucleic acid). On the other hand, fixatives have to provide maximal retention of cellular target RNA and to maintain the morphological details of the tissue structure *(10,11)*. Unfortunately, no universal fixation protocol useful for all substrates has yet been described.

Table 1
Procedures Used for Different *In Situ* Hybridization Approaches in the Cardiovascular Field

Target	Tissue	Fixation procedure	Probe type	Probe labeling	Ref.
Iroquois homeobox genes family mRNA	Embedded mouse embryos (whole mount)	Formaldehyde (4%)	Riboprobe	Digoxigenin	*21*
ICAM-1, TGF-β, TNF-α mRNA	Rat heart sections (cryostat)	Formaldehyde (4%)	cDNA random primed	^{35}S isotope	*25*
apoA-I mRNA	Mouse heart sections (paraffin)	Paraformaldehyde (4%)	Riboprobe	Psoralen-biotin	*24*
Abcg2 mRNA	Mouse embryos sections (paraffin)	Paraformaldehyde (4%)	Riboprobe	^{35}S isotope	*27*
HCN4 mRNA	Mouse embryos sections (paraffin)	Paraformaldehyde (4%)	Riboprobe	^{35}S isotope	*28*
SGLT1 mRNA	Human heart sections (paraffin)	VWR International procedure	Riboprobe	Digoxigenin	*29*
Melanocortin-4 receptor mRNA	Rat heart sections (cryostat)	Paraformaldehyde (4%)	Riboprobe	^{33}P isotope	*30*
Titin mRNA	Rat heart sections (cryostat)	Paraformaldehyde (4%)	Riboprobe	Biotin	*31*
Protozoan parasite detection	Human heart sections (cryostat)	Paraformaldehyde (4%)	Double-strand DNA	Digoxigenin	*26*
Enterovirus RNA	Human valve tissue sections (paraffin)	Formalin	Olignucleotide	Digoxigenin	*32*
RGS5 mRNA	Macaque aorta sections (cryostat)	Formalin (10%)	Riboprobe	^{35}S isotope	*33*
EDGs and p63RhoGEF mRNAs	Human heart sections (paraffin)	Novagen procedure	Riboprobe	^{35}S isotope	*6,7*

Fixatives like acetic acid-alcohol mixtures (i.e., formalin-acetic acid-alcohol) provide the best probe penetration but may permit the loss of RNA from tissue *(12)*. Glutaraldehyde allows the best RNA retention and tissue morphology, but because of extensive protein crosslinking, the probe penetrance and the target accessibility are low *(13)*. To increase penetrance, it is possible to treat the tissue with detergents (i.e., saponin, sodium dodecyl sulfate, Triton, Tween) that permeabilize the membranes by extracting the lipids or to use proteinase K, a nonspecific endopeptidase that removes proteins surrounding the target sequence. Protease incubation has to be carefully monitored since overdigestion may lead to the loss of DNA and RNA. The balance between tissue structure preservation and target accessibility also has to be taken into account when one is using 4% paraformaldehyde (PAF), which is so far the most commonly used solution (*see* **Table 1**). The protocol that we recommend provides a good compromise between sensitivity and morphological structure maintenance.

3. Nature of the probe: the hybridization step can be performed essentially using four different types of probes: riboprobes, synthetic oligonucleotides, single-stranded DNA, or double-stranded DNA; each has its advantages and disadvantages. Researchers must consider the most appropriate probe to use for their experiments.

As the strength of the bonds between the probe and the target sequence plays an important role in hybridization sensitivity, riboprobes have the advantage of generating RNA-RNA interactions that are more stable than DNA-RNA hybrids. The high sensitivity of riboprobes has led most investigators (including us) to use this approach (*see* **Table 1**). Similarly to oligonucleotides and single-strand DNA probes, transcribing only one strand to synthesize the antisense probe prevents reannealing of complementary strands during hybridization. The main disadvantage is that these probes are very sensitive to RNases and therefore special attention has to be focused on RNAse-free conditions during probe synthesis. To allow good penetration of the RNA probe in tissue, it is recommended to use plasmid insert sequences with sizes in the range of 200 to 500 bp. Alternatively, RNA probes longer than 600 bp could be submitted to alkaline hydrolysis to obtain respectable size probes *(14,15)*.

Designed from perfectly known sequences, oligonucleotides are automatically produced by DNA synthesizers from labeled nucleotides. The advantages of such oligoprobes are that (1) no molecular biology expertise is required for synthesis, (2) the molecules generated are resistant to RNases, (3) the sequence can be ideally designed to represent the correct GC percentage and to avoid secondary structures, (4) because of their small size, generally from 20 to 40 pairs, they can easily penetrate the cells or tissues of interest, and (5) they are single stranded, therefore excluding the possibility of self-hybridization *(16)*. The low specific activity of oligonucleotide probes is a common feature that represents one disadvantage. Likewise, owing to their small size, they cover less of the target than longer DNA or RNA probes. These two drawbacks often lead to lower hybridization signals. Additional disadvantages come from the fact that the hybrids are less stable because of the short length of the hybridized sequence.

The use of single-stranded DNA probes has similar advantages to the oligonucleotide probes regarding stability and the absence of probe self-hybridization. However, generating single-stranded DNA probes requires technically complex procedures. They can be produced by reverse transcription of RNA or by amplified primer extension of a PCR-generated fragment in the presence of a single antisense primer *(17)*.

In contrast, the major advantage of double-stranded DNA is that probe preparation is relatively easy to perform (classic RT-PCR or enzymatic digestion of a high quantity of plasmid), leading to high specific activity during the labeling processes (nick translation, random primed) *(18)*. The major disadvantage of double-stranded DNA is that during the hybridization step the complementary strand competes with the target sequence, which can strongly decrease the sensitivity. For this reason this last approach is principally used to localize transcripts or DNA abundantly represented in tissue sections such as virus *(19,20)*.

4. Probe labeling: it is important to determine how to achieve the best probe labeling in order to obtain the highest level of resolution under the strongest stringency conditions. Although recent nonradioactive *in situ* hybridization protocols (i.e., fluorescent, digoxigenin label) have been developed using riboprobes that produce quick results, avoid the handling of hazardous materials, and seem to exhibit good sensitivity *(21–24)*, we have chosen and recommend to label probes with [^{35}S]UTP since this methodology has been shown to be typically more sensitive and allows one to perform quantitative studies *(14,25)*.

5. Hybridization and washing: an accessory, single prehybridization procedure may be carried out before the hybridization step in order to limit nonspecific hybridization and reduce background staining. The prehybridization solution is composed of all the components of the hybridization solution minus the probe. In most of our experiments, we do not prehybridize when previous experiments have shown that it is not necessary and has no influence on the final results. During hybridization, there are several parameters that influence the specificity and sensitivity of the procedure, including the probe, formamide and salt concentrations in the hybridization buffer, hybridization temperature, and pH. Hybridization of a labeled probe to the target sequence is defined by hydrogen bonding and hydrophobic interactions in equilibrium. This phenomenon is represented by the mean thermal stability of hybrids (T_m), which corresponds to the temperature at which 50% of the target sequences are dissociated/associated. This value varies with the length of the probe, the G/C content of the sequence, the number of mismatches between the target and the probe, and the salt and formamide concentrations in the hybridization buffer. Formamide, which is an organic solvent, reduces the thermal stability of the bonds by disrupting hydrogen bonds and therefore allows hybridization to be carried out at a lower temperature. Salts (generally monovalent cations: sodium ions) interact mainly with the phosphate groups of the nucleic acids and neutralize the negative or repulsive nature of the nucleic acid strand negative charge. These factors could be adapted by the investigators depending on the nature of the probe (type, length, and sequence) and the relative amount and location of the transcripts being targeted.

Following hybridization, washes are applied to remove unbound probe or probe that is not hybridized to a specific target. Generally, serial washes at increasing temperatures and/or decreasing salt concentrations should be carried out close to the stringency condition of the hybridization procedure. The stringency of the washes can be manipulated by varying the formamide concentration, salt concentration, and temperature. For correct sensitivity and low background, it is necessary to optimize the washing conditions according to the nature of the tissue, the type of probe, and the hybridization procedure followed. Note that it is generally recommended to hybridize stringently rather than wash stringently.

6. Detection step: for a radioactive approach, film and emulsion autoradiographies are the two detection methods generally performed in parallel. Autoradiographic film detection is most useful for quick detection in experiments, although emulsion autoradiography combined with counterstaining allows one to detect hybridization signal precisely at the single cell level with longer exposure.

    Most of the nonradioactive detection procedures require the use of enzymes or fluorescence-conjugated antibodies. The nature of the epitope recognized by antibodies depends on the type of probe labeling (i.e., digoxigenin, biotin). Probe-target hybrids are usually detected with an alkaline-phosphatase-conjugated antibody either by a color reaction or by a chemiluminescence reaction *(21,26)*. Of the many alkaline phosphatase substrates that can be used, CDP-Star or CSPD are adapted for chemiluminescence reactions and NBT/BCIP for color reactions. When one is using such nonradioactive approaches, it is important to check for the presence of endogenous enzymes in tissue and to neutralize them to avoid a very high background. In addition to such labels, companies have developed a range of fluorescent dyes that can be either coupled directly to the nucleotides or conjugated to antibodies. The advantage of these fluorescent dyes used in the Fluorescence *in situ* hybridization approach is that two or more different probes can be visualized at one time in the same experiment.

7. Control procedures: whatever the procedure followed (radioactive or nonradioactive probe, RNA or DNA probe, and so on), the *in situ* hybridization protocol must include negative and positive controls to ensure that the detected signal results from specific hybridization. Sense probe/oligonucleotide hybridization is the common negative control used in parallel with antisense riboprobe/oligonucleotide on adjacent sections (**Figs. 5** and **6**). For a more sophisticated control, a competition between labeled and unlabeled probes can be performed. Pretreating sections with unlabeled probe should in fact lead to the inaccessibility of the target for the labeled probe. When probing for mRNA, RNase could also be applied onto the tissue before hybridization to determine whether labeling results from RNA/RNA hybrids. Regarding the positive control, it is of particular interest to test the sequence used for probe synthesis in a preliminary Northern blot analysis. The procedure that we usually follow ensures that the probe binds to the correct transcript target at the exact molecular size (**Fig. 1**). Another positive control would be to hybridize probe onto tissues or fixed cell types known to

express the transcript of interest abundantly. A negative result with such a control should reflect a problem encountered during probe synthesis. Finally, to check for the quality of the tissue, one can hybridize tissue with probes specific to housekeeping sequences (actin), which are always expressed constituently. Absence of signal should indicate target/tissue degradation.

References

1. Pardue, M. L. and Gall, J. G. (1969) Molecular hybridization of radioactive DNA to the DNA of cytological preparations. *Proc. Natl. Acad. Sci. USA* **64,** 600–604.
2. John, H. A., Birnstiel, M. L., and Jones, K. W. (1969) RNA-DNA hybrids at the cytological level. *Nature* **223,** 582–587.
3. Harper, M. E. and Marselle, L. M. (1986) In situ hybridization—application to gene localization and RNA detection. *Cancer Genet. Cytogenet.* **19,** 73–80.
4. Steenman, M., Chen, Y. W., Le Cunff, M., et al. (2003) Transcriptomal analysis of failing and nonfailing human hearts. *Physiol. Genomics* **12,** 97–112.
5. Mangi, A. A., Glueck, S. B., and Pratt, R.E. (2002) Getting to the heart of the matter: focus on "microarray analysis of global changes in gene expression during cardiac myocyte differentiation". *Physiol. Genomics* **9,** 131–133.
6. Souchet, M., Portales-Casamar, E., Mazurais, D., et al. (2002) Human p63RhoGEF, a novel RhoA-specific guanine nucleotide exchange factor, is localized in cardiac sarcomere. *J. Cell. Sci.* **115,** 629–640.
7. Mazurais, D., Robert, P., Gout, B., Berrebi-Bertrand, I., Laville, M. P., and Calmels, T. (2002) Cell type-specific localization of human cardiac S1P receptors. *J. Histochem. Cytochem.* **50,** 661–670.
8. Simmons, D. M., Arriza, J. L., and Swanson, L. W. (1989) A complete protocol for *in situ* hybridization of messenger RNAs in brain and other tissues with radio-labelled single-stranded RNA probes. *J. Histotechnol.* **12,** 169–181.
9. Hayashi, S., Gillam, I. C., Delaney, A. D., and Tener, G. M. (1978) Acetylation of chromosome squashes of *Drosophila melanogaster* decreases the background in autoradiographs from hybridization with [^{125}I]-labeled RNA. *J. Histochem. Cytochem.* **26,** 677–679.
10. Tecott, L. H., Eberwine, J. H., Barchas, J. D., and Valentino, K. L. (1987) Methodological considerations in the utilization of *in situ* hybridization, in In Situ *Hybridization: Applications to Neurobiology* (Valentino, K. L., Eberwine, J. H., and Barchas, J. D., eds.), Oxford University Press, New York.
11. Hofler, H. (1990) Principles of *in situ* hybridization, in In Situ *Hybridization: Principle and Practice* (Polak, J. M. and McGee, J. O'D., eds.), Oxford University Press, Oxford.
12. Thompson, C. H. and Rose, B. R. (1991) Deleterious effects of formalin/acetic acid/alcohol (FAA) fixation on the detection of HPV DNA by in situ hybridization and the polymerase chain reaction. *Pathology* **23,** 327–330.
13. Uehara, F., Ohba, N., Nakashima, Y., Yanagita, T., Ozawa, M., and Muramatsu, T. (1993) A fixative suitable for in situ hybridization histochemistry. *J. Histochem. Cytochem.* **41,** 947–953.

14. Winzer-Serhan, U. H., Broide, R. S., Chen, Y., and Leslie, F. M. (1999) Highly sensitive radioactive in situ hybridization using full length hydrolyzed riboprobes to detect alpha 2 adrenoceptor subtype mRNAs in adult and developing rat brain. *Brain Res. Protoc.* **3,** 229–241.
15. Yang, H., Wanner, I. B., Roper, S. D., and Chaudhari, N. (1999) An optimized method for in situ hybridization with signal amplification that allows the detection of rare mRNAs. *J. Histochem. Cytochem.* **47,** 431–446.
16. Crabb, I. D., Hughes, S. S., Hicks, D. G., Puzas, J. E., Tsao, G. J., and Rosier, R. N. (1992) Non radioactive in situ hybridization using digoxigenin-labelled oligonucleotides. Applications to musculoskeletal tissues. *Am. J. Pathol.* **141,** 579–589.
17. Konat, G. W. (1996) Generation of high efficiency ssDNA hybridization probes by linear polymerase chain reaction (LPCR). *Scanning Microsc. Suppl.* **10,** 57–60.
18. Saeki, K., Mishima, K., Horiuchi, K., et al. (1993) Detection of low copy numbers of Epstein-Barr virus by in situ hybridization using nonradioisotopic probes prepared by the polymerase chain reaction. *Diagn. Mol. Pathol.* **2,** 108–115.
19. Summers, J., Jilbert, A. R., Yang, W., et al. (2003) Hepatocyte turnover during resolution of a transient hepadnaviral infection. *Proc. Natl. Acad. Sci. USA* **100,** 11652–11659.
20. Lehr, E. and Brown, D. R. (2003) Infection with the oncogenic human papillomavirus type 59 alters protein components of the cornified cell envelope. *Virology* **309,** 53–60.
21. Moorman, A. F., Houweling, A. C., de Boer, P. A., and Christoffels, V. M. (2001) Sensitive nonradioactive detection of mRNA in tissue sections: novel application of the whole-mount in situ hybridization protocol. *J. Histochem. Cytochem.* **49,** 1–8.
22. Ishikawa, K., Azuma, S., Ikawa, S., et al. (2003) Cloning and characterization of *Xenopus laevis* drg2, a member of the developmentally regulated GTP-binding protein subfamily. *Gene* **322,** 105–112.
23. Katz, S. G., Williams, A., Yang, J., et al. (2003) Endothelial lineage-mediated loss of the GATA cofactor Friend of GATA 1 impairs cardiac development. *Proc. Natl. Acad. Sci. USA* **100,** 14030–14035.
24. Baroukh, N., Lopez, C. E., Saleh, M. C., et al. (2004) Expression and secretion of human apolipoprotein A-I in the heart. *FEBS Lett.* **557,** 39–44.
25. Sun, Y., Zhang, J., Lu, L., Bedigian, M. P., Robinson, A. D., and Weber, K. T. (2004) Tissue angiotensin II in the regulation of inflammatory and fibrogenic components of repair in the rat heart. *J. Lab. Clin. Med.* **143,** 41–51.
26. Elias, F. E., Vigliano, C. A., Laguens, R. P., Levin, M. J., and Berek, C. (2003) Analysis of the presence of *Trypanosoma cruzi* in the heart tissue of three patients with chronic Chagas' heart disease. *Am. J. Trop. Med. Hyg.* **68,** 242–247.
27. Martin, C. M., Meeson, A. P., Robertson, S. M., et al. (2004) Persistent expression of the ATP-binding cassette transporter, Abcg2, identifies cardiac SP cells in the developing and adult heart. *Dev. Biol.* **265,** 262–275.
28. Stieber, J., Herrmann, S., Feil, S., et al. (2003) The hyperpolarization-activated channel HCN4 is required for the generation of pacemaker action potentials in the embryonic heart. *Proc. Natl. Acad. Sci. USA* **100,** 15235–15240.

29. Zhou, L., Cryan, E. V., D'Andrea, M. R., Belkowski, S., Conway, B. R., and Demarest, K. T. (2003) Human cardiomyocytes express high level of Na+/glucose cotransporter 1 (SGLT1). *J. Cell. Biochem.* **90,** 339–346.
30. Mountjoy, K. G., Jenny Wu, C. S., Dumont, L. M., and Wild, J. M. (2003) Melanocortin-4 receptor messenger ribonucleic acid expression in rat cardiorespiratory, musculoskeletal, and integumentary systems. *Endocrinology* **144,** 5488–5496.
31. Warren, C. M., Jordan, M. C., Roos, K. P., Krzesinski, P. R., and Greaser, M. L. (2003) Titin isoform expression in normal and hypertensive myocardium. *Cardiovasc. Res.* **59,** 86–94.
32. Li, Y., Pan, Z., Ji, Y., Peng, T., Archard, L. C., and Zhang, H. (2002) Enterovirus replication in valvular tissue from patients with chronic rheumatic heart disease. *Eur. Heart J.* **23,** 567–573.
33. Adams, L. D., Geary, R. L., McManus, B., and Schwartz, S. M. (2000) A comparison of aorta and vena cava medial message expression by cDNA array analysis identifies a set of 68 consistently differentially expressed genes, all in aortic media. *Circ. Res.* **87,** 623–631.

III

CARDIAC GENE REGULATION
PROMOTER CHARACTERIZATION IN THE MYOCARDIUM

10

Characterization of *cis*-Regulatory Elements and Transcription Factor Binding

Gel Mobility Shift Assay

Jim Jung-Ching Lin, Shaun E. Grosskurth,
Shannon M. Harlan, Elisabeth A. Gustafson-Wagner, and Qin Wang

Summary

To understand how cardiac gene expression is regulated, the identification and characterization of *cis*-regulatory elements and their *trans*-acting factors by gel mobility shift assay (GMSA) or gel retardation assay are essential and common steps. In addition to providing a general protocol for GMSA, this chapter describes some applications of this assay to characterize cardiac-specific and ubiquitous *trans*-acting factors bound to regulatory elements [novel TCTG(G/C) direct repeat and A/T-rich region] of the rat *cardiac troponin T* promoter. In GMSA, the specificity of the binding of trans-acting factor to labeled DNA probe should be verified by the addition of unlabeled probe in the reaction mixture. The migratory property of DNA-protein complexes formed by protein extracts prepared from different tissues can be compared to determine the tissue specificity of *trans*-acting factors. GMSA, coupled with specific antibody to *trans*-acting factor (antibody supershift assay), is used to identify proteins present in the DNA-protein complex. The gel-shift competition assay with an unlabeled probe containing a slightly different sequence is a powerful technique used to assess the sequence specificity and relative binding affinity of a DNA-protein interaction. GMSA with SDS-PAGE fractionated proteins allows for the determination of the apparent molecular mass of bound *trans*-acting factor.

Key Words: Gel mobility shift assay (GMSA); gel retardation assay; antibody supershift assay; gel-shift competition assay; cardiac troponin T promoter; D module; F module; TCTG(G/C) direct repeat; cardiac-specific trans-acting factor; A/T-rich region; MEF2-like motif; HMG2; nondenaturing polyacrylamide gel electrophoresis; SDS-PAGE (polyacrylamide gel electrophoresis); heparin-agarose column chromatography.

From: *Methods in Molecular Biology, vol. 366: Cardiac Gene Expression: Methods and Protocols*
Edited by: J. Zhang and G. Rokosh © Humana Press Inc., Totowa, NJ

1. Introduction

The identification and characterization of *cis*-regulatory elements and *trans*-acting factors is essential for understanding the control of cardiac gene expression. Promoter studies on cardiac-restricted genes have suggested that the activation of these genes is likely controlled through interactions of a combination of numerous cardiac-restricted and ubiquitous transcriptional factors. Additionally, transcriptional repressors may also be present in noncardiac tissues and may play a role in reinforcing this cardiac specificity during differentiation. The gel mobility shift assay (GMSA) provides an approach to characterize the interactions between *cis*-regulatory elements and *trans*-acting factors that regulate cardiac gene expression. The fundamental principle behind this assay is that the binding of a protein to a DNA fragment, which has been radiolabeled for ease in analysis, will reduce the mobility of this DNA fragment when it is electrophoresed on a nondenaturing polyacrylamide gel. This differential mobility allows the complex to be distinguished from the unbound probe. This technique can be performed with whole cell extracts from cardiac tissue, as well as nuclear extracts from cardiac tissue, to investigate binding of *trans*-acting factors to the promoter regions of cardiac genes. This assay can further be applied to confirm interactions of suspected transcription factors with a regulatory region of the promoter.

This chapter discusses a specific application of the GMSA to characterize the binding of cardiac-specific and ubiquitous *trans*-acting factors to the regulatory region of rat *cardiac troponin T (cTnT)* promoter. Expression of the rat/mouse *cTnT* gene is clearly detectable in the lateral plate mesoderm of the heart forming fields at embryonic day 7.5 (E7.5), and in the linear heart tube at E8.0 *(1,2)*. At E10.5, *cTnT* is strongly expressed in the heart and weakly expressed in a few of the most posterior somites. However, no skeletal muscle *TnT* gene expression is seen at this stage of development, suggesting that the expressed cTnT protein may be functional in these skeletal muscle precursors. At later fetal stages, transcription of the *cTnT* gene is specifically repressed in developing skeletal muscle. The cTnT proteins detected in the skeletal muscles decline progressively and disappear 2 wk after birth. In the adult, *cTnT* is expressed only in the cardiomyocytes. This expression pattern implies both temporal and spatial regulation of *cTnT* expression and suggests a tightly regulated expression profile. Using reporter assays, a *cTnT* proximal promoter region –497 bp from the transcriptional start site has been identified to drive cardiac-specific expression specifically in cultured cardiomyocytes and to recapitulate the endogenous *cTnT* expression pattern in transgenic mice *(1–3)*. This indicates that the essential regulatory elements for driving cardiac specificity are contained within this –497 bp promoter (**Fig. 1A**).

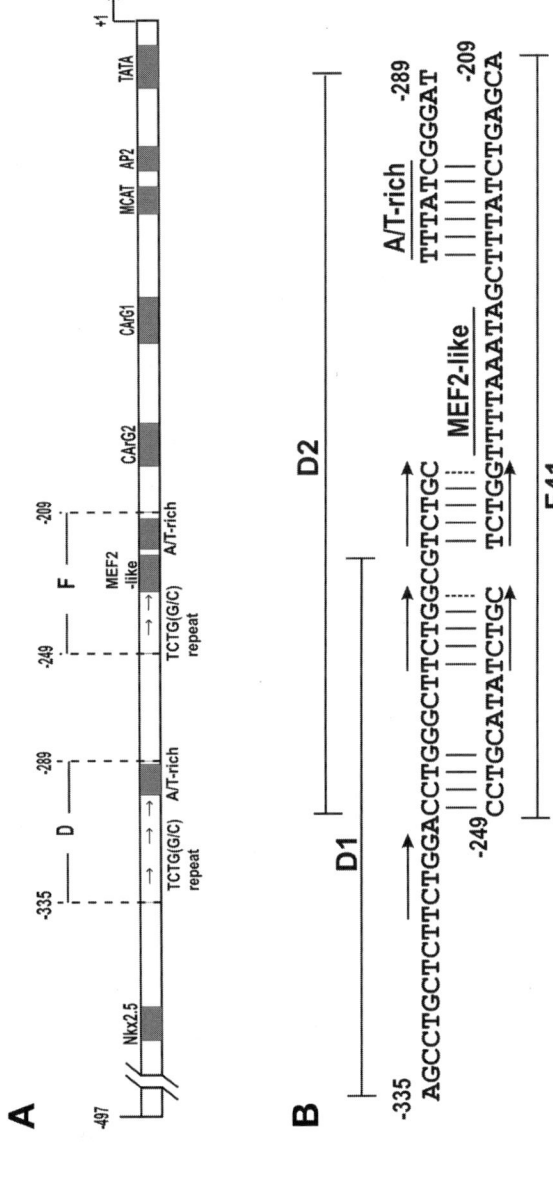

Fig. 1. Rat cardiac troponin T (*cTnT*) promoter. (**A**) Diagram of rat −497 bp *cTnT* promoter. Both D and F modules have a TCTG(G/C) direct repeat indicated by arrows and an A/T-rich site. In addition, the F module contains an A/T-rich core sequence within a consensus MEF2 site (MEF2-like motif). (**B**) Sequence comparison between D (D1 and D2, −335 bp to −289 bp) and F (F41, −249 bp to −209 bp) modules. The vertical lines represent identical sequence. Gaps are introduced to obtain maximal alignment. The TCTG(G/C) direct repeats in D1, D2, and F41 are indicated by arrows.

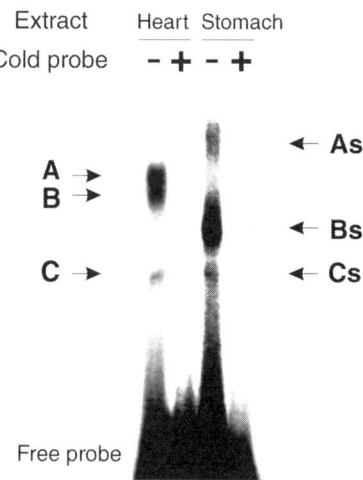

Fig. 2. Gel mobility shift assays (GMSAs) with the F41 probe. Protein extract (10 µg) prepared from heart or stomach was incubated with ^{32}P-labeled F41 probe in each reaction. For examination of the specificity, a 100-fold excess of the unlabeled F41 probe (cold probe) was added to the reaction mixture before incubation with the labeled probe. Complexes A and B formed by heart extracts appeared to have different mobility from complexes formed by stomach extracts (As and Bs). In contrast, the fast-migrating complexes (C and Cs) formed by heart and stomach extracts had the same mobility, suggesting the involvement of a ubiquitous factor in this complex.

Detailed evaluation of this –497 bp proximal promoter region uncovered two sequence homologous modules, module D and module F, which each contain TCTG(G/C) direct repeating units as well as an A/T-rich site (**Fig. 1B**). In addition, module F contains a myocyte-specific enhancer factor 2 (MEF2)-like motif, which is also an A/T-rich region. Analysis with deletion and religation mutant promoters suggests that module D serves as an enhancer to increase promoter activity but cannot totally substitute for the function of module F *(4)*. GMSA and competition GMSA with probes (F41, D1, and D2) containing direct repeats and A/T-rich sites of modules D and F were used to demonstrate that the same protein factors bound to these *cis*-regulatory elements. Using application of the GMSA, the 42-kDa proteins were identified to bind the direct repeat regions of this promoter in a cardiac tissue-specific manner (**Fig. 2**). Furthermore, a 25-kDa ubiquitous protein was identified to bind the A/T-rich site and was later shown to be a high mobility group 2 (HMG2) family protein *(4)*. Such application of the GMSA provides an effective tool for investigating the details of the complex regulatory process through which *cTnT* directs cardiac-specific expression.

Gel Mobility Shift Assay

In this chapter, we describe in detail the preparation of total protein extracts from cardiac tissue, the preparation of nondenaturing polyacrylamide gel and labeled probe, the DNA-protein binding reaction in the GMSA, and the specific applications of this technique.

2. Materials

2.1. Protein Preparation

1. 20 g of Each tissue (fresh or frozen): heart, liver, stomach, and skeletal muscle. Frozen rat tissues can be purchased from Pel-Freez Biologicals, Rogers, AR; www.pelfreez-bio.com.
2. Liquid nitrogen.
3. Mortar and pestle.
4. Lysis buffer: 15 mM HEPES, pH 7.6, 100 mM KCl, 5 mM MgCl$_2$, 1 mM dithiothreitol (DTT), 1 mM phenylmethylsulfonyl fluoride (PMSF), and 10% glycerol.
5. 400 mM NaCl extraction buffer: 15 mM HEPES, pH 7.6, 400 mM NaCl, 5 mM MgCl$_2$, 1 mM DTT, 1 mM PMSF, and 10% glycerol.
6. Heparin-agarose column: 2.5 mL prepacked column (Sigma, St. Louis, MO, cat. no. HEP-1-5).
7. 1 M NaCl extraction buffer: 15 mM HEPES, pH 7.6, 1 M NaCl, 5 mM MgCl$_2$, 1 mM DTT, 1 mM PMSF, and 10% glycerol.
8. 1X GMSA binding buffer: 20 mM HEPES, pH 7.6, 100 mM KCl, 5 mM MgCl$_2$, 0.2 mM EDTA, 0.5 mM DTT, and 10% glycerol.
9. Dialysis tubing: mol. wt. cutoff 12 to 14,000 or Spectra/por #1 mol. wt. cutoff 6 to 8,000 (Spectrum Medical Industries, Los Angeles, CA).

2.2. DNA Probe Preparation

The length of the DNA fragment can vary between 20 and 300 bp for production of a probe. The longer the fragment, the more likely it is that multiple binding sites are present, and nonspecific binding may occur.

1. Restriction enzyme-digested, or polymerase chain reaction (PCR)-amplified, end-labeled DNA fragments or annealed, end-labeled oligonucleotides (commercially synthesized) can be used as probes for GMSA.
2. [α-^{32}P]dNTP (>3000 Ci/mmol) for Klenow fill-in, end-labeling or [γ-^{32}P]ATP (>3000 Ci/mmol) for T4 polynucleotide kinase end-labeling.
3. Klenow fragment of *E. coli* DNA polymerase I, restriction enzymes, T4 polynucleotide kinase (Promega, Madison, WI or New England Biolabs, Beverly, MA).
4. 10X T4 polynucleotide kinase buffer: 700 mM Tris-HCl, pH 7.6, 100 mM MgCl$_2$, and 50 mM DTT.
5. 10X Klenow buffer: 100 mM Tris-HCl, pH 7.5, 50 mM MgCl$_2$, and 75 mM DTT.
6. 0.5 mM each of dNTPs.
7. Razor blade.
8. Qiaex II gel extraction kit (Qiagen, Valencia, CA).
9. Micro Bio-spin 6 column (Bio-Rad, UK).

2.3. Polyacrylamide Gel Preparation

1. 0.1% Sodium dodecyl sulfate (SDS) solution for washing glass plates.
2. 95% Ethanol for cleaning plates.
3. 1% Agarose for sealing plates when pouring the acrylamide gel.
4. 5X Tris-glycine electrophoresis buffer stock: 60.56 g (0.25 M) Tris base, 285.4 g (1.9 M) glycine, 7.44 g (10 mM) EDTA disodium salt, dihydrate, and deionized, distilled H$_2$O to 2 L. Check that the pH is 8.5. Store for several months at room temperature.
5. 30% Acrylamide.
6. 1% Bis-acrylamide.
7. 50% Glycerol.
8. TEMED (N,N,N',N'-tetramethylenediamine).
9. 10% Ammonium persulfate (freshly prepared).
10. Whatman 3MM filter paper.
11. Plastic wrap.
12. Gel dryer (Bio-Rad, UK).
13. AR X-ray film, intensifying screen and cassette or PhosphorImager (Molecular Dynamics).

2.4. DNA-Protein Binding Reaction

1. 2X GMSA binding buffer: 40 mM HEPES, pH 7.6, 200 mM KCl, 10 mM MgCl$_2$, 0.4 mM EDTA, 1 mM DTT, 20% glycerol.
2. 1 µg/µL poly(dI-dC)·poly(dI-dC).
3. 10X Gel loading dye: 0.25% bromophenol blue in 1X GMSA binding buffer.

3. Methods

3.1. Preparation of Protein Extracts

The preparation of the protein extracts from frozen adult rat tissues such as heart, liver, stomach, and skeletal tissues for use in GMSA is described in **Subheadings 3.1.1.** and **3.1.2.** This includes (1) a detailed description for preparing total protein extracts and (2) a brief description of the preparation of nuclear extracts.

3.1.1. Preparation of Total Protein Extracts

1. Grind 20 g of frozen adult rat heart, skeletal muscle, liver, or stomach (*see* **Note 1**) in liquid nitrogen using a mortar and pestle.
2. Incubate the pulverized tissue in 200 mL of lysis buffer at 4°C for 30 min while gently stirring.
3. Centrifuge lysate at 8000g at 4°C for 20 min.
4. Save a small amount of the supernatant for later footprinting or GMSA analysis to make sure that the supernatant does not contain any binding activity. Resuspend the pellet with 400 mM NaCl extraction buffer. Incubate at 4°C for 30 min.

5. Centrifuge the mixture at 8000g at 4°C for 30 min and collect the supernatant. Repeat centrifugation, if needed. Save a small amount of the supernatant for later footprinting or GMSA analysis.
6. Apply the supernatant to a 2.5-mL heparin-agarose column preequilibriated with 20 mL of 400 mM NaCl extraction buffer (*see* **Note 2**).
7. Wash the column with 10 mL of 400 mM NaCl extraction buffer.
8. Elute the bound proteins with 12 mL of 1 M NaCl extraction buffer.
9. Dialyze the eluent with two changes (2 L each) of 1X GMSA binding buffer for at least 4 h.
10. Check the DNase I footprinting or GMSA activity of the extracts before and after heparin-agarose chromatography to compare the specific activity of *trans*-acting proteins in each fraction. Based on our studies, the resulting protein extracts after heparin-agarose column purification contain binding activity for many known transcription factor-binding motifs, such as the AP2 site, TATA site, CArG box, M-CAT site, MEF2-like motif, and others, as well as unknown factor-binding sites *(3,4)*.
11. Determine the protein concentration by the Bradford method *(5)* or the Bio-Rad Protein Assay (Bio-Rad, UK).
12. Aliquot and store the protein extracts at –70°C.

3.1.2. Preparation of Nuclear Extracts

The preparation of nuclear extract requires isolation of nuclei from tissues or cultured cells. To reduce the contribution of nuclear extract by cell types other than muscle cells, primary cardiomyocytes can first be isolated, cultured, and subsequently used for the isolation of nuclei *(6–8)*. In striated muscle tissues, large quantities of myofibrils interfere with the isolation of nuclei. Therefore, a special step to relax the muscle tissue is included before homogenization of tissues *(9)*.

1. Nuclei from the homogenated tissues/cells are pelleted through a sucrose cushion by centrifugation to separate the cytoplasmic contents from the nuclei.
2. The protein extraction from the collected nuclei is generally carried out by the careful dropwise addition of a high-salt buffer into the resuspended nuclei, as described in detail elsewhere *(10)*.
3. The high-salt extracts are subsequently clarified by centrifugation, and the supernatant is saved.
4. This supernatant now contains nuclear extracts and is dialyzed into GMSA binding buffer.
5. The nuclear extracts can then be aliquoted and stored in liquid nitrogen or at –70°C.

3.2. Preparation of Nondenaturing Polyacrylamide Gel

1. Wash glass plates thoroughly with 0.1% SDS solution, rinse well with deionized water, and dry (*see* **Note 3**).
2. Clean the plate with 95% ethanol.

3. Assemble two plates with 1.5-mm spacers on the sides and bottom and clip plates together securely to prepare for casting the gel.
4. Seal the sides and bottom of the plates with warm 1% agarose, and allow the agarose to solidify.
5. Prepare a low-ionic-strength gel solution (*see* **Note 4**). For a 6% native polyacrylamide gel, combine: 4.6 mL 5X Tris-glycine electrophoresis buffer stock, 4.6 mL 30% acrylamide, 3.0 mL 1% *bis*-acrylamide, 1.15 mL 50% glycerol, 22 µL TEMED, 9.43 mL H_2O, and 220 µL 10% ammonium persulfate, for a total of 23 mL. This amount can be scaled appropriately depending on the size of gel needed.
6. Pour the gel solution into the space between the plates, insert the comb into the gel, and allow gel to polymerize completely for approx 30 min.
7. Remove bottom spacer and attach plates to electrophoresis tank. Fill the lower and upper reservoirs of the tank with 1X Tris-glycine electrophoresis buffer (*see* **Note 5**). Remove the comb, and use a bent-needle syringe to remove air bubbles trapped between the plates below the gel, and flush out and straighten the wells.
8. Prerun the gel for 30 to 60 min at 100 V.

3.3. Preparation of Labeled DNA Probe

3.3.1. Preparation of Restriction Fragment Probes or PCR-Amplified DNA Probes

Prior to constructing the radiolabeled probe, the DNA fragment of interest must be selected (*see* **Note 6**).

1. Isolate a DNA fragment containing the binding sites of interest using a standard restriction enzyme digestion (*see* **Note 7**). Restriction endonucleases that leave 5' overhangs are preferable for use in the subsequent end-labeling step.
2. End-label the restriction fragment by a Klenow fill-in reaction (*see* **Note 8**): 25 µL of 1X Klenow buffer containing 1 to 50 pmol of DNA fragment, 150 µCi of a suitable [α-^{32}P]dNTP (>3000 Ci/mmol), 40 µM each of the other three unlabeled dNTPs, and 10 U of the Klenow fragment of *E. coli* DNA polymerase I. Incubate for 15 min at room temperature. Then chase with the unlabeled form of the fourth dNTP and incubate for an additional 5 min.
3. Separate the desired labeled restriction fragment from the plasmid using gel electrophoresis. Identify the labeled DNA fragment by briefly exposing this electrophoresed gel to X-ray film. Excise the identified radiolabeled band from the gel with a razor blade and elute it by the Qiaex II gel extraction kit (Qiagen).
4. If PCR is used to amplify the region of interest, separate the PCR product by gel electrophoresis. Excise the DNA band from the gel and elute it by the Qiaex II gel extraction kit (Qiagen). The isolated DNA fragment is subsequently end-labeled using T4 polynucleotide kinase as described in **Subheading 3.3.2.**

3.3.2. Preparation of Oligonucleotide Probes

1. Synthesize commercially complementary oligonucleotides (*see* **Note 9**).
2. Anneal them to generate the double-stranded oligonucleotide probe containing the specific binding site of interest.
3. End-label the probe using T4 polynucleotide kinase in the following reaction conditions: 30 µL of 1X T4 polynucleotide kinase buffer containing dephosphorylated DNA fragments (1–50 pmol of PCR-amplified or double-stranded oligonucleotide probe), 150 µCi of [γ-^{32}P]ATP (6000 Ci/mmol), and 20 U of T4 polynucleotide kinase, 37°C for 1 h.

   If the double-stranded oligonucleotides are designed to have a 5' overhang, they can be labeled using Klenow fragment as previously described.
4. Remove unincorporated [γ-^{32}P]ATP by passing the reaction mixture through a micro Bio-spin 6 column (Bio-Rad, UK).
5. Collect the flow through fraction and determine the radioactivity.

3.4. Gel Mobility Shift Assay

Numerous conditions can affect DNA-protein binding *(11)*, and thus many parameters of the binding reaction can be altered to obtain optimal conditions for the DNA-protein interaction of interest. A range of conditions varying in amounts of DNA probe and binding proteins, ionic strength, pH, divalent ion, temperature, glycerol content, and nonionic detergent content, as well as including carrier proteins and/or carrier DNA, may have to be tested to identify a potential interaction. The following is a GMSA protocol that has been used effectively in identifying the *cis*-regulatory elements and *trans*-acting factor bindings on the *cTnT* proximal promoter *(2–4)* (*see* **Note 10**).

3.4.1. Reaction Setup

To show the specificity of the binding reaction, adequate positive and negative controls must be established. It is also important to show specificity using a competition assay by including specific unlabeled probe as well as nonspecific unlabeled probe.

1. Negative controls include using the labeled DNA probe of interest with no protein added or with a protein that should not bind the probe.
2. A positive control would be a known DNA-protein interaction, which may not always be available.
3. If competition assays are to be performed, conditions should be optimized so that 20 to 30% of the probe is bound. Specific competitors (specific unlabeled probes or self-probes) are used at increasing molar excess of the labeled probe to show the specificity of the binding reaction (*see* **Note 11**).
4. Nonspecific competitors include an unlabeled DNA fragment that is different from the sequence of interest in the binding reaction. Nonspecific competitors

are usually used at increasing molar excess of the labeled probe. These reactions show the specificity of the desired DNA fragment with the protein. Nonspecific competitors can also be used to identify the exact nucleotide sequence needed for binding. A probe with a sequence similar to the fragment of interest, but with an altered nucleotide(s) incorporated, can be used to show exactly which nucleotides are necessary for binding.

3.4.2. Binding Reaction

1. While the gel is prerunning, add the following reagents to a 1.5-mL microcentrifuge tube in the order listed: 5000 cpm labeled probe (0.1–0.5 ng), 1 µL of 1 µg/µL carrier DNA poly(dI-dC).poly(dI-dC), 10 µL of 2X GMSA binding buffer, protein/nuclear extracts (10–20 µg protein extract or 5–25 ng purified protein). Adjust total volume to 20 µL with deionized distilled water.
2. Mix by tapping, and incubate at room temperature for 30 min.

3.4.3. Running the Gel

1. Add 2 µL of 10X gel loading buffer to each reaction.
2. Immediately load each reaction on a 6% nondenaturing polyacrylamide gel, which has been prerun in 1X Tris-glycine buffer for 30 min to 1 h.
3. Run the gel at 100 V for 1 to 1.5 h until the tracking dye is near the bottom of the gel. The tracking dye (bromophenol blue) migrates roughly at the same position as a 70-bp DNA probe. For probes <70 bp, the dye should not be run to the bottom of the gel (*see* **Note 12**).

3.4.4. Drying the Gel

1. Remove the gel from the gel tank and carefully remove side spacers.
2. Slowly pry the glass plates apart, allowing the gel to remain on one plate.
3. Lay the glass plate with the gel attached on the bench (gel side up) and cover the gel with a sheet of Whatman 3MM filter paper.
4. Flip the gel sandwich so that the filter paper is on bottom and glass plate is on top.
5. Remove the glass plate.
6. Cover the gel with plastic wrap and dry under a vacuum at 80°C for 90 min.
7. Autoradiograph the dried gel on X-ray film with an intensifying screen at –70°C or by PhosphorImager and ImageQuant software (Molecular Dynamics).

Figure 2 illustrates a typical example of GMSA results.

3.5. Applications

The GMSA has been adapted to be used for many additional applications; such as the antibody supershift assay, gel-shift competition assay, and GMSA with SDS-PAGE gel fractionated proteins. These applications are discussed in **Subheadings 3.5.1.** to **3.5.3.**

3.5.1. Antibody Supershift Assay

The antibody supershift assay is an adapted variant of the GMSA in which antibodies are used to identify proteins present in the DNA-protein complex. This variant is limited because the researcher must have prior knowledge of what the binding protein may be, in order to test with accurate antibodies. Owing to this limitation, this technique would not be suitable for identifying novel proteins. If the antibody is included in the GMSA, three outcomes are possible in regard to the mobility of the DNA-protein complex: (1) no effect on the mobility if the protein recognized by the antibody is not involved in complex formation, (2) no complex formation and the probe runs like free probe if the protein is necessary for complex formation and the antibody interferes with the DNA-interacting sites of the protein, or (3) reduced mobility (supershift) if the protein that forms the complex is recognized by the antibody and the antibody does not perturb DNA-protein complex formation (*see* **Note 13**).

When one is investigating the regulatory mechanisms of the rat *cTnT* *(2)* and the rat *cardiac ventricular myosin light-chain 2* (*MLC-2v*) *(12,13)*, the antibody supershift assay has been successfully used to demonstrate the absence or presence of specific transcription factors within DNA-protein complexes. Previously, MEF2A and HF1b transcription factors had both been shown to bind to the MEF2 consensus sequence and regulate transcription of cardiac-restricted genes *(12,14)*. When the antibody supershift assay was used to investigate whether MEF2A or HF1b was present at the MEF2-like site within the rat *cTnT* promoter, neither supershift nor effect on the DNA-protein complexes was observed. This result strongly suggests that neither MEF2A nor HF1b is present in the complex formed by cardiac proteins and DNA probe containing the MEF2-like site *(2)*. Previously, a 250-bp *MLC-2v* promoter fragment containing three conserved regulatory elements (HF1a, HF1b/MEF2, and HF3) has been shown to confer ventricular-specific expression of a *lacZ* reporter gene *(7,13,15)*. However, evidence also suggests that cardiac-specific expression of *MLC-2v* requires combinatorial interactions between various elements located within the 250-bp fragment, which led to the discovery of a novel upstream stimulating factor (USF)-binding site named MLE1 *(13)*. When using the antibody supershift assay, USF antibodies supershifted both endogenous MLE1 and the HF1a radiolabeled probes, suggesting that USF can bind specifically to the MLE1 site and the HF-1a site within the *MLC-2v* promoter *(13)*.

3.5.2. Gel-Shift Competition Assays

Gel-shift competition assay is a powerful technique used to asses the sequence specificity of a DNA-protein interaction, especially since most protein extracts contain both specific and nonspecific DNA binding proteins.

When one is assessing a specific DNA-protein interaction, an unlabeled competitor DNA probe with the same sequence as the protein binding site can be used to sequester/compete the protein away from labeled probe. As a control for binding specificity, a nonspecific competitor probe must be used to show that the unlabeled nonspecific probe cannot sequester/compete the protein away from the labeled probe. Generally, a nonspecific competitor can be any fragment of unrelated sequence with the same size and configuration at the ends as the labeled probe, but ideally a nonspecific competitor should be identical to the labeled probe except for a key mutation in the binding site known to disrupt protein binding.

When we investigated the proteins that interact with the D and F modules of the rat *cTnT* promoter, competition GMSAs were used to demonstrate that the same proteins could bind to different *cTnT* promoter modules with differential affinities *(4)*. Using F41, D1, or D2 as labeled probes, GMSAs were carried out with the absence (probe alone) and presence of an increased amount of non-self-competitors as well as self-competitors. When assays were performed with the F41 probe (**Fig. 3A**), both D1 and D2 competitors were able to compete off complexes A and B, but only D2 could compete off complex C because D1 contained no A/T-rich site (*see* **Fig. 1B** for sequence difference). When assays were performed with the D1 probe (**Fig. 3B**), only D2 but not F41 competitor could effectively compete off complexes A and B. When assays were performed with the D2 probe (**Fig. 3C**), neither D1 nor F41 competitors could compete off the complexes formed by D2. These results together suggest that the TCTG(G/C) direct repeats within these different probes have different binding affinities for the same protein factors and that the relative affinity is D2>D1>F41.

3.5.3. GMSA with SDS-PAGE Gel Fractionated Proteins

When we investigated the proteins that interact with the D and F modules of the rat *cTnT* promoter, GMSAs were used to identify the proteins involved in binding to the TCTG(G/C) direct repeat and A/T-rich site *(4)*. First, F41 probed GMSAs using protein extracts from heart and stomach formed an identical complex C, suggesting the involvement of a ubiquitous factor. In contrast, tissue-specific factors are responsible for the formation of complexes A and B (minus lanes in **Fig. 2**). Competitive GMSAs using 100-fold molar excess of unlabeled F41 demonstrated that the DNA-protein interactions were specific for all complexes that were formed from both heart and stomach protein extracts (plus lanes in **Fig. 2**). To characterize the proteins involved in complex formation further, protein extracts from adult rat hearts were fractionated by SDS-PAGE into many fractions, eluted, renatured *(16)*, and then incubated

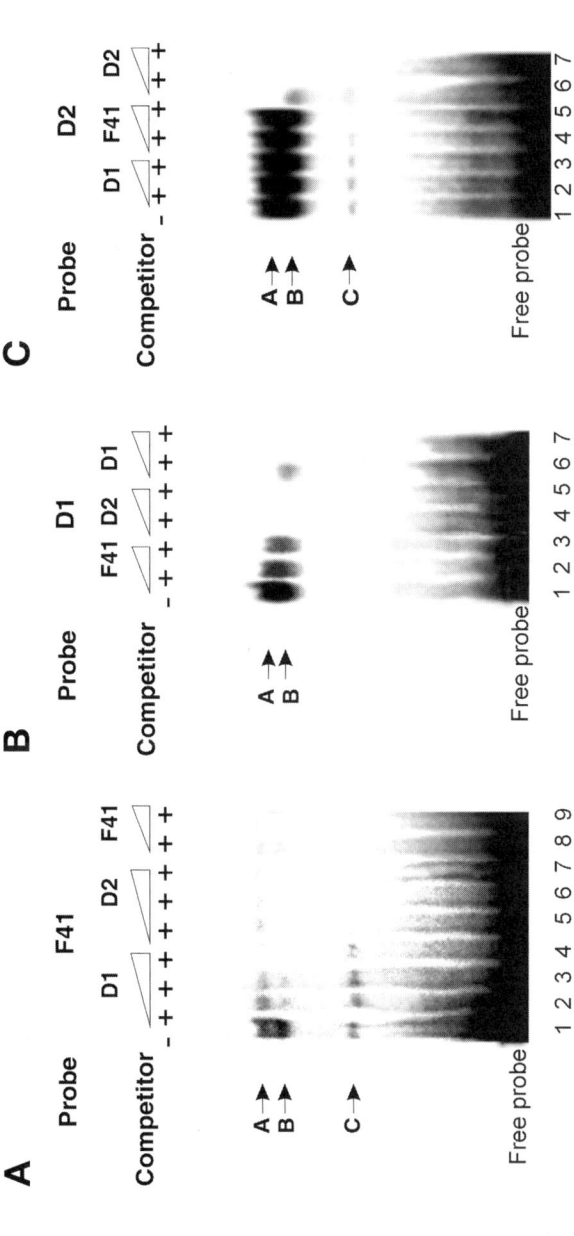

Fig. 3. Competition GMSAs using the F41, D1, and D2 probes. (A) GMSA with labeled F41 probe (lane 1). Competition assays were carried out with 100-, 200-, and 400-fold molar excess of D1 (lanes 2, 3, and 4, respectively), D2 (lanes 5, 6, and 7, respectively) and 100- and 200-fold molar excess of self-competitor (lanes 8 and 9, respectively). Both D1 and D2 can compete off complexes A and B, whereas only D2 can compete off complex C. (B) GMSA with labeled D1 probe (lane 1). A 100- and 200-fold molar excess of F41 (lanes 2 and 3, respectively), D2 (lanes 4 and 5, respectively), and D1 itself (lanes 6 and 7, respectively) served as competitors. D2 but not F41 effectively competes off complexes A and B formed by the D1 probe. (C) GMSA with labeled D2 probe (lane 1). Unlabeled D1 (lanes 2 and 3), F41 (lanes 4 and 5), and D2 itself (lanes 6 and 7) at 100- and 200-fold molar excess were used as competitors. Neither D1 nor F41 can compete off complexes A and B formed by D2.

with either F41 probe (**Fig. 4A**) or D1 probe (**Fig. 4B**) for the subsequent GMSAs (*see* **Note 14**).

1. Briefly, 50 µg of cardiac protein extracts were heated at 37°C for 5 min in SDS-PAGE loading buffer and subsequently separated on a 12.5% SDS-PAGE gel.
2. The gel was then sliced into 23 pieces, each of which was then mashed and incubated in 240 µL of elution/renaturation buffer (1% Triton X-100, 20 m*M* HEPES, pH 7.6, 1 m*M* EDTA, 100 m*M* NaCl, 5 mg/mL bovine serum albumin, 2 m*M* DTT, and 1 m*M* PMSF) at 37°C for 3 h and then 4°C overnight.
3. The eluent was separated from gel residues by centrifugation at 12,000*g* for 10 min at 4°C and stored at –70°C before use.
4. Twelve microliters of eluent from each fraction was then used for GMSAs.

Compared with protein extracts before gel fractionation as seen in lane C of **Fig. 4**, the 42-kDa proteins in fraction 10 were shown to form complexes A and B with either the F41 or D1 probe. However, the 25-kDa protein found in fraction 15 could form complex C with the F41 probe. Furthermore, when the D2 probe was used in GMSAs with these gel-fractionated proteins, the same results were obtained as when F41 was used as a probe (data not shown). Since both F41 and D2, but not D1, contain the A/T-rich site in addition to the TCTG(G/C) direct repeat, these results further confirmed that the 25-kDa protein bound to an A/T-rich site, whereas the 42-kDa proteins recognized the TCTG(G/C) direct repeat. The appreciation of this GMSA technique allowed for the identification of a 25-kDa ubiquitous protein bound to the A/T-rich site. This protein was subsequently shown to be an HMG2 protein *(4)*.

4. Notes

1. When one is characterizing the *cis*-regulating elements and the *trans*-activating factors of the rat *cTnT* gene, protein extracts are taken from multiple tissues for the following reasons. First, multiple tissue types may use different transacting factors to either enhance or suppress the *cTnT* expression. Both cardiac and skeletal muscles express *cTnT* during embryonic and fetal stages *(1,17–22)*, but *cTnT* expression is restricted and upregulated in cardiac muscle during later fetal stages and adult *(23)*. Furthermore, rat c*TnT* expression has also been identified in regenerating adult skeletal muscle after cold injury and in mature skeletal muscle fibers after denervation *(19)*. To understand the mechanism regulating *cTnT* expression and whether there is a mechanism specific to cardiac muscle, extracts are made from cardiac and skeletal muscle for further examination. Second, since cardiac tissue primarily contains cardiac muscle cells and fibroblasts, liver tissue is used as a control for the nonmuscle cell contamination in the cardiac tissue extracts. Since cardiac and skeletal muscles are both striated muscle types, stomach tissue is used as a control to establish that *cTnT* is not expressed in smooth muscle.

A GMSA using F41 as a probe

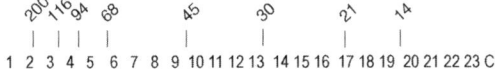
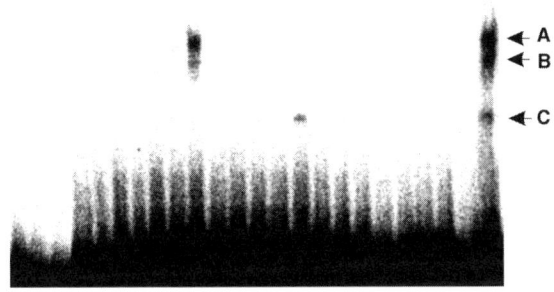

B GMSA using D1 as a probe

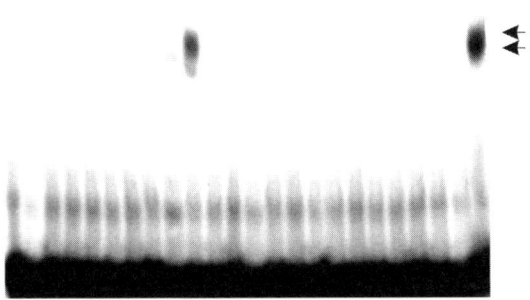

Fig. 4. GMSAs of renatured proteins after separation by SDS-PAGE. Protein extracts from adult rat hearts were fractionated by SDS-PAGE into 23 fractions. Proteins in each fraction were eluted, renatured, and incubated with labeled F41 (**A**) or D1 (**B**). Lane C in each panel represents GMSA with protein extracts before gel fractionation. Proteins with 42 kDa found in fraction 10 are capable of forming complexes A and B with F41 or D1 probe. On the other hand, protein with 25 kDa in fraction 15 forms complex C only with F41 but not with D1.

2. If the lysate is too thick to go through the heparin-agarose column by gravity, the lysate loaded column can be centrifuged using a low-speed setting (2–3000g) at 4°C.
3. The plates must be thoroughly rinsed to remove all detergent, as detergent can interfere with the binding reaction.
4. Typically gels in the range of 4 to 6% acrylamide are best for this application. A 6% polyacrylamide gel worked well under these outlined conditions but will be variable depending on the sizes of the DNA-protein complexes.
5. The stability of the DNA-protein complex is also affected by the choice of buffer used for running the polyacrylamide gel, and this parameter must be optimized for individual binding reactions. Generally, Tris-glycine (as recommended by this protocol) works well, but alternatively Tris-acetate (TAE), and Tris-borate (TBE) electrophoresis buffers can be used. The buffer contained in the gel and the electrophoresis buffer should be the same.
6. Although the length of the DNA fragment can vary, our lab has had success with probes ranging from 16 to 92 bp *(3,4)*. The longer the fragment, the more likely it is that multiple binding sites are present, which could result in nonspecific binding *(11)*.

    Consensus sequence databases provide a useful tool for identifying transcription factor binding sites, which can then be used for determination of a DNA fragment to use in probe construction. Examples of such sites are available to the public (www.gene-regulation.com or www.cbrc.jp/research/db/TFSEARCH.html). Additionally, DNase I footprinting can also be employed to identify a binding region of interest more specifically for use in probe generation.
7. Digestion of 1 µg of plasmid DNA will yield sufficient restriction fragment for use in 200 binding reactions.
8. It is most convenient to choose restriction digest sites that leave 5' overhangs, as these can be filled in using the Klenow fragment and an appropriate [α-^{32}P]dNTP and other nonlabeled dNTPs and thereby radiolabeled.
9. Oligonucleotide probes are generally designed to be conveniently labeled and are typically 20 to 40 nucleotides in length *(11)*.
10. Various components can be added to the binding reaction to optimize or achieve the desired DNA-protein interaction. The inclusion of a carrier protein, such as BSA, in the binding reaction can increase the stability of some complexes during electrophoresis. When one is using protein extracts, increasing the salt concentration to 50 to 100 mM can reduce the number of nonspecific DNA-protein interactions *(24)*. In contrast, when one is using purified proteins, lower salt concentrations can be used. A final component to consider is the amount of carrier DNA used. At a given protein concentration, too little carrier DNA will cause the entire probe to be bound and not enter the gel. If too much carrier DNA is added, none of the probe will be bound *(11)*. The sequence of the carrier DNA should not resemble the DNA probe used.

    Various parameters of the binding reaction can also be adjusted to optimize the DNA-protein interaction. An important parameter is the temperature at which the reaction takes place. Most binding reactions are performed at 30°C, yet optimal temperature ranges can be from room temperature to 37°C *(24)*. A second variable is the incubation time of the binding reaction. Most reactions will reach equilibrium within 30 min; therefore incubations are typically done from 10 to 30 min *(24)*. Another important parameter to consider is pH. Various DNA-protein interactions are dependent on pH. Most reactions will be stable at pH 7.9, yet interactions need to be empirically determined at ranges of 6.5 to 8.5 *(24)*.

11. Typical amounts of competitor in these assays are 50X, 100X, and 200X molar excess relative to the labeled probe *(11)*.
12. While you are running the gel, it is important that the plates do not get too warm, as denaturation can occur, which interferes with protein binding. The voltage must be decreased if the plates become too warm, or, alternatively, the gel can also be run in a cold room if the gel is run at higher voltages.
13. The amount of antibody used in the supershift assays should be the minimal amount needed to produce a detectable effect. Possible complications in antibody supershifts occur if the antibody recognition site overlaps with the DNA binding domain.
14. Although it is typical to heat-denature protein extracts at 100°C prior to running them on a SDS-PAGE protein gel, it is important to note that this treatment may change the protein structure and alter the DNA-protein complex migration in the gel.

Acknowledgments

We thank Rebecca Reiter, Jenny L.-C. Lin, and Kibby Wall for routine technique help. This work was supported by NIH grants HL72910 and HL75015. S.E.G., S.M.J., and E.A.G-W. contributed equally to this study.

References

1. Wang, Q., Reiter, R. S., Huang, Q.-Q., Jin, J.-P., and Lin, J. J.-C. (2001) Comparative studies on the expression patterns of three troponin T genes during mouse development. *Anat. Rec.* **263,** 72–84.
2. Wang, Q., Sigmund, C. D., and Lin, J. J.-C. (2000) Identification of *cis* elements in the *cardiac troponin T* gene conferring specific expression in cardiac muscle of transgenic mice. *Circ. Res.* **86,** 478–484.
3. Wang, G., Yeh, H.-I., and Lin, J. J.-C. (1994) Characterization of cis-regulating elements and trans-activating factors of the rat cardiac troponin T gene. *J. Biol. Chem.* **269,** 30595–30603.
4. Wang, Q., Lin, J. L.-C., and Lin, J.J.-C. (2002) A novel TCTG(G/C) direct repeat and an A/T-rich HMG2-binding site control the expression of the rat *cardiac troponin T* gene. *J. Mol. Cell. Cardiol.* **34,** 1667–1679.
5. Bradford, M. M. (1976) A rapid and sensitive method for the quantitation of microgram quantities of protein utilizing the principle of protein-dye binding. *Anal. Biochem.* **72,** 248–254.

6. Chen, C. Y. and Schwartz, R. J. (1996) Recruitment of the tinman homolog Nkx-2.5 by serum response factor activates *cardiac α-actin* gene transcription. *Mol. Cell. Biol.* **16,** 6372–6384.
7. Navankasattusas, S., Zhu, H., Garcia, A. V., Evans, S. M., and Chien, K. R. (1992) A ubiquitous factor (HF-1a) and a distinct muscle factor (HF-1b/MEF-2) form an E-box-independent pathway for cardiac muscle gene expression. *Mol. Cell. Biol.* **12,** 1469–1479.
8. Zou, Y. and Chien, K. R. (1995) EFIa/YB-1 is a component of cardiac HF-1A binding activity and positively regulates transcription of the myosin light-chain 2v gene. *Mol. Cell. Biol.* **15,** 2972–2982.
9. Mar, J. H. and Ordahl, C. P. (1990) M-CAT binding factor, a novel *trans*-acting factor governing muscle-specific transcription. *Mol. Cell. Biol.* **10,** 4271–4283.
10. Abmayr, S. M. and Workman, J. L. (1998) Preparation of nuclear and cytoplasmic extracts from mammalian cells, in *Current Protocols in Molecular Biology* (Ausubel, F. M., Brent, R., Kingston, R. E., et al., eds.), John Wiley & Sons, New York, pp. 12.11.11–12.11.19.
11. Buratowski, S. and Chodosh, L. A. (1998) Mobility shift DNA-binding assay using gel electrophoresis. in *Current Protocols in Molecular Biology* (Ausubel, F. M., Brent, R., Kingston, R. E., et al., eds.), John Wiley & Sons, New York, pp. 12.12.11–12.12.11.
12. Zhu, H., Nguyen, V. T., Brown, A. B., et al. (1993) A novel, tissue-restricted zinc finger protein (HF-1b) binds to the cardiac regulatory element (HF-1b/MEF-2) within the rat *myosin light chain-2* gene. *Mol. Cell. Biol.* **13,** 4432–4444.
13. Navankasattusas, S., Sawadogo, M., van Bilsen, M., Dang, C. V., and Chien, K. R. (1994) The basic helix-loop-helix protein upstream stimulating factor regulates the cardiac *ventricular myosin light-chain 2* gene via independent *cis* regulatory elements. *Mol. Cell. Biol.* **14,** 7331–7339.
14. Yu, Y.-T., Breitbart, R. E., Smoot, L. B., Lee, Y., Mahdavi, V., and Nadal-Ginard, B. (1992) Human myocyte-specific enhancer factor 2 comprises a group of tissue-restricted MADS box transcription factors. *Genes Dev.* **6,** 1783–1798.
15. Lee, K. J., Hickey, R., Zhu, H., and Chien, K. R. (1994) Positive regulatory elements (HF-1a and HF-1b) and a novel negative regulatory element (HF-3) mediate ventricular muscle-specific expression of myosin light-chain 2 luciferase fusion genes in transgenic mice. *Mol. Cell. Biol.* **14,** 1220–1229.
16. Ossipow, V., Laemmli, U. K., and Schibler, U. (1993) A simple method to renature DNA-binding proteins separated by SDS-polyacrylamide gel electrophoresis. *Nucleic Acids Res.* **21,** 6040–6041.
17. Cooper, T. A. and Ordahl, C. P. (1984) A single *cardiac troponin T* gene regulated by different programs in cardiac and skeletal muscle development. *Science* **226,** 979–982.
18. Mar, J. H., Antin, P. B., Cooper, T. A., and Ordahl, C. P. (1988) Analysis of the upstream regions governing expression of the chicken *cardiac troponin T* gene in embryonic cardiac and skeletal muscle cells. *J. Cell Biol.* **107,** 573–585.

19. Saggin, L., Gorza, L., Ausoni, S., and Schiaffino, S. (1990) Cardiac troponin T in developing, regenerating and denervated rat skeletal muscle. *Development* **110,** 547–554.
20. Swiderski, R. E. and Solursh, M. (1990) Precocious appearance of *cardiac troponin T* pre-mRNAs during early avian embryonic skeletal muscle development in ovo. *Dev. Biol.* **140,** 73–82.
21. Sutherland, C. J., Elsom, V. L., Gordon, M. L., Dunwoodie, S. U., and Hardeman, E. C. (1991) Coordination of skeletal muscle gene expression occurs late in mammalian development. *Dev. Biol.* **146,** 167–178.
22. Jin, J.-P., Huang, Q.-Q., Yeh, H.-I., and Lin, J. J.-C. (1992) Complete nucleotide sequence and structural organization of rat *cardiac troponin T* gene. A single gene generates embryonic and adult isoforms via developmentally regulated alternative splicing. *J. Mol. Biol.* **227,** 1269–1276.
23. Long, C. S. and Ordahl, C. P. (1988) Transcriptional repression of an embryo-specific muscle gene. *Dev. Biol.* **127,** 228–234.
24. Tymms, M. J. (2000) *Transcription Factor Protocols.* Humana Press, Totowa, NJ.

11

Mapping Transcriptional Start Sites and *In Silico* DNA Footprinting

Martin E. Cullen and Paul J. R. Barton

Summary

Determination of a gene's transcriptional start site underlies the identification of the proximal promoter region and thus facilitates the subsequent analysis of components controlling its expression, namely, *cis*-acting regulatory elements and their cognate binding proteins. It also enables assembly of meaningful reporter constructs to examine promoter function in different cellular contexts. In this chapter, basic protocols for two experimental approaches to transcriptional start site determination are described: primer extension analysis and the ribonuclease protection assay. Consideration is also given to RNA sources, RNA purification, and primer design. The explosion in genomic DNA and mRNA sequence information derived from genomic sequencing projects, expressed sequence tags and microarrays, combined with *in silico* analysis, such as automated sequence annotation and gene identification algorithms, now provides an alternative source of detailed information on gene structure and expression. Two approaches to the *in silico* identification of transcription factor binding sites are described.

Key Words: Primer extension; RNase protection assay; bioinformatics; phylogenetic footprinting.

1. Introduction

Accurate determination of the transcriptional start site underlies the identification of the proximal promoter and subsequent analysis of the *cis*-acting regulatory elements of a gene. The transcriptional start site is defined as the most 5' nucleotide of the gene that is transcribed into RNA and is generally annotated "+1." It is extremely important that the correct start site be identified when detailed gene regulation studies are to be undertaken. The gene promoter, the region responsible for regulating gene transcription, either alone or in concert with more distal enhancer elements, is defined as the region immediately upstream of the transcription start site. Accurate definition of the start site therefore sets the scene for identification of the *cis*-acting elements and their cognate binding proteins. It also allows for meaningful construction of promoter-reporter constructs for examining the function of the promoter. Erroneous identification would lead to considerable difficulties in interpretation of any subsequent studies and would likely result in misleading data.

When one is deciding how to establish the transcriptional start site, it is also important to include a broad view of the gene and its expression. Many genes use a single transcriptional start site, but it may be that the gene under investigation has more than one start site or that alternative start sites are used in different contexts, for example, for different tissue-specific or developmental patterns of expression *(1,2)*. Two basic experimental approaches have remained the mainstay of start site determination over the years since their establishment in the late 1970s. These are (1) primer extension analysis, to determine accurately the length of messenger RNA (mRNA), which lies upstream (5') of a given point, and (2) nuclease protection assay, to confirm colinearity of this upstream mRNA sequence with the corresponding genomic DNA. These methods have been refined and adapted over the years in response to new reagents and techniques, for example, by using polymerase chain reaction (PCR)-based techniques including 5' rapid amplification of cDNA ends (RACE) in place of primer extension, and by replacement of the original S1 nuclease protection assay with a ribonuclease protection assay (RPA; *see* **Subheading 3.2.**). Nonetheless, in essence the approach and rationale have remained the same. With the rapid increase in genomic DNA and mRNA sequence information such as that derived from expressed sequence tags (ESTs), it is now often possible to derive quite detailed information on gene structure and expression through *in silico* analysis. However, being certain of the start site before embarking on detailed promoter studies is essential and should be backed up with experimental confirmation of such *in silico*-derived data.

2. Materials

1. Trizol (Invitrogen).
2. RNeasy RNA purification kit (Qiagen).
3. Ultra Turrax T8 homogenizer (VWR).
4. Microtube pestle (Anachem).
5. QIAShredder (Qiagen).
6. RNaseIn (Promega) or RNaseOUT (Invitrogen).
7. Oligonucleotide primer.
8. [γ-32P]ATP (>5000 Ci/mmol 10 mCi/mL) (Amersham).
9. T4 polynucleotide kinase.
10. 10X Polynucleotide kinase buffer: 0.5 M Tris-HCl pH 7.5, 0.1 M MgCl$_2$, 50 mM DTT, 1 mM spermidine, 1 mM ETDA.

11. Water bath.
12. STE buffer: 100 mM NaCl, 20 mM Tris-HCl, pH 7.5, 10 mM EDTA.
13. NucTrap purification column (Stratagene).
14. 10X RT buffer: 0.5 M KCl, 0.5 M Tris-HCl, pH 7.5, 50 mM MgCl$_2$, 0.1 M DTT, 10 mM spermidine.
15. Heating block.
16. 10X dNTP mix: 5 mM each of dATP, dCTP, dGTP, and dTTP.
17. AMV reverse transcriptase enzyme.
18. Polyacrylamide DNA sequencing gel system.
19. RPA kits (e.g., Ambion, BD Biosciences).

3. Methods

In this chapter we describe basic methods for primer extension (*see* **Subheading 3.1.**) and ribonuclease protection assays (*see* **Subheading 3.2.**) for the experimental determination of the transcription start site and discuss aspects of *in silico* sequence analysis (*see* **Subheading 3.3.**) that are of benefit in predicting transcription factor binding sites within the proximal promoter of a gene.

3.1. Primer Extension Analysis

The basic principle behind primer extension analysis is straightforward. A gene-specific complementary labeled DNA "primer" is annealed to its target mRNA in vitro and used to initiate DNA synthesis by the enzyme reverse transcriptase (a viral enzyme that has the ability to synthesize a DNA copy of the target RNA). As with all nucleic acid synthesis, extension proceeds in a 5' to 3' direction with respect to the synthesized complementary strand. Thus, the synthesized DNA will extend from the primer towards the 5' end of the mRNA template, and the size of the extended product will determine the length of the RNA that lies upstream (5') of the primer. Mapping this information onto the genomic sequence allows the location of the start site of transcription to be determined. Note, however, that the start site can only be defined if mRNA and genomic sequences are identical over the entire length of the extended product. If this is not known from available sequence data, it needs to be proved using RPA (*see* **Subheading 3.2.** below).

3.1.1. RNA

In order for the primer extension assay to be meaningful, it is essential that suitable RNA be available. For the reasons described in **Subheading 1.**, this must be from the tissue or cell type(s) in which expression is being analyzed. It is equally important that the RNA be of high quality, as fragmented or partly degraded RNA will result in premature termination of the reverse transcriptase reaction, a truncated product, and inaccurate or inconsistent start site determination. RNA can be made in-house using a number of commercially available kits. For myocardium we use tissue homogenization in Trizol (Invitrogen) followed by RNA extraction using RNeasy (Qiagen). Adult myocardium can be difficult to homogenize efficiently; when tissue samples are limited, for example, from human endomyocardial biopsies, homogenization with a microtube pestle (e.g., Anachem, cat. no. 749520-0090), followed by a handheld Ultra Turrax T8 homogenizer (VWR) fitted with a 5-mm probe, as well as the use of a QIAShredder (Qiagen), can significantly improve RNA yield (*3*). Alternatively, RNA can be purchased from a variety of suppliers for several species including human (e.g., Clontech, Stratagene, Origene) if tissue is not available in-house.

All RNA should be assessed for quality by spectrophotometry and agarose gel electrophoresis before proceeding. For most primer extension analysis, total RNA is usually sufficient. However, if the transcript is of low abundance it may be necessary to use "mRNA-enriched" RNA. mRNA typically comprises only a few percent of the total RNA content of a cell, and purified mRNA should therefore result in 10- to 20-fold increase in signal. It should be noted, however, that most mRNA purification is based on the use of polyA$^+$ selection. That is, mRNAs are purified on the basis of trapping the 3' end of the polyadenylated molecule. When transcripts are long, or when there may be a degree of RNA damage, such selection will result in underrepresentation of the 5' end of the mRNA. Again, many commercial suppliers offer polyA$^+$-purified RNA, albeit at a premium.

3.1.2. Primer

At the start of analysis a suitable primer must be designed. This must be complementary to the mRNA being analyzed, and so sequence data for the corresponding gene and transcript are needed. Early methods used cloned cDNA or genomic DNA fragments as primers for reverse transcription; typically these involved the use of strand separation techniques to isolate labeled primer from a DNA duplex. More straightforward and far more commonly used currently are oligonucleotide primers synthesized either in-house or to order through a suitable commercial supplier. When one is designing a suitable oligonucleotide primer, several factors should be considered. Short oligonucleotide primers of 10 nucleotides can give sufficient specificity and successful extension, but it is generally more efficient overall to use longer ones. Typically aim for 20 to 25 nucleotides. As with all oligonucleotides used for hybridization, regions of unusual sequence such as high GC or AT content, potential stem-loops, or repetitive sequence should be avoided. Primer design programs are available to help define ideal sequences, although generally careful visual inspection is sufficient.

Finally, locate the position of the primer so that the extended product is long enough to measure easily but short enough so as to maximize the synthesis of a full-length extension product by reverse transcriptase. Typically aim to be in the region of 50 to 150 nucleotides from the putative start site. Products of less than 50 nucleotides can be problematic, as they can be mistaken for artifactual bands resulting from "primer–dimer" effects or bands resulting from reverse transcriptase stalling early after initiation. When a suitable primer has been defined, it is worthwhile conducting a BLAST analysis to check for inadvertent sequence identity with other known transcripts. As a negative control to confirm specificity of product formation, an identical primer extension reaction should be carried out in the absence of the target mRNA template. Ideally, this should be done using a similar RNA (total or polyA$^+$ as appropriate) when the specific mRNA being examined is not present (e.g., RNA from an alternative tissue, different developmental time point, and so on).

3.1.3. Basic Protocol

Note that all materials, reagents, and solutions must be high grade and free from any contaminating RNase activity. RNase inhibitors such as RNaseIn (Promega) or RNaseOUT (Invitrogen) can be added as a precaution, or reagents can be bought in the form of a preoptimized primer extension kit (e.g., Promega).

3.1.3.1. PRIMER PREPARATION

1. Label oligonucleotide by incubation at 37°C for 1 h in a 10 µL mixture containing: 2 pmol oligonucleotide, 1 µL 10X polynucleotide kinase buffer, 3 µL [γ-^{32}P]ATP (>5000 Ci/mmol, 10 mCi/mL), H$_2$O, and T4 polynucleotide kinase.
2. Purify labeled oligonucleotide either by diluting the above reaction mix in 150 µL of STE buffer and passing it through a 2 mL G-25 sephadex column equilibrated using STE buffer (either dropwise, monitoring the peak of excluded labeled oligonucleotide using a hand-held Geiger monitor, or by gentle centrifugation of the column, e.g., 1000g × 5 min), or by using a commercial purification column such as NucTrap (Stratagene), following the manufacturer's instructions.
3. Determine the radioactive count of the resultant primers, which should be in the region of 50,000 to 100,000 cpm/µL in a total of approx 150 µL (using Cerenkov counting, i.e., without scintillant).

3.1.3.2. ANNEALING

1. Mix 2 µL of labeled primer from above with *either* 10 to 50 µg of total RNA *or* 1 to 2 µg polyA$^+$-enriched RNA, ethanol-precipitate, and dry pellet. (To precipitate, add 1/10th vol of 3 *M* sodium acetate, pH 5.0, and 2.5 vol ethanol. Chill to –20°C for at least 30 min and centrifuge to pellet.)

2. Resuspend pellet in 10 µL of 1X RT buffer.
3. Anneal, either by incubation at 60°C for 1 h or by placing the tube in a heating block at 80°C, which is then allowed to cool gradually to room temperature.
4. When cool, briefly spin tube in a microfuge to ensure that any condensation is brought to the bottom of the tube.

3.1.3.3. EXTENSION AND ANALYSIS

1. Extension is achieved by further addition of 2 µL 10X dNTP mix, 1 µL 10X RT buffer, AMV reverse transcriptase enzyme, and H_2O to a combined total volume of 20 µL, followed by incubation at 42°C for 1 h (NB: Incubation temperature can be increased to diminish the effects of potential interference from RNA secondary structure formation if required.)
2. Purify nucleic acids from reaction mix above by phenol-chloroform extraction, followed by ethanol precipitation.
3. Analyze product on a standard denaturing 8% polyacrylamide sequencing gel using standard gel loading buffers and heat denaturing of the sample before loading.

3.1.4. Use of DNA Sequencing Size Markers

Size determination of the extended product can be made by reference to any suitable size marker. For short products, it is often the case that the most accurate size determination is from a DNA sequencing ladder run on the same gel and providing a size ladder in single nucleotide steps. When a cloned copy of the gene is available, it is usual to use a sequencing reaction derived from the target gene, using the same oligonucleotide as for primer extension. In this way, when primer extension product and DNA sequencing reaction are run alongside each other it is possible to "read" the position of the start site directly. It is important to note, however, that primer extension can only determine the length of the extended product and gives no information on colinearity of product with genomic DNA. This must be proved by combining the primer extension assay with mRNA sequence data and/or a nuclease protection assay.

3.2. Ribonuclease Protection Assay (RPA)

The basic principle behind the RPA rests on the ability of certain ribonucleases (RNases A, T1, T2, and I) to digest selectively only single-stranded RNA, thereby distinguishing free RNA from RNA duplexed with either DNA or RNA. For start site determination, a labeled RNA probe derived by in vitro transcription of cloned genomic DNA, which encompasses the suspected start site, is annealed with its target mRNA. Unhybridized probe regions are removed using ribonucleases that will digest single-stranded but not double-stranded regions, and the size of the resulting protected molecule is determined.

Therefore, what the nuclease protection assay is determining is the length of the hybridized probe.

It is essential to note that using this method it is impossible to distinguish between the start site and an internal intron-exon boundary within the gene. Whereas primer extension analysis will determine the total distance from the annealed primer to the 5' end of the mRNA irrespective of gene structure, RPA determines the length of contiguous hybridization between probe and mRNA. The presence of an intron would give apparently discordant results between primer extension and RPA analysis, with the RPA analysis indicating a shorter contiguous region of hybridization than the length of transcript predicted from primer extension. For genes for which there is essentially complete sequence information on both gene and transcript, there is little need to demonstrate colinearity experimentally, as this will be self-evident from sequence alignment. When there are doubts as to exon usage or when sequence data are absent or incomplete, nuclease protection remains the assay of choice for determining colinearity.

3.2.1. Basic Protocol

The same precautions should be exercised for ribonuclease protection assays as for primer extension (*see* **Subheading 3.1.2.**) with regard to RNA source and quality and the use of RNase-free equipment and solutions. Many commercial companies now market RPA kits (e.g., Ambion, BD Biosciences), although they are usually marketed for use in RNA quantification studies. There is little merit in deriving the multiple buffers and components required to generate riboprobes from first principles, as commercial kits both avoid the need for time-consuming optimizing of reagents and are guaranteed RNase-free. A detailed methodology is therefore not given here, as this will vary depending on the system being used. Instead, an overview of the RPA method is given as it relates to start site determination.

In order to conduct RPA, genomic DNA containing the presumptive start site should be cloned in a suitable plasmid vector. This allows in vitro transcriptional synthesis of a radiolabeled single-stranded RNA probe, typically from T3, T7, or SP6 promoters located next to the multiple cloning site. The cloned fragment should be oriented such that in vitro transcription results in the production of an antisense RNA probe of about 100 to 300 nucleotides when the plasmid is linearized using a restriction enzyme site located upstream of the start site. To avoid complicating factors arising from intron sequences, the construct should ideally be made so that the 3' end of the cloned fragment is located within the first exon, as the use of probes that contain more than one exon can lead to results that are difficult to interpret owing to the generation of multiple individual protected fragments.

An RNA probe generated as just described will be complementary to the mRNA and will run through (part of) exon 1 and out into the gene promoter region of the gene. The probe can be labeled using ^{32}P or a suitable nonisotopic system such as those based on biotin according to the RPA kit manufacturer's instructions. The probe is purified and annealed with mRNA, and unhybridized regions are digested using single-strand specific RNAses. The resulting product is purified, denatured, and run on a standard denaturing 8% polyacrylamide sequencing gel with suitable size markers (see above). Initial titration of RNase concentration may be required to find conditions under which digestion is driven to completion. Control tracks should include samples treated identically, except when either no RNase is added or when no hybridizing DNA is added, to test for probe integrity and digestion artifacts, respectively.

3.3. Footprinting: An In Silico Approach

This section will outline some of the approaches that researchers have used to identify functional transcription factor binding sites (TFBS) and the changes from wet lab work to an *in silico* approach that have occurred recently. A recent review of bioinformatic approaches to identify regulatory elements provides more detailed information and includes a table of online resources *(4)*. It should be noted that an *in silico* approach can also be used to determine the 5' end of transcription, as discussed above. Databases such as the Eukaryotic Promoter Database (EPD; *[5]*), a curated reference collection of experimentally verified promoter sequences, are now being expanded by *in silico* "primer extension," based on information provided by 5' ESTs. In a complementary approach, algorithms that detect candidate promoters, based on a combination of transcription start site prediction and detection of CpG dinucleotide imbalance ("CpG islands"), have been useful in annotating putative promoters in the completely sequenced genomes currently interpreted via genome browsers such as Ensembl *(6)* or the UCSC Genome Browser *(7)*.

Standard bench experiments to establish the presence of bound transcription factors (TFs) on a promoter utilize one of several approaches, including electrophoretic mobility shift assays (EMSAs), chromatin immunoprecipitation (ChIP), and DNA footprinting. EMSA typically involves sequence-specific binding of TFs in crude nuclear extracts to defined end-labeled DNA fragments, followed by electrophoretic separation on nondenaturing gels *(8–10)*. Bound complexes are revealed by a mobility shift, relative to unbound DNA fragment. With this method, the ability of a given promoter region to bind specific TFs can be tested, with the identity of the bound factor determined using antibodies and the sequence specificity of binding revealed by mutagenesis of the DNA fragment or of competitor DNA. ChIP *(11)* involves treating living cells or tissue with formaldehyde to crosslink TFs and other DNA-binding proteins to their cognate sites. Antibodies directed against specific TFs are used to immunoprecipitate crosslinked DNA, and PCR analysis of the immunoprecipitate is used to detect the presence of any given promoter or other *cis*-element, thereby determining whether specific TFs are bound to the promoter in vivo. ChIP has been used to provide a snapshot of native chromatin structure and TFs bound to genes in different functional states.

3.3.1. Experimental Footprinting

Footprinting methods are based on the ability of bound protein to protect DNA from enzymatic or chemical cleavage (thereby leaving a "footprint") and can be separated into in vitro and in vivo methods. In vitro footprinting methods *(12,13)* first require the equilibration of an end-labeled DNA fragment containing the sequence of interest with a crude nuclear protein extract, usually in a titrated/dilution series. DNA is then cleaved by a variety of agents, including enzymes such as DNase I (also titrated in a dilution series) and specific restriction enzymes or chemical agents such as $KMnO_4$, dimethyl sulfate (DMS), Fe-EDTA *(14)*, and methidium-propyl-EDTA (MPE *[15]*). Comparison of cleaved DNA with similarly treated naked DNA separated on denaturing sequencing gels (followed by autoradiography) identifies protected regions and hence the footprints of bound proteins. However, to obtain a useful signal-to-noise ratio, it may be necessary to use abnormally high and potentially nonphysiological stoichiometries of extract to DNA.

In vivo footprinting methods probably more accurately represent the true status of promoter occupancy, but they are difficult to achieve and require powerful amplification techniques *(16,17)*. Such methods include the use of DMS, which rapidly crosses cell membranes and methylates exposed guanines, which are then cleaved by piperidine. Cleaved genomic DNA is recovered and amplified by ligation-mediated PCR. In this method, gene-specific primers define one end of the sequence of interest and allow synthesis of double-stranded product, which is then ligated to a defined linker sequence, thereby allowing exponential PCR amplification of the intervening DNA. Comparison of amplified fragments separated on a sequencing gel with DMS-treated naked genomic DNA allows the identification of protected guanines, and hence the presence of bound protein. Starting with a few micrograms of mammalian genomic DNA, footprints on single-copy genes can be detected following overnight autoradiography.

These approaches, which do not require any *a priori* information about the sequences bound by TFs, have gone hand-in-hand with DNA sequencing in an effort to identify TFBS within promoters and enhancers. With the rapid increase in discovery of new TFBS, online databases of experimentally determined sites have allowed researchers to search for putative sites within specific DNA sequences (e.g., TRANSFAC *[18]*), TRRD *[19]*, or COMPEL *[20]*). Data from footprinting, as well as other in vitro methods that select TFBS from a library of random oligomers, including CAST (cyclic amplification and selection of targets; *[21]*) and SELEX (systematic evolution of ligands by exponential amplification; *[22]*), have allowed the derivation of consensus binding sites for a variety of transcription factors, which can then be used to search promoter sequences for putative binding sites.

However, the short length, degeneracy, and flexible spacing of TFBS leads to an unacceptably high rate of false positives when a simple string search is performed. For example, searching the complete human genome with a consensus sequence for MyoD suggests approx 10^6 potential binding sites, of which only 10^3 are likely functional *(4)*. Using position weight matrices *(23)*, which take account of the individual contributions of different nucleotides in making a binding site, provides a modest improvement and adds a quantitative measure of the similarity a sequence has to a consensus TFBS. Nonetheless, these approaches assume that each nucleotide of a recognition site acts independently and do not take account of other contextual effects, such as chromatin or the combinatorial effects of transcription factor binding partners, which are well known to be of key importance to functional TF binding in living cells. Other methods include the use of neural networks trained on sets of known binding sites to predict putative sites *(24)*.

3.3.2. In Silico *Approaches to Footprinting*

Prior to the availability of the complete human genome sequence, sequence analysis to discover potential regulatory sequences required the curation of candidate sequences, such as muscle-specific gene promoters *(25)*, or other reference collections, such as the EPD *(5)*. By searching muscle-specific gene promoter alignments with known consensus sites, it became clear that true positive muscle regulatory TFBS could be more easily distinguished from false positives if more than one site or class of TF was included in comparisons and if spacing between sites was not fixed during searches. In a complementary approach (and an early application of phylogenetic footprinting—*see* below), comparison of noncoding regulatory sequences from distantly related organisms was used to discover evolutionarily conserved sequences upstream of orthologous genes. Correlation of conserved sequences with those found by comparing coregulated muscle-specific genes would support the possibility that true, functional regulatory sequences had been found. Although this piecemeal approach has been successful in identifying *cis*-regulatory modules in muscle *(25)* and liver *(26)*, it relies on curated assembly of a data set.

With the recent advent of two major technical advances—complete genome sequencing and transcriptional profiling by microarray—new *in silico* approaches have been made possible. These approaches take two forms. The first, so-called phylogenetic footprinting *(27,28)*, uses orthologous genes (obtained from COGS *[29]* or HomoloGene *[30]*, for example) and takes an evolutionary approach. Comparison of the promoters of orthologous genes from evolutionarily distantly related genomes can reveal TFBS subject to evolutionary constraint and likely to indicate true functional sites. To a certain extent this approach relies on the assumption that orthologous genes have retained identical transcriptional regulation, and hence the same upstream regulatory sequences and associated transcription factors, so some thought must be given to the selection of appropriate evolutionarily distant species. In earlier models, algorithms relied on an initial local or global alignment of promoter sequences, followed by inspection to detect conserved elements. However, using these methods to create alignments can be confounded by the flexible spacing and orientation of TFBS that may be allowed as promoter sequences diverge from a common ancestor. Recently, bioinformatic methods have become more sophisticated. CONREAL *(31)*, for example, uses biologically relevant information in the shape of potential TFBS to establish anchors between orthologous sequences and to guide a potentially more relevant promoter sequence alignment.

The second approach relies on mining the plethora of data provided by transcriptional profiling experiments that utilize microarray technology *(32)*. Assuming that coregulated expression of a set of genes is likely to be at least partly determined by shared sets of TFs, clustering algorithms identify candidate sets of genes whose promoters are likely to be enriched for that TF's binding sites. Extraction of upstream sequences, i.e., promoters, now made more feasible with the advent of completely sequenced genomes, provides a set of sequences that can be subjected to search algorithms that identify *ab initio* overrepresented sequence motifs. Cross-referencing these results with known consensus binding sites, such as those in the TRANSFAC or JASPAR *(33)* databases, can establish correlations of overrepresented sequences with biological data established in the laboratory. For example, Li et al. recently used a microarray approach to define a set of insulin-like growth factor 1 (IGF-1)-responsive genes in cultured cardiac myocytes *(34)*. Computational analysis revealed overrepresented motifs, including the TFBS for Sp1, the importance of which was confirmed by EMSAs and reporter gene assays. The process of compiling the upstream regulatory regions of differentially expressed sets of genes, examining them for overrepresented motifs, and extracting networks of TF interactions has recently become highly automated and available via a www interface *(35)*.

Identification of TFBS using an *in silico* approach is fast becoming a useful first step in exploring potential regulatory modules in genes of interest, helping to focus wet lab studies on a more tractable set of sequences. Given the fast pace of complete genome sequencing and developments in bioinformatics, however, a new generation of emergent technologies is quickly bringing researchers nearer to the goal of being able to describe entire transcriptional regulatory networks *in silico*.

Acknowledgments

We gratefully acknowledge the financial support of the Magdi Yacoub Institute.

References

1. Nicholas, S. B., Yang, W., Lee, S. L., Zhu, H., Philipson, K. D., and Lytton, J. (1998) Alternative promoters and cardiac muscle cell-specific expression of the Na+/Ca2+ exchanger gene. *Am. J. Physiol.* **274**, H217–H232.
2. Toutant, M., Gauthier-Rouviere, C., Fiszman, M. Y., and Lemonnier, M. (1994) Promoter elements and transcriptional control of the chicken tropomyosin gene [corrected]. *Nucleic Acids Res.* **22**, 1838–1845.
3. Felkin, L. E., Taegtmeyer, A. B., and Barton, P. J. R. (2006) Real-time quantitative PCR in cardiac transplant research, in *Transplantation Immunology: Methods and Protocols* (Hornick, P. and Rose, M. L., eds.), Humana Press, Totowa, NJ, pp. 305–330.
4. Wasserman, W. W. and Sandelin, A. (2004) Applied bioinformatics for the identification of regulatory elements. *Nat. Rev. Genet.* **5**, 276–287.
5. Schmid, C. D., Praz, V., Delorenzi, M., Perier, R., and Bucher, P. (2004) The Eukaryotic Promoter Database EPD: the impact of in silico primer extension. *Nucleic Acids Res.* **32**, D82–D85.
6. Ureta-Vidal, A., Ettwiller, L., and Birney, E. (2003) Comparative genomics: genome-wide analysis in metazoan eukaryotes. *Nat. Rev. Genet.* **4**, 251–262.
7. Karolchik, D., Baertsch, R., Diekhans, M., et al. (2003) The UCSC Genome Browser Database. *Nucleic Acids Res.* **31**, 51–54.
8. Strauss, F. and Varshavsky, A. (1984) A protein binds to a satellite DNA repeat at three specific sites that would be brought into mutual proximity by DNA folding in the nucleosome. *Cell* **37**, 889–901.
9. Carthew, R. W., Chodosh, L. A., and Sharp, P. A. (1985) An RNA polymerase II transcription factor binds to an upstream element in the adenovirus major late promoter. *Cell* **43**, 439–448.
10. Singh, H., Sen, R., Baltimore, D., and Sharp, P. A. (1986) A nuclear factor that binds to a conserved sequence motif in transcriptional control elements of immunoglobulin genes. *Nature* **319**, 154–158.
11. Johnson, K. D. and Bresnick, E. H. (2002) Dissecting long-range transcriptional mechanisms by chromatin immunoprecipitation. *Methods* **26**, 27–36.
12. Galas, D. J. and Schmitz, A. (1978) DNAse footprinting: a simple method for the detection of protein-DNA binding specificity. *Nucleic Acids Res.* **5**, 3157–3170.

13. Lin, K. C. and Shiuan, D. (1995) A simple method for DNaseI footprinting analysis. *J. Biochem. Biophys. Methods* **30**, 85–89.
14. Tullius, T. D., Dombroski, B. A., Churchill, M. E., and Kam, L. (1987) Hydroxyl radical footprinting: a high-resolution method for mapping protein-DNA contacts. *Methods Enzymol.* **155**, 537–558.
15. Hertzberg, R. P. and Dervan, P. B. (1984) Cleavage of DNA with methidium-propyl-EDTA-iron(II): reaction conditions and product analyses. *Biochemistry* **23**, 3934–3945.
16. Mueller, P. R. and Wold, B. (1989) In vivo footprinting of a muscle specific enhancer by ligation mediated PCR. *Science* **246**, 780–786.
17. Strauss, E. C. and Orkin, S. H. (1999) Guanine-adenine ligation-mediated polymerase chain reaction in vivo footprinting. *Methods Enzymol.* **304**, 572–584.
18. Matys, V., Fricke, E., Geffers, R., et al. (2003) TRANSFAC: transcriptional regulation, from patterns to profiles. *Nucleic Acids Res.* **31**, 374–378.
19. Kolchanov, N. A., Ignatieva, E. V., Ananko, E. A., et al. (2002) Transcription Regulatory Regions Database (TRRD): its status in 2002. *Nucleic Acids Res.* **30**, 312–317.
20. Kel-Margoulis, O. V., Kel, A. E., Reuter, I., Deineko, I. V., and Wingender, E. (2002) TRANSCompel: a database on composite regulatory elements in eukaryotic genes. *Nucleic Acids Res.* **30**, 332–334.
21. Wright, W. E., Binder, M., and Funk, W. (1991) Cyclic amplification and selection of targets (CASTing) for the myogenin consensus binding site. *Mol. Cell Biol.* **11**, 4104–4110.
22. Pollock, R. and Treisman, R. (1990) A sensitive method for the determination of protein-DNA binding specificities. *Nucleic Acids Res.* **18**, 6197–6204.
23. Frech, K., Herrmann, G., and Werner, T. (1993) Computer-assisted prediction, classification, and delimitation of protein binding sites in nucleic acids. *Nucleic Acids Res.* **21**, 1655–1664.
24. Workman, C. T. and Stormo, G. D. (2000) ANN-Spec: a method for discovering transcription factor binding sites with improved specificity. *Pac. Symp. Biocomput.* 467–478.
25. Wasserman, W. W. and Fickett, J. W. (1998) Identification of regulatory regions which confer muscle-specific gene expression. *J. Mol. Biol.* **278**, 167–181.
26. Krivan, W. and Wasserman, W. W. (2001) A predictive model for regulatory sequences directing liver-specific transcription. *Genome Res.* **11**, 1559–1566.
27. Aparicio, S., Morrison, A., Gould, A., et al. (1995) Detecting conserved regulatory elements with the model genome of the Japanese puffer fish, *Fugu rubripes*. *Proc. Natl. Acad. Sci. USA* **92**, 1684–1688.
28. Lenhard, B., Sandelin, A., Mendoza, L., Engstrom, P., Jareborg, N., and Wasserman, W. W. (2003) Identification of conserved regulatory elements by comparative genome analysis. *J. Biol.* **2**, 13.
29. Tatusov, R. L., Fedorova, N. D., Jackson, J. D., et al. (2003) The COG database: an updated version includes eukaryotes. *BMC Bioinformatics* **4**, 41.

30. Wheeler, D. L., Church, D. M., Edgar, R., et al. (2004) Database resources of the National Center for Biotechnology Information: update. *Nucleic Acids Res.* **32,** D35–D40.
31. Berezikov, E., Guryev, V., Plasterk, R. H. A., and Cuppen, E. (2004) CONREAL: Conserved Regulatory Elements Anchored Alignment algorithm for identification of transcription factor binding sites by phylogenetic footprinting. *Genome Res.* **14,** 170–178.
32. Werner, T. (2001) Target gene identification from expression array data by promoter analysis. *Biomol. Eng* **17,** 87–94.
33. Sandelin, A., Alkema, W., Engstrom, P., Wasserman, W. W., and Lenhard, B. (2004) JASPAR: an open-access database for eukaryotic transcription factor binding profiles. *Nucleic Acids Res.* **32,** D91–D94.
34. Li, T., Chen, Y. H., Liu, T. J., et al. (2003) Using DNA microarray to identify Sp1 as a transcriptional regulatory element of insulin-like growth factor 1 in cardiac muscle cells. *Circ. Res.* **93,** 1202–1209.
35. Vadigepalli, R., Chakravarthula, P., Zak, D. E., Schwaber, J. S., and Gonye, G. E. (2003) PAINT: a promoter analysis and interaction network generation tool for gene regulatory network identification. *OMICS* **7,** 235–252.

12

Characterization of Cardiac Gene Promoter Activity

Reporter Constructs and Heterologous Promoter Studies

Hsiao-Huei Chen and Alexandre F. R. Stewart

Summary

Cardiac gene promoter analysis remains an integral method in molecular cardiology and continues to provide novel insights into the transcriptional mechanisms that regulate gene expression in the myocardium. Initial studies focused on the regulated expression of contractile genes, since their transcripts are abundant and their cDNAs were among the first to be cloned. More recent studies have focused on the promoters of genes expressed at much lower levels, including those that encode ion channels, signaling proteins, and the cardiac transcription factors. The standard approach to analyze myocardial gene promoters has been to transfect reporter plasmids into cultured neonatal rat cardiac myocytes. This approach has the unique advantage of allowing the exploration of different signaling mechanisms by supplementing culture media with different agonists and inhibitors. In addition, *cis*-elements that control gene expression under different physiological stresses have been further characterized in the context of heterologous promoters to demonstrate their "stand-alone" functional properties in the absence of confounding influences from other *cis*-elements and their cognate transcription factors. Here we illustrate the characterization of cardiac gene promoter activity using reporter constructs and heterologous promoter studies in cultured cardiac myocytes.

Key Words: Cardiac myocytes; gene promoter; *cis*-elements; reporter gene; heterologous promoter.

1. Introduction

The β-myosin heavy chain gene has one of the best studied cardiac promoters *(1,2)*. In rodent hearts, β-myosin heavy chain is expressed during the fetal period of cardiac growth but is downregulated shortly after birth with the concomitant upregulation of α-myosin heavy chain *(3,4)*. The hypertrophic myocardium upregulates the expression of the β-myosin heavy chain and other fetal genes in a stereotypical response to injury or increased load *(5)*. β-Myosin heavy chain gene expression can also be upregulated in cultured neonatal rat cardiac myocytes in response to α_1-adrenergic activation *(6)*. Using truncated constructs of the β-myosin heavy chain promoter, Simpson and coworkers have identified minimal sequences required for the α_1-adrenergic upregulation of promoter activity *(7,8)*. These studies typify a minimalist approach to mapping regulatory elements controlling cardiac gene expression. Robbins and coworkers challenged the validity of this minimalist approach by testing similar mutant constructs of large β-myosin promoter fragments in transgenic mice *(9)*. In the context of larger promoter fragments, individual *cis*-elements were shown to be dispensable, but when several were mutated together, promoter activity was markedly compromised.

These elegant studies confirmed the utility of the minimalist approach, while demonstrating the complexity of combinatorial interactions at the β-myosin heavy chain promoter. Our understanding of transcriptional regulation of the β-myosin heavy chain promoter is far from complete. As newly identified cardiac transcription factors are characterized, their input on transcriptional control is being tested and is yielding new insight into the mechanisms controlling cardiac gene transcription *(10,11)*. The ability to test individual *cis*-elements outside the context of a muscle-specific promoter has allowed the further characterization of combinatorial interactions and synergy between different cardiac transcription factors *(12)*.

2. Materials
2.1. Reporter Constructs

1. Promoterless luciferase reporter vector (e.g., pGL3 from Promega) for cardiac promoter analysis.
2. Minimal eukaryotic expression vector (e.g., herpes simplex virus thymidine kinase promoter truncated at 46 bp upstream of a luciferase reporter).
3. Sequence-specific oligonucleotides (bearing 5' restriction enzyme sequences to subclone into promoterless luciferase reporter) to amplify the promoter fragment to be tested.
4. Double-stranded oligonucleotides corresponding to the *cis*-elements to be tested in heterologous promoter studies. (Oligonucleotides should be designed to carry partial restriction enzyme sequences at each end to facilitate end-to-end multimerization when ligated into a heterologous promoter.)
5. Restriction enzymes.
6. DNA ligase.
7. *Taq* DNA polymerase.
8. Competent bacteria for plasmid vector transformation and large-scale preparation.
9. Plasmid purification kit (Qiagen).

2.2. Media and Reagents for Cell Culture

Reagents are obtained from Invitrogen and Sigma, and all solutions are filter-sterilized.

1. HBSS-G: Hanks' balanced salt solution + 10 g/L glucose. Make at least 200 mL by adding 20 mL of 10% glucose to 180 mL HBSS. Prewarm to 37°C.
2. HBSS-G-Hep: to 50 mL HBSS-G, add 500 µL heparin solution (heparin sodium for injection, 1000 USP U/mL stored at 4°C).
3. Basic medium: MEM + vitamin B_{12} + fungizone + penicillin-streptomycin.
4. MEM25: basic medium + 25% bovine calf serum (BCS).
5. MEM10: basic medium + 10% BCS.
6. MEM5: basic medium + 5% BCS + 5-bromo-2-deoxyuridine (100 µM).
7. TIBSA: MEM + 5 µg/mL transferrin + 1 µg/mL insulin + 1 nM sodium selenite + 25 µg/mL ascorbic acid.
8. 10X Enzyme solution (prepare fresh, and keep at 4°C): 10 mL HBSS-G, 50 mg Worthington collagenase type 2, and 500 µL DNAse I (0.4% solution prepared fresh).

2.3. Solutions for Transfection

1. 1 M HEPES (pH 7.5).
2. 2X HBS (HEPES-buffered saline): 8.2 g NaCl, 5.95 g HEPES, 0.15 g Na_2HPO_4 in 500 mL water, pH 6.95 to 7.1.
3. TCEB (transfected cell extraction buffer): 0.1 M KH_2PO_4 (from 1 M stock, pH 7.4), 2 mM EDTA, 1 mM dithiothreitol (DTT).
4. 10X ATP (0.1 M, pH 7.0), stored as frozen aliquots at –20°C.
5. 10X luciferase buffer (250 mM glycyl-glycine, 100 mM $MgCl_2$, pH 7.8 with NaOH). For 1X luciferase buffer, mix 2.5 mL of freshly thawed 10X ATP with 1.25 mL 10X luciferase buffer and add water to 20 mL.
6. D-Luciferin (Acros Chemicals, sodium salt): 0.3 mg/mL, in water.
7. Ca^{2+}- and Mg^{2+}-free phosphate-buffered saline (PBS; Invitrogen).

3. Methods

3.1. Selection and Construction of Cardiac Promoter/Reporters

Choosing a promoter fragment to test in cultured myocytes requires knowledge of the identity of the first exon and the start site of transcription. Once the start site of transcription has been characterized (by primer extension and RNase protection analysis), promoter fragments of various sizes that include the start site of transcription are amplified using polymerase chain reaction (PCR) and sequence-specific primers. Amplified promoter fragments are then ligated upstream of an appropriate reporter gene (e.g., firefly luciferase). Several promoterless luciferase plasmids are available commercially (e.g., pGL3 basic from Promega lacks eukaryotic promoter and enhancer sequences, allowing maximum flexibility in cloning putative regulatory sequences).

Using the rat β-myosin heavy chain as an example, selecting the promoter fragment to test in cultured myocytes influences not only the outcome of the experiment but also the complexity of interpretation (**Fig. 1A**).

3.2. Construction of Heterologous Promoters to Characterize "Stand-Alone" cis-Elements and Transcription Factor Interaction

Once *cis*-regulatory sequences are identified in cardiac promoters, it is often necessary to understand how these sequences affect the activity of a promoter on their own or when they are juxtaposed to other sequences of interest (**Fig. 1B**). In this case, heterologous promoters (often consisting of a "minimal" TATA binding site) can be used to test whether a given *cis*-element, with its associated transcription factor, is sufficient to activate transcription in cardiac myocytes. In addition, adding multiple copies of the *cis*-element in tandem, and altering spacing between elements, can be informative if cooperative binding mediates transcriptional activity. An example of this approach comes from studies of *cis*-elements that confer cardiac-specific expression to a heterologous promoter.

3.3. Cardiac Myocyte Culture

Cardiac myocytes are cultured in media buffered with Hanks' balanced salts in a 1% CO_2 water-saturated incubator. The low CO_2 is critical for efficient cardiac myocyte transfection using the calcium phosphate precipitation method. This approach is also relatively inexpensive compared with lipid-based transfection reagents.

1. To each of two 10-cm dishes, add cold HBSS-G-Hep and keep on ice.
2. Count and weigh the neonates.
3. Take 1- to 2-d-old neonates, roll in 70% ethanol, and dry on a paper towel.
4. Decapitate with scissors and allow the head to fall into liquid nitrogen. Perform midline sternotomy, apply light pressure to pop out the heart, hold the apex with scissors, cut out heart at the base, and place heart in a 10-cm plate with HBSS-G-Hep.
5. Remove all hearts as above and dissect away the atria, placing dissected hearts in a second 10-cm plate with HBSS-G-Hep. (Atria can be digested separately for atrial myocyte cultures.)
6. Cut hearts in half.
7. Collect heart pieces into a 15-mL polypropylene tube.
8. Aspirate supernatant and wash hearts with 10 mL cold HBSS-G-Hep.
9. Repeat **step 8** twice.
10. Wash once with warm HBSS-G.
11. Add 4.5 mL of prewarmed HBSS-G and 0.5 mL of 10X enzyme solution.
12. Rotate at 37°C in the 15-mL tube for 15 min.

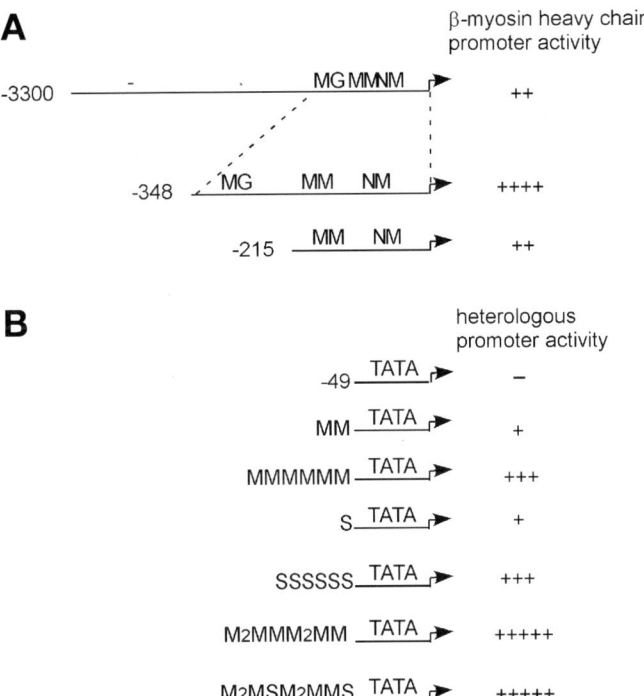

Fig. 1. Construction of a cardiac promoter/luciferase reporter (**A**) and heterologous promoter/luciferase reporter (**B**). In (**A**), the rat β-myosin heavy chain promoter is shown as an example. Inclusion of various promoter fragment lengths (determined by the location of sequence-specific primers used in PCR amplification) will determine the activity of the promoter in cardiac myocyte cultures. cis-regulatory sequences that are known to affect β-myosin heavy chain promoter activity include several MCAT elements (M), a GATA site (G), and an NFAT site. A distal putative negative regulatory element is indicated by a minus (–) sign. Relative luciferase activities are from prior reports *(11,15)* and unpublished results. (**B**) The herpes simplex virus thymidine kinase promoter/luciferase reporter can be used as a heterologous (noncardiac muscle) minimal promoter/reporter to test multimerized copies of different cis-elements or combinations thereof. The example shows multiple copies of cis-elements active in cardiac muscle, and their relative luciferase activities are representative *(12)*. This approach allows qualitative assessment of cooperative binding and interactions between transcription factors. TATA, the TATA-binding protein consensus; M, MCAT sites; G, GATA sites; SRF, serum response factor sites; M2, myocyte enhancer factor 2 binding sites.

13. Collect liquid with a large-bore sterile plastic pipet and triturate (pipet up and down) three to four times.
14. Allow large fragments to settle to the tip of the pipet and expel back into the tube.

15. Put remaining liquid with cells into a 50-mL tube containing 10 mL MEM25 on ice.
16. Add 5 mL of warm HBSS-G to heart pieces, rinse, and add to cells of **step 15**.
17. Repeat **steps 11 to 16** seven times until digestion is complete.
18. After each cell collection:
    a. Add 100 µL DNAse I solution (0.4%).
    b. Spin cells at 30 g for 7 min at 4°C.
    c. Resuspend myocytes in 10 mL MEM10 and store on ice until all aliquots are harvested.
19. When dissociation is complete, pool all MEM10 samples and pellet as above.
20. Resuspend in plating medium (MEM5).
21. Pass cells through a 40-µm nylon mesh.
22. Preplate on 10-cm dishes for 2 h (one dish per eight rat pups).
23. Remove and save floating cardiac myocytes and wash plate gently with MEM5 to dislodge weakly attached myocytes.
24. Plate myocytes at 3×10^5 cells per cm^2 in a six-well plate.
25. Next day, change the medium for fresh MEM5 or for TIBSA if serum-free cultures are desired.

3.4. Transfection of Cardiac Myocytes and Reporter Gene Assay

1. On the morning of the first day in culture, wash away floating dead cells with fresh medium, and then 1 h prior to transfection, replace medium with MEM5 containing 30 m*M* HEPES, pH 7.5.
2. The calcium phosphate DNA precipitate is prepared as follows:
    a. To a 15-mL polystyrene tube add 275 µL of 2X HBS.
    b. To a 15-mL polypropylene tube add 6 µg of a luciferase reporter together with 34 µg of carrier DNA (typically pBluescript plasmid) in 220 µL of water.
    c. Add 55 µL of 2.5 M CaCl$_2$ to the DNA solution and mix well.
3. Add the DNA/CaCl$_2$ solution dropwise to the 2X HBS solution while shaking tube on vortex mixer set to low.
4. Let the precipitate stand for 15 min and add 250 µL in each of two 35-mm wells (allowing for duplicate transfections).
5. Discard remaining solution (typically less than 50 µL).
6. Incubate myocytes for 2 to 4 h, and then rinse with prewarmed media once with MEM5, and once again with MEM5 with 30 m*M* HEPES pH 7.5.
7. Return cells to incubator for 15 min, and change the medium to MEM5 or to TIBSA for serum-free culture.
8. One or 2 d after transfection, rinse myocytes in Ca^{2+}- and Mg^{2+}-free PBS and harvest directly in 500 µL of TCEB.
9. Freeze cells in a dry ice bath and thaw quickly three times, mix with a vortex, and pellet cell debris in a microcentrifuge.
10. Luciferase reporter assays are typically carried out with 20 to 50 µL of supernatant cell extract. In clear luminometer tubes, place 750 µL luciferase buffer, and

add 20 µL of cell extract and 100 µL of 0.3 mg/mL D-luciferin (either manually or with an automatic injector).
11. Read light emission in a luminometer.

4. Notes

1. The pGL3-Promoter vector (Promega), which contains an SV40 promoter upstream of the luciferase gene, is not recommended for heterologous promoter studies of *cis*-regulatory elements since the SV40 promoter fragment (vector sequence 48-250) contains numerous *cis*-elements in addition to the TATA element that will confound studies of individual or multimerized cardiac *cis*-elements. Instead, a synthetic promoter containing only a TATA element or the thymidine kinase –49 promoter fragment shown in **Fig. 1B** is preferable.
2. In determining sequences minimally required for the β-myosin heavy chain promoter to respond to α_1-adrenergic stimulation in cardiac myocytes, Simpson's group found that a promoter truncated 215 bp upstream from the transcriptional start site was sufficient. With site-directed mutagenesis, individual sequences were identified that correspond to consensus TEF-1-binding MCAT sites. The MCAT element, 5'-CATTCCT-3', is a muscle-specific CAT sequence first described by Ordahl and associates that confers high levels of promoter activity in striated muscles *(13)*. On the other hand, Kitsis and coworkers examined sequences between –303 to –197 relative to the start site and identified a GATA element to be important for hypertrophy-induced upregulation of the β-myosin heavy promoter directly injected into the rat myocardium following aortic coarctation. Importantly, mutation of two TEF-1-binding sequences within this region of the promoter had apparently no effect on the hypertrophy-induced upregulation of the promoter, suggesting that TEF-1 factors do not participate in the response *(14)*. However, Farrance and coworkers demonstrated that several additional TEF-1 binding sites are located within this region of the proximal promoter and are required for α_1-adrenergic induced upregulation *(11)*. Thus, although GATA and its associated transcription factor GATA4 doubtless play a part in cardiac gene regulation, dismissing a role for TEF-1 factors in the hypertrophic activation of the β-myosin heavy chain promoter based on the ablation of two of four functional TEF-1 sites is probably erroneous. Studies from our own laboratory have implicated members of the TEF-1 family of transcription factors including RTEF-1 (TEAD4) and DTEF-1 (TEAD3), in mediating the α_1-adrenergic response in cardiac myocytes. RTEF-1 overexpression augments by twofold the α_1-adrenergic induction of both a minimal (–215 bp) and a much longer (–3300 bp) β-myosin heavy chain promoter *(15)*. This result suggests that TEF-1 factors act not only in the context of a minimal promoter, but also within the context of a much more complex set of regulatory sequences.
3. When one is testing for different promoter construct activities, inclusion of a second reporter plasmid (e.g., Renilla luciferase or β-galactosidase) may be required to normalize for differences in transfection efficiencies. However, when one is testing the effects of various transcription factors on a cardiac promoter by

cotransfection with expression plasmids, expression of internal control reporter plasmids will also be affected by the transcription factors tested, thereby rendering internal control reporter plasmids useless. Week-to-week variability in transfection efficiencies is typically not large, particularly if myocytes are cultured at the same density and myocyte viability is similar. Each primary cardiac myocyte preparation and transfection should be considered as a single experiment. Weekly or biweekly cardiac myocyte cultures are required to achieve sufficient numbers (typically 6 to 10 experiments) to attain significance.

Acknowledgments

This work was supported by an Operating Grant from the Canadian Institutes of Health Research (145322) to Alexandre Stewart.

References

1. Shimizu, N., Prior, G., Umeda, P. K., and Zak, R. (1992) *cis*-acting elements responsible for muscle-specific expression of the myosin heavy chain beta gene. *Nucleic Acids Res.* **20,** 1793–1799.
2. Thompson, W. R., Nadal-Ginard, B., and Mahdavi, V. (1991) A MyoD1-independent muscle-specific enhancer controls the expression of the β-myosin heavy chain gene in skeletal and cardiac muscle cells. *J. Biol. Chem.* **266,** 22678–22688.
3. Lompré, A.-M., Mercadier, J.-J., Wisnewski, C., et al. (1981) Species and age-dependent changes in the relative amounts of cardiac isoenzymes in mammals. *Dev. Biol.* **84,** 286–290.
4. de Groot, I. J. M., Lamers, W. H., and Moorman, A. F. M. (1989) Isomyosin expression patterns during rat heart morphogenesis: an immunohistochemical study. *Anat. Rec.* **224,** 365–373.
5. Schwartz, K., Boheler, K. R., de la Bastie, D., et al. (1992) Switches in cardiac muscle gene expression as a result of pressure and volume overload. *Am. J. Physiol.* **262,** R364–R369.
6. Waspe, L. E., Ordahl, C. P., and Simpson, P. C. (1990) The cardiac β-myosin heavy chain isogene is induced selectively in α_1-adrenergic receptor stimulated hypertrophy of cultured rat heart myocytes. *J. Clin. Invest.* **85,** 1206–1214.
7. Kariya, K., Farrance, I. K., and Simpson, P. C. (1993) Transcriptional enhancer factor-1 in cardiac myocytes interacts with an α_1-adrenergic- and β-protein kinase C-inducible element in the rat β-myosin heavy chain promoter. *J. Biol. Chem.* **268,** 26658–26662.
8. Kariya, K., Karns, L. R., and Simpson, P. C. (1994) An enhancer core element mediates stimulation of the rat β-myosin heavy chain promoter by an α_1-adrenergic agonist and activated β-protein kinase C in hypertrophy of cardiac myocytes. *J. Biol. Chem.* **269,** 3775–3782.
9. Knotts, S., Rindt, H., Neumann, J., et al. (1994) In vivo regulation of the mouse beta myosin heavy chain gene. *J. Biol. Chem.* **269,** 31275–31282.

10. Stewart, A. F. R., Suzow, J., Kubota, T., et al. (1998) Transcription factor RTEF-1 mediates α_1-adrenergic reactivation of the fetal gene program in cardiac myocytes. *Circ. Res.* **83,** 43–49.
11. McLean, B. G., Lee, K. S., Simpson, P. C., et al. (2003) Basal and alpha(1)-adrenergic-induced activity of minimal rat betaMHC promoters in cardiac myocytes requires multiple TEF-1 but not NFAT binding sites. *J. Mol. Cell Cardiol.* **35,** 461–471.
12. Li, X., Eastman, E. M., Schwartz, R. J., et al. (1999) Synthetic muscle promoters: activities exceeding naturally occurring regulatory sequences. *Nat. Biotechnol.* **17,** 241–245.
13. Mar, J. H. and Ordahl, C. P. (1988) A conserved CATTCCT motif is required for skeletal muscle-specific activity of the cardiac troponin T gene promoter. *Proc. Natl. Acad. Sci. USA* **85,** 6404–6408.
14. Hasegawa, K., Lee, S. J., Jobe, S. M., et al. (1997) Cis-acting sequences that mediate induction of beta-myosin heavy chain gene expression during left ventricular hypertrophy due to aortic constriction. *Circulation* **96,** 3943–3953.
15. Ueyama, T., Zhu, C., Valenzuela, Y. M., et al. (2000) Identification of the functional domain in the transcription factor RTEF-1 that mediates α_1-adrenergic signaling in hypertrophied cardiac myocytes. *J. Biol. Chem.* **275,** 17476–17480.

IV

In Silico Assessment of Regulatory *cis*-Elements and Gene Regulation

13

Comparative Genomics
A Tool to Functionally Annotate Human DNA

Jan-Fang Cheng, James R. Priest, and Len A. Pennacchio

Summary

The availability of an increasing number of vertebrate genomes has enabled comparative methods to infer functional sequences based on evolutionary constraint. Although this has proved powerful for gene identification, significant progress has also been made in uncovering gene regulatory sequences such as distant acting transcriptional enhancers. These pursuits have led to the development of a variety of valuable databases and resources that should serve as a routine toolbox for biological discovery.

Key Words: Comparative genomics; gene regulation; enhancer; transcription; databases.

1. Introduction

Advances in DNA sequencing technologies over the past two decades have resulted in the exponential growth of sequence information available in public databases. None is more impressive than the recent completion of the 3 billion nucleotides of the human genome *(1,2)*. In addition, there is a growing list of draft genomes from numerous additional vertebrates, and various computational tools have been developed to compare whole genomes to facilitate the identification of DNA sequences conserved over evolutionary time *(3–5)*. When the human genome is compared with that of other vertebrates, particularly mammals, the most highly conserved sequences are protein-encoding genes *(6–8)*. In addition, sequence comparisons in noncoding DNA are increasingly being leveraged to identify conserved regulatory elements responsible for the temporal and spatial patterns of gene expression *(9–12)*. In this chapter we describe some of the basic comparative methodologies for researchers without detailed experience in obtaining genomic sequence data and applying bioinformatic tools for comparative analysis.

2. Public Databases

The most straightforward method for obtaining a genomic sequence is through preexisting public databases. The databases described here are limited to animals in the Phylum Chordata owing to their particular relevance in annotating the human genome through comparative genomics. Currently, a handful of chordate genomes have been sequenced, and much larger numbers will be available in the near future *(13)*. More details are provided here of several chordates whose genomes have been assembled (**Table 1**) and are ready for global comparative analysis.

Of the chordates, the human genome was the first to be completed to a high degree of accuracy. With the exception of a small number of difficult to clone gaps, the human euchromatic DNA sequence has been determined through an international effort *(2)*. A handbook describing the assembly and annotation process is also available through the National Center for Biotechnology Information (NCBI) *(14)*. Today, three large centers provide web-accessible annotation to the human genome: the NCBI, the University of California at Santa Cruz (UCSC), and the Sanger Center. In addition to exon annotation across the entire genome, these browsers contain a tremendous amount of additional annotations for features such as repetitive DNA, expressed sequence tags (ESTs), CpG islands, and single-nucleotide polymorphisms (SNPs). Updates to the human genome sequence occur on a semiregular basis, so it is important to specify the source of the sequence data used for analysis. The examples given in this chapter use the human Build 34 sequence, which is based on an assembly dated, July 2003.

The house mouse (*Mus musculus*) was the first nonhuman chordate genome to be sequenced and assembled and was followed by two pufferfish (*Takifugu rubripes* and *Tetraodon nigroviridis*), the rat (*Rattus norvegicus*), the zebrafish (*Danio rerio*), the sea squirt (*Ciona intestinalis*), and recently the first nonhuman primate, the chimpanzee (*Pan troglodytes*). Currently the assemblies of these chordate genomes exist as drafts (**Table 1**), in that they are lower quality than that of the human genome owing to the significant human and monetary resources necessary to complete the final 5 to 10% of sequence coverage. Additional assembled chordate genomes are also available and among others include the rhesus macaque, dog, chicken, cow, and frog.

3. Precomputed Genomes

3.1. Comparative Genomic Browsers

The easiest method for obtaining human-based comparative sequence data is through precomputed whole-genome alignments. These resources are accessible through web-based browsers and provide a wealth of user-friendly fea-

Table 1
Selected Web Sites for Gathering Assembled Sequences and Genomic Information of Chordates

Institution	Web site
NCBI	http://www.ncbi.nlm.nih.gov/genomes/leuks.cgi
UCSC	http://genome.ucsc.edu/cgi-bin/hgGateway
EMBL	http://www.ensembl.org/index.html
JGI	http://genome.jgi-psf.org/euk_home.html

tures and options for the biomedical scientific community. Precomputed alignments relieve users from obtaining and annotating their own sequences and waiting for the computationally intensive process of preparing alignments. A variety of algorithms have been developed to prepare these alignments, each with their advantages and disadvantages. **Table 2** lists existing web-servers dedicated solely to precomputed alignments.

For many users, the familiar environment of the UCSC Genome Browser *(15,16)* provides a good introduction to comparative genomics. Similar to the browser's other annotation fields, comparative genomic information is presented as adjustable "tracks" listed under the heading "Comparative Genomics." Highly conserved sequence tracks are displayed as blocks, with the length and shading of the blocks indicative of the size and level of homology between human and a variety of organisms including mouse, rat, and pufferfish (fugu). Furthermore, human and rodent conservation is ranked by an "L-score," which is plotted on a separate track. This unique scoring system examines conservation in the context of the local genomic interval, rather than a simple percent identity or length. A conserved sequence receives a higher score when found in a region of low conservation or, conversely, receives a lower score when found in a highly conserved region. This rationale is based on the observation that neutral rates of DNA sequence change are highly variable throughout the mammalian genome *(8)*. Thus, the L-score assumes that conservation in regions in which more change has occurred is more likely to be functional than conservation in slowly evolving intervals.

As dedicated portals for comparative genomic information, the VISTA Genome Browser *(17,18)* and ECR Browser *(19)* are easier to use than the UCSC Browser. The fundamental difference between the two sites is found in the algorithms used to compute the alignments, as the ECR Browser uses a BlastZ local alignment *(20)* whereas VISTA relies on LAGAN *(21)* and AVID *(22)* for global alignments. The algorithmic differences between the two have been discussed at length and recently tested in detail on simulated datasets

Table 2
Web-Based Browsers Containing Pre-Computed Comparative Genomic Alignments

	URL	Alignment algorithm	Reference
VISTA Genome Browser	http://pipeline.lbl.gov/	LAGAN, AVID, MLAGAN	*17,21,22*
UCSC Browser	http://genome.ucsc.edu/	BlastZ, Multiz	*20*
ECR Browser	http://ecrbrowser.dcode.org/	BlastZ	*20*

(23). Each web-interface allows users the option to access alignments by gene name or chromosomal coordinates. Upon viewing the alignments within the browser, the user may navigate, zoom in and out, and adjust the organisms being compared in addition to a number of smaller features unique to each site.

3.2. Custom Comparison with Whole Genomes

In addition to preprocessed whole-genome comparative data, several additional tools allow for any sequence to be compared with previously assembled and annotated genomes. For a sequence originating from an organism that has yet be sequenced or annotated, it is informative to examine that sequence in the context of the annotated human or mouse genomes. At the UCSC Browser, any sequence may be aligned to the available genomes using BLAT *(24)*, a tool designed to detect quickly large regions of extremely similar sequence. However, for comparisons between evolutionarily distant organisms, slower, more sensitive algorithms are necessary to produce a useful alignment. Both the VISTA and ECR Browsers provide the ability to compare any sequence with a given genome with an appropriately sensitive algorithm. Sequence files can be acquired from in-house sequencing projects or from sequence databases such as GenBank, and this will be discussed in detail in a later section.

4. Comparative Sequence Analysis with User-Obtained Sequence Resources

Although precomputed alignments are powerful, accessible, and simple, their sensitivity and flexibility are somewhat restricted. The whole-genome browsers described provide brute force alignments that are accurate for the most part, but biologically inappropriate alignments and missing alignments are commonplace owing to the size of such analyses. Here we describe how to carry out multiple-species comparative sequence analysis with user-obtained sequence resources. Sequence data may come from existing databases (*see* **Subheading 2.**) or be generated in the investigator's laboratory. Regardless, the basic substrate for comparative sequence analysis in most cases is orthologous sequences from two or more species.

4.1. Finding Additional Species for Comparative Analysis

In the current era of genomics, new sequence data are continuously being generated and released to large web-accessible databases. For the knowledgeable user, there is a growing amount of sequence data for many genes of interest suitable for comparative sequence analysis. In addition to Genbank, many sequence resources exist for specific organisms in the form of EST sequencing projects and the individual sequencing reads of whole-genome shotgun sequence, commonly known as traces. Many of these sequences, such as those for cow and dog, are currently searchable through the BLAST website *(25)* at NCBI under Genomes, in the trace archive *(26)*, or at organism-specific websites maintained by genome sequencing centers such as those listed in **Table 1**.

The type of sequence needed for comparative analysis is large contiguous sequence blocks (commonly referred to as contigs) that contain detailed structural information such as 5' and 3' boundaries of genes, the order of exons, and, importantly, the availability of neighboring noncoding sequences. Such sequence features provide the opportunity to identify *cis*-regulatory elements such as promoters, enhancers, silencers, and repressors, in the neighborhood of the target gene. Nevertheless, genomic sequences in public databases are often found as individual reads, long sequence scaffolds made up of several contigs bridged by defined gaps, or completely finished sequences, all depending on the status of a particular sequencing project. Therefore it is important to understand the quality of the sequence before using it in the analysis.

4.2. Target Sequencing of Selected Species

If the sequence of a particular species is not available, it is possible to obtain specific sequence information through targeted sequencing. Access to sequencing resources is available to the scientific community in the form of sequence libraries, sequence reagents, and sequencing services. Although the whole-genome sequence of an organism may not be available for comparative analysis, a genomic library from which a sequence of interest can be identified and extracted may be available *(27)*. Often, genome-sequencing centers provide sequencing services for projects of merit at little or no cost to the investigator. Two prominent examples include the Department or Energy's Joint Genome Institute *(28)* and the National Institutes of Health's (NIH's) National Human Genome Research Institute *(29)*.

4.3. Comparative Sequence Analysis Tools

Our discussion of resources and tools until this point has been limited to comparisons between distantly related organisms (i.e., human/mouse). Such comparisons contribute to identifying the general functional classes of DNA elements common across long evolutionary time periods including protein coding exons, noncoding RNAs, and gene regulatory elements. Implicit in this approach is the expectation that DNA sequences conserved between two organisms are responsible for their similarities. However, in other cases investigators may be interested in biology, which is restricted to a closely related set of organisms such as primates. For example, when compared with other mammals, there are many biological characteristics that are unique to primates that will not be accessible through human-mouse comparisons. A prime example is the apo(a) gene, which is only present in primates *(30)*. Human and mouse comparisons are therefore impossible because no orthologous mouse sequences exists. Thus, different strategies have been developed to utilize comparative genomics for basic vertebrate functions versus those that are lineage specific.

Conceptually, the way to gain comparative insight into a uniquely primate gene is to compare many primates and add up the cumulative differences, a recently developed approach named *phylogenetic shadowing (30)*. In contrast to tools designed for comparisons over long evolutionary distances, phylogenetic shadowing amplifies the evolutionary distance between closely related species by comparing not two but rather many sequences simultaneously. This approach has successfully identified functional elements specific to primates and will be discussed in further detail in **Subheading 4.3.2.**

4.3.1. Available Tools and Web-Servers for Comparing Sequences of Distantly Related Species

If users are interested in biological functions common across large evolutionary distances, the aligning of DNA sequences from distant evolutionarily related species is a logical approach to identify functionally constrained regions. Numerous sequence alignment tools have been developed to align long stretches of DNA. Some of the publicly available tools (Avid, BlastZ, Chaos, ClustalW, DiAlign, Lagan, Needle, and WABA) have been tested for the relative sensitivity and specificity in finding short blocks of conserved sequences from sequences spanning a range of divergence *(23)*. Researchers should consider the performance assessment in order to select the proper algorithm for their sequence comparison, and in most cases comparisons should be performed with multiple algorithms.

Another critical aspect of comparative sequence analysis is visualization of the alignment results. There are a few publicly available servers that provide not only tools to align sequences but also graphical interfaces to display sequence alignment data in a meaningful way. We describe three commonly used sequence comparison servers.

1. The VISTA server *(17,18)* aligns two or more sequences using AVID, LAGAN, and Shuffle-LAGAN *(21,22)*, calculates percent identify over a specified window length at each base pair, and then draws a continuous curve to display levels of sequence similarity along the base sequence. One of the submitted sequences is designated by the user as the base sequence, for which an annotation file describing the locations of known coding exons, untranslated exons, and any other user defined feature of interest is also provided. An integrated tool named rVISTA *(31)* is also available through the VISTA server as an option to detect putative transcription factor binding sites in the conserved sequence domains. An example is described in **Subheading 5.**

2. PipMaker *(32,33)* and MultiPipMaker *(34)* use the alignments generated by BlastZ and display the level of sequence similarity as horizontal dots/bars.
3. zPicture and Multi-zPicture *(35)* also use BlastZ alignments but provide more options such as uploading UCSC Genome Browser sequences and annotation files and automatic submission of alignments to rVISTA processing.

4.3.2. Available Tools and Web-Servers for Comparing Sequences of Closely Related Species

If the researcher is interested in a biology feature limited to a restricted phylogenetic lineage, phylogenetic shadowing *(30)* is a logical strategy. This approach is based on the simple premise that if the differences between two closely related species do not reveal regions of conservation, sufficient sequence variation can be accumulated by combining the differences found in a number of closely related species. The first step is to identify the most informative minimum number of species needed to achieve the discriminative power for a shadowing comparison (i.e., the level of sequence variations to distinguish conserved from not conserved). In the human and mouse comparison, the genome-wide average substitution per site is 0.47 *(8)*, which seems to have sufficient discriminative power to detect constrained conservation. The discriminative power of a given set of closely related species is directly related to the overall number of accumulated mutations in that family tree. In primates, the minimum number of species required to achieve greater than 0.3 substitution per site is between 4 and 7 *(30)*. This level of variation is recommended based on the ability to detect conservation and the fast diminishing benefit of sequencing more species. Naturally, a subset of species with higher sequence divergences would require a fewer number of sequenced species to reach an informative comparison.

There are two servers providing on-line "phylogenetic shadowing" analysis:

1. RankVISTA *(36)* uses MLAGAN *(21)* or MAVID *(37)* to generate multiple sequence alignments from a set of orthologous sequences in the Multi-FASTA format and then identifies variation using the likelihood ratio of a fast-versus slow-mutation rate at each nucleotide position of the base sequence *(30)*. A VISTA-like plot is generated to show regions that are resist to accumulating variation. A multiple sequence alignment file is also provided to show sequence conservation at the nucleotide level.
2. eShadow *(38)* uses ClustalW to produce multiple sequence alignments and calculates accumulated sequence variation in a different fashion *(39)*. In addition, eShadow implements two statistical approaches: Hidden Markov Model Islands (HMMI) and Divergence Threshold (DT) to distinguish between putative functional verse neutrally evolving regions.

5. Example of Using Public Data to Perform Multispecies Sequence Comparison

We present two examples (here and in **Subheading 6.**) to illustrate how researchers can use public databases and tools to perform comparative sequence analysis. These analyses use only a selection of the available databases (UCSC Genome Browser and the JGI frog and pufferfish databases) and web-based tools (VISTA, rVISTA, and RankVISTA) described previously. Readers are encouraged to explore other comparative genomic databases (VISTA Genome Browser and ECR Browser) and web-servers (PipMaker and eShadow), each with their own advantages and disadvantages.

A complex pattern of expression makes human fibroblast growth factor-8 (*FGF8*) a good example for comparative sequence analysis of temporal and spatial gene expression. *FGF8* is primarily an embryonic epithelial growth factor that interacts with FGF tyrosine kinase receptors to mediate the development of liver *(40)*, brain *(41,42)*, and limb *(43–45)*; it is overexpressed in prostate and breast cancers *(46)*. The diverse roles of this growth factor in mammalian organogenesis and its misexpression in neoplastic tissue make *FGF8* an interesting target for investigation. As the example shows, comparative sequence analysis quickly and cleanly identifies many *cis*-regulatory elements that have already been experimentally shown to regulate *FGF8*, in addition to other elements without currently defined functions.

5.1. Collection of Available Orthologous Sequences

The human genomic sequence and the associated information such as intron/exon structure can be obtained from the UCSC Genome Browser *(16)* by using the key word "FGF8." This gene spans 5578 bases of genomic DNA and consists of six exons (1a, 1b, 1c, 1d, 2, and 3). In the July 2003 assembly (an equivalent of the GenBank Build 34), the *FGF8* gene resides between nucleotides 103,194,667 and 103,200,294 of chromosome 10. One can download any amount of flanking regions by resetting the viewing window to a specified pair of nucleotide coordinates. Here we use the same amount of the 5' and 3' flanking regions as the gene size to compare with other animals (i.e., chr10:103,189,089–103,205,822). As shown in **Fig. 1**, the amount of human sequence used in the comparison is 16,734 bp in length. A gene annotation file describing the location of known exons and/or predicted exons in this region can be extracted from the information embedded in the "Tables" section of the UCSC Genome Browser. To use that information, the nucleotide coordinate of the base sequence must start with "1." Alternatively, the pair of nucleotide coordinates can be submitted to http://pga.lbl.gov/Tools/annotation.html to produce an annotation file that can be used directly in the VISTA server for comparative sequence analysis.

The "Comparative Genomics" section of the UCSC Genome Browser lists the available species whose genomes have been aligned with the human genome. The mouse and rat sequences of the *FGF8* genes can be obtained from the corresponding mouse and rat tracks, which point to the corresponding Genome Browser where sequences can be downloaded. The mouse orthologous sequence downloaded for this particular analysis is located at chr19:45,870,580 to 45,886,816 of the October 2003 assembly, and the rat sequence is located at chr1:251,207,050 to 251,222,621 of the June 2003 assembly.

The assembled sequence contigs of frog (*Xenopus tropicalis*) and pufferfish (*Takifugu rubripes*) are available at GenBank (NCBI) and the Joint Genome Institute web sites, respectively (**Table 1**). Sequence scaffolds containing the Fgf8 orthologs can be found by tblastn (use protein sequence to search against translated nucleotide sequence) search using the human FGF8 protein sequence (NCBI: NP_149353). The frog and pufferfish scaffolds identified in the tblastn searches are then individually blasted against the human genome to confirm that they are indeed the best match to human FGF8 and not a related protein found somewhere else in the genome sequence. In this example, we download the frog scaffold_2522 from the assembly v1.0 and the pufferfish scaffold_778 from the release v3.0 for the purpose of comparative analysis.

5.2. Detection of Evolutionarily Conserved Sequences

The obtained sequences and annotation files are then submitted to the multi-VISTA server *(47)* to generate a multiple pairwise sequence conservation plot (or VISTA plot), as shown in **Fig. 1**. In addition to the conserved coding exons, the VISTA plot also reveals several conserved noncoding elements including two located at the 5' flanking region (2.7 kb and immediately upstream of the transcription start site), at least three in the introns, and three more at the 3' flanking region. It is important to note that all sequences except human in this comparison contain sequence gaps, and so a lack of sequence homology may result from the lack of sequence information. Gaps occur frequently in draft genome sequences and in this particular example explain the missing conservation at exon five in the human/frog comparison.

The window-size and percent identity parameters that define conservation within the alignment have a profound effect on the VISTA results. It is important to run the same comparison using a different window size and thereby alter the sensitivity of the alignment. As shown in the lower panels of **Fig. 1**, the conserved sequence elements become more prominent when a smaller window size is applied. Some conserved elements, particularly those in distantly related species, may be limited to sequences shorter than the default window size of 100 bp. By reducing the window size to 20 bp, the resultant VISTA plot

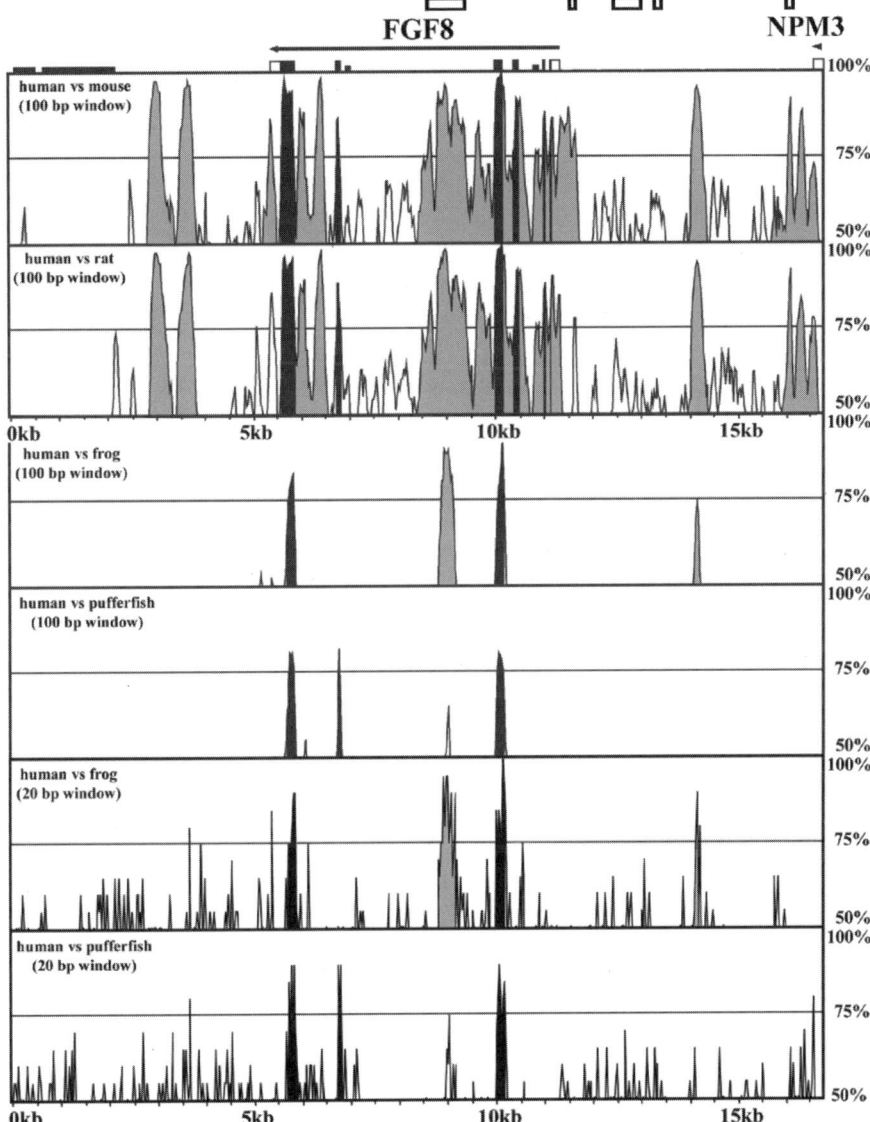

Fig. 1. Example of multiple species sequence comparison in the region of FGF8 gene. The *x*-axis represents the human sequence spanning from nucleotides 103,189,089 to 103,205,822 of chromosome 10 based on the NCBI Build 34. The *y*-axis indicates the pairwise percent identities calculated between the specified pair of organisms and within the specified window sizes in each panel. Horizontal arrows indicate the direction of transcription. Boxes above the arrow position the regulatory sequences whose function has been confirmed experimentally. The locations of coding exons and untranslated regions (UTRs) are shown as black and white rectangles above the identity profile, respectively. Dark and gray peaks represent sequences with greater than 75% identity and over 100 nucleotides in the coding and noncoding regions, respectively.

reveals regions of conserved sequence that failed to register in less sensitive comparisons at the default window size.

5.3. Detection of Transcription Factor Binding Sites

The rVISTA program is given as an option when submitting sequences for a VISTA plot. When this option is selected, the users need to select a set of transcription factor binding sites and the cutoff level for sequence similarity of the binding matrix to display. For this example, we arbitrarily selected six transcription factor binding sites (AP2, CREB, EGR1, EN1, PBX1, and SP1) and an 80% sequence similarity cutoff to generate an rVISTA plot, shown in **Fig. 2**. The program filters out the binding sites that are not part of the conserved sequences (**Fig. 2A**). However, it is possible that this filtering step may remove true transcription factor binding sites that are not located in the conserved domains of the genome. However, rVISTA is useful for prioritizing putatively functional binding sites located within a conserved element before laborious experimental analyses. **Figure 2B** shows the narrowing of one of the conserved sequences through multiple species comparison. The sequences of six conserved Engrail (EN-1) transcription factor binding sites can also be obtained from the server (**Fig. 2C**).

5.4. Implications of the FGF8 Comparison Results

As already indicated, several noncoding genomic fragments surrounding the FGF8 gene have experimentally determined regulatory activities *(48–50)*. These fragments with functions in transcriptional regulation include a 0.2-kb fragment with promoter activity immediately upstream of the transcriptional start site, a 0.8-kb fragment with enhancer activity in embryonic cells residing in the large intron, a 0.6-kb fragment with androgen induction activity that is located between –996 and –1,613 bp, and two independent retinoid response elements residing at –4.7 kb and –1.9 kb. Comparative sequence analysis of multiple species immediately detects the 0.2-kb and 0.8-kb elements and the element located at –4.7 kb (**Fig. 1**). However, the other two functional elements are not detected in our analysis, indicating that a handful of comparisons will detect only a handful of the functional elements. Conversely, there are at least six evolutionarily conserved regions, one residing at –2.7 kb, two residing in the last intron, and three residing at the 3' flanking region, that lack any experimental evidence for regulatory activity. Strikingly, the –2.7-kb element has been conserved for the nearly 400 million years of evolutionary time that separate the divergence of humans and frogs. Such conservation suggests a high likelihood of functional significance. Not every conserved sequence is necessarily involved in gene regulation; DNA retains a host of other functions including chromosomal architecture, chromatin remodeling, and nuclear matrix interactions, for which few sequence elements have been well defined. However, in the search for gene-regulatory activity, the identification of conserved sequences via sequence comparison remains an easy first step in prioritizing candidate regions for further study before painstaking experimental effort.

6. Example of Performing Primate Sequence Comparison

For our second example, we analyze the *APOA1* (Apolipoprotein A1) gene and its flanking genomic interval to illustrate the detection of conserved sequences within the primate lineage. ApoAI is the major component of high-density lipoprotein (HDL) cholesterol and is relatively abundant in plasma. The liver and small intestine are the main sites of *APOA1* expression, and the protein product is known to promote efflux of cholesterol from cells and is thought to play an important protective action against the accumulation of platelet thrombi at sites of vascular damage. Not only does *APOAI* play an important role in cardiovascular biology, but it is one of a few genes that has been sequenced in several primates, which makes it an ideal candidate for a phylogenetic shadowing analysis. We expect that several primate genomes will be sequenced in the next few years, and the approach described here will frequently be used to identify primate-specific sequence conservation with potentially primate specific functions.

6.1. Collection of Primate Sequences

Eight nonhuman primate genomic sequences containing the *APOA1* gene are found in GenBank by a tblastn search using the human APOA1 protein sequence (NCBI: AAQ91811), including chimpanzee (AC113242), gibbon (AC146472), baboon (AC145521), colobine monkey (AC148228), Dusky titi (AC144989), marmoset (AC145529), squirrel monkey (AC146293), and owl monkey (AC146499). In order to contrast a shadowing approach of closely related species to comparisons with more distant relatives, we also collect the rabbit (AC118580) and mouse (AC116503) genomic sequences for VISTA analysis. The human *APOA1* sequence and the corresponding gene annotation files are available from the UCSC Genome Browser, as described in **Subhead-**
  Here we use the July 2003 assembly of the human nucleotide coordinate, chr11:116,242,253 to 116,247,859, as the base sequence.

6.2. Primate Shadowing of the APOA1 Gene Interval

In addition to the annotation file for the base sequence, a multi-FASTA file is required for input into the RankVISTA web-server (see http://genome.lbl.gov/vista/rankVISTA.shtml for a description) *(36)*. After submission, the server returns links to several files including a multiple sequence alignment

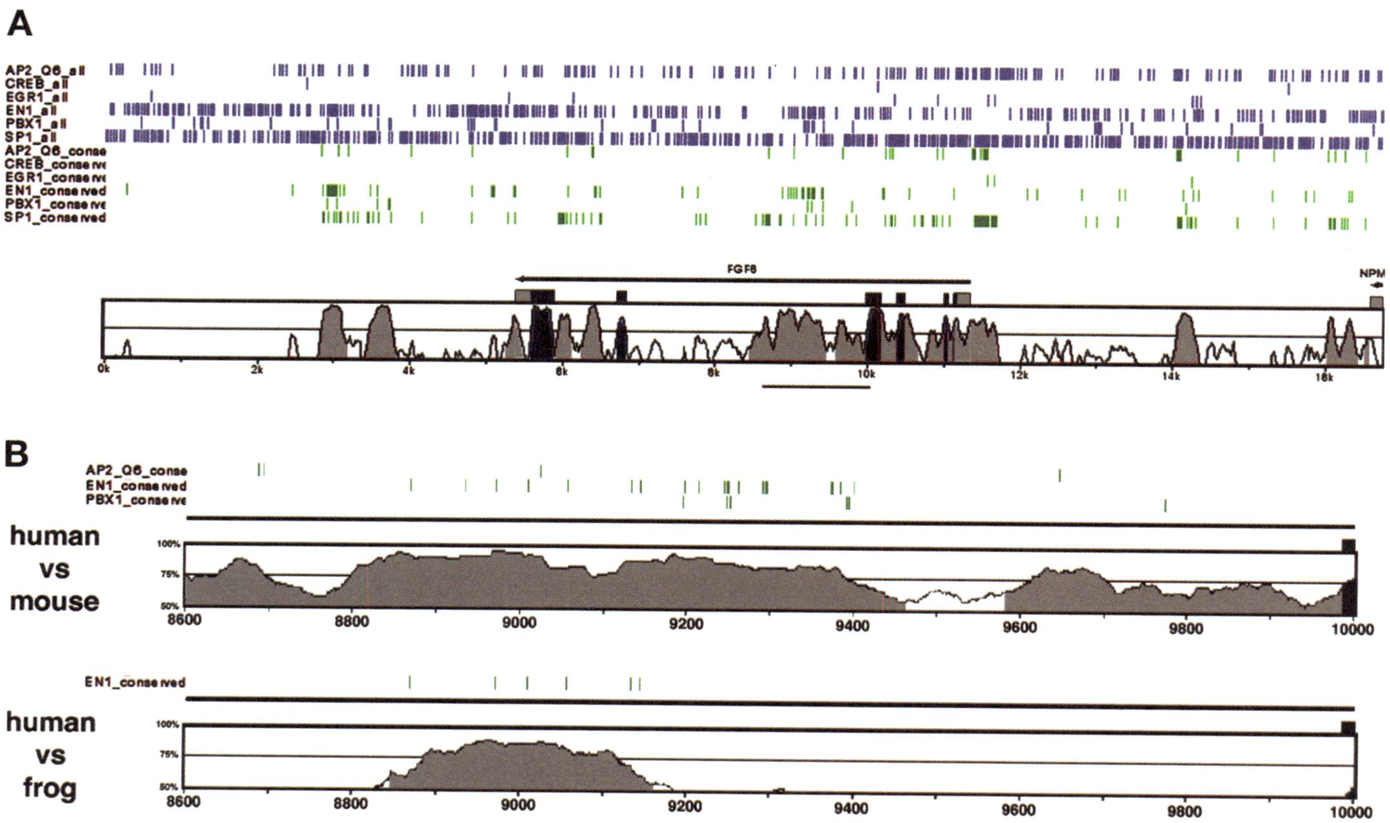
A
AP2_Q6_all
CREB_all
EGR1_all
EN1_all
PBX1_all
SP1_all
AP2_Q6_conse
CREB_conserve
EGR1_conserve
EN1_conserved
PBX1_conserve
SP1_conserved
FGF8
NPM
0k
2k
4k
6k
8k
10k
12k
14k
16k
B
AP2_Q6_conse
EN1_conserved
PBX1_conserve
human vs mouse
100%
75%
50%
8600
8800
9000
9200
9400
9600
9800
10000
EN1_conserved
human vs frog
100%
75%
50%
8600
8800
9000
9200
9400
9600
9800
10000

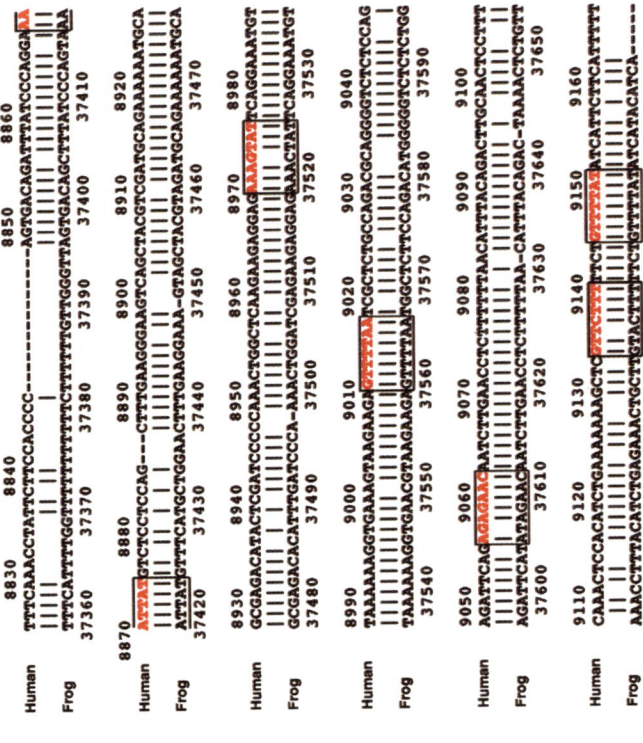

Fig. 2. Example of rVISTA comparison in searching for transcriptional factor binding sites. (**A**) Graphical display of the rVISTA output indicates the locations of six transcription factor binding sites (AP2, CREB, EGR1, EN1, PBX1, and SP1) as short vertical lines above the human/mouse VISTA plot. The top six rows are binding sites found in the 16,734 bp of the human genomic interval, and the lower six rows are binding sites found in the conserved regions of the human and mouse sequences. (**B**) A comparison of transcription factor binding sites found in the human/mouse and human/frog conserved regions shows that only six of the conserved human/mouse EN1 binding sites are present in the human/frog analysis. The human/mouse rVISTA plot is a magnified view of the underlined area in . (**C**) The human/frog sequence alignment in the conserved region of (**B**) shows the six EN1 binding sites, indicated by boxes.

adjust the *X* (nucleotide position of the base sequence) and *Y* (the likelihood ratio of fast and slow mutation rates) scales to capture the regions of increased conservation. The resultant Shadow plot is shown in **Fig. 3B**. Pairwise VISTA plots of human/rabbit and human/mouse sequences in the orthologous region are generated following the procedure described in **Subheading 5.2.** and shown in **Fig. 3A**. The coding exons as well as a previously identified liver enhancer located within 256 bp of the 5' flanking region *(51)* are conserved in both the VISTA plots and the Shadow plot, as well as additional noncoding conserved sequences uniquely identified in the Shadow plot. As in the *FGF8* analysis, comparisons identify known gene-regulatory sequences in addition to conserved regions without known function. The biological relevance of these additional slowly evolving regions can only be speculated on, but current studies suggests they warrant further exploration (D. Boffelli and C. Weer, unpublished data).

6.3. Strategies for Successful Phylogenetic Shadowing

6.3.1. Species Selection

Appropriate selection of closely related species is critical for carrying out the sequencing necessary for phylogenetic shadowing in a cost-effective way. We have previously determined *(30)* that although sequencing a large number (>15) of nonhuman primate species reveals the most powerful information, the sequence of five carefully chosen species that maximize phylogenetic diversity captures most of the information revealed in the larger data set. As illustrated in **Fig. 4**, the species at the most distant branches from human on the primate evolutionary tree capture the most diversity.

Fig. 3. *(opposite page)* VISTA and phylogenetic shadowing comparison of the *APOA1* gene interval. (**A**) Pairwise VISTA comparisons of human/rabbit and human/mouse sequences reveal conserved sequence information. In contrast, the human/chimpanzee VISTA comparison shows a high degree of sequence homology in the entire region. The *x*-axis represents the human *APOA1* sequence spanning from nucleotides 116,242,253 to 116,247,859 of chromosome 11 based on the NCBI Build 34. The definitions of the *y*-axis, horizontal arrows, rectangles, and peaks are the same as that in the legend for **Fig. 1**. The vertical arrows point to the experimentally confirmed liver enhancer located 5' proximal of the transcription start site. (**B**) The Shadower plot of human and eight nonhuman primates reveals more conserved regions than the distant comparisons in (**A**). The *x*-axis represents the same human sequence used in the VISTA comparison. The *y*-axis indicates the log likelihood ratio at that position. A lower ratio indicates a higher degree of constraint on mutability of that site.

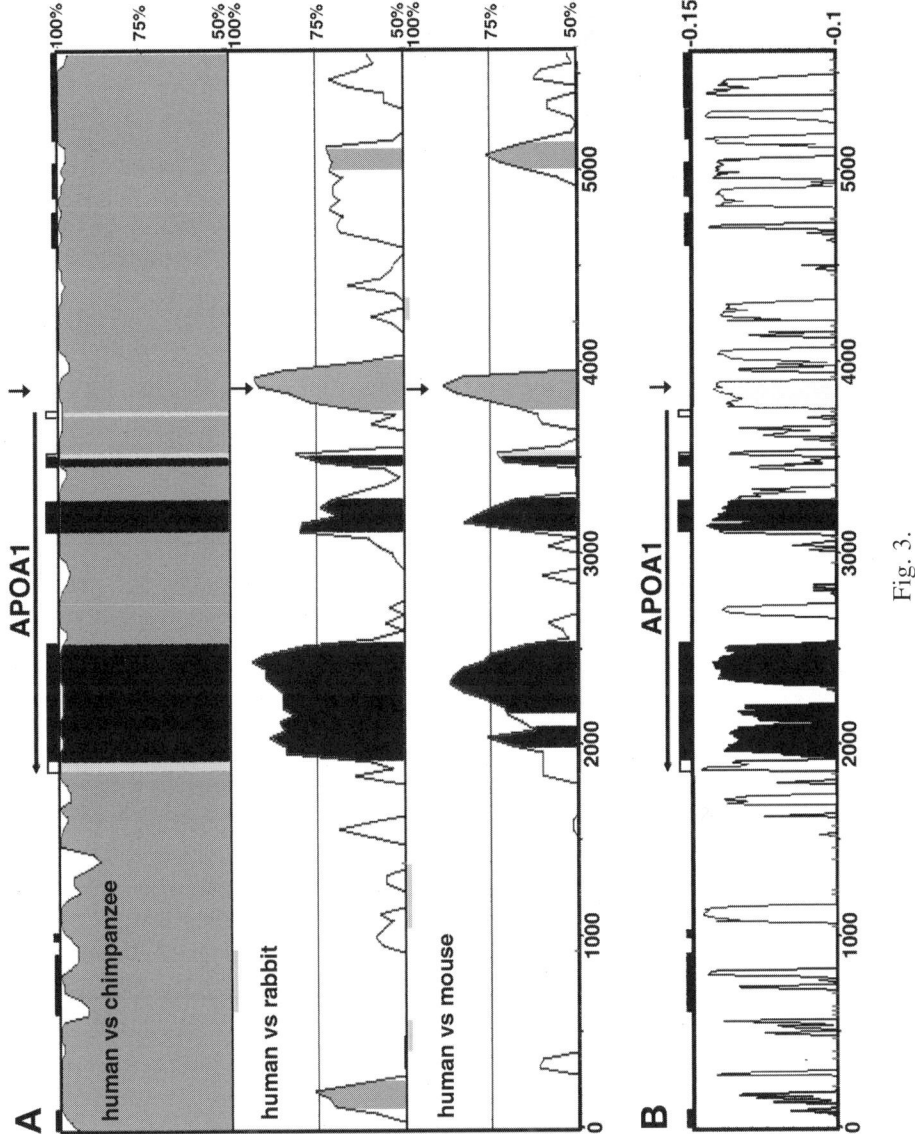

Fig. 3.

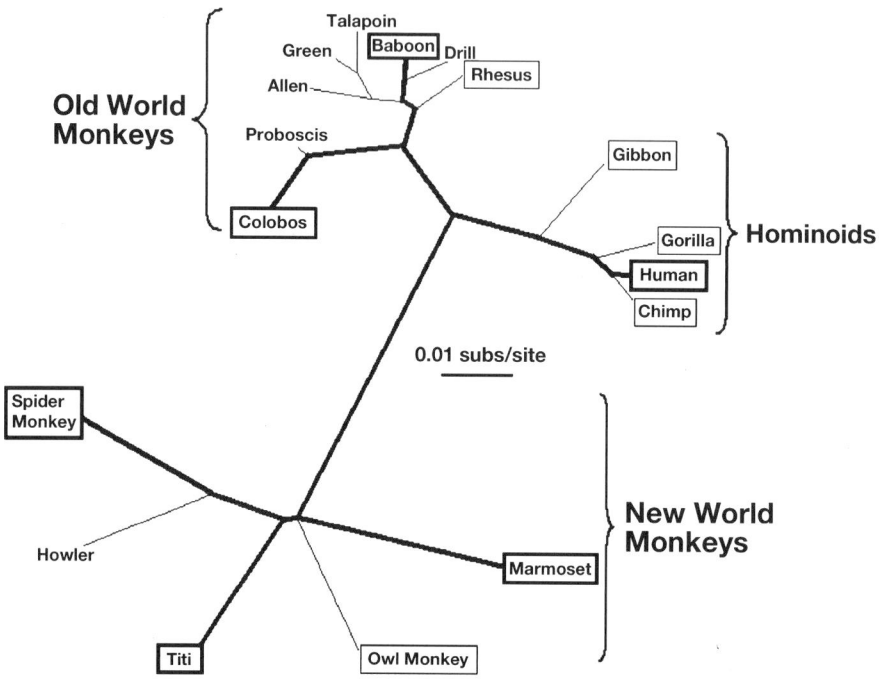

Fig. 4. Selection of evolutionarily distant primates for phylogenetic shadowing studies. An evolutionary tree of primates based on the divergence of the LXRA gene sequence. The length of branches represents levels of divergence calculated as base substitutions per site. BAC libraries are available for the boxed species. Five non-human primates in thick boxes represent the most diverse group of primates.

6.3.2. Species-Specific Biology

Comparing closely related species may result in a large amount of background conservation owing to insufficient divergent time. Specific criteria in selecting a gene for comparison will increase the potential for insight into species-specific biology. For example, selecting genes with preserved regulatory function in a set of closely related species but diverged in others may reduce the number of candidate conserved sequences to those specific to a subspecies lineage.

6.3.3. Sequence Quality

Sequences with gaps are allowed in the shadowing analysis. However, in an overall comparison, missing data from one species are indistinguishable from a large amount of diversity within that species; therefore missing data can be a source of biologically relevant or irrelevant information, and gaps warrant careful manual examination.

One major goal in the field of genomics is to reduce sequencing costs so that genomes of many organisms can be compared *(52)*. As current funding levels stand, the number of sequenced primate genomes will only increase in the foreseeable future, one day allowing a whole-genome shadowing analysis. Such a global analysis is expected to lead to a better understanding of the conserved sequences that differentiate primates from other mammals.

7. Final Remarks

Since the discovery of the double helix, researchers have been comparing DNA and protein sequences across species to study the molecular processes of evolution and to gain insights into the functional regions of proteins. Globin sequence comparisons are a classic example of a comparative study that was carried out over two decades ago *(53)*. The phylogenetic relationships among all globin genes were determined based on comparison of over 120 hemoglobin and 60 myoglobin protein sequences from vertebrates ranging from fish to human and 11 monomeric globin sequences from invertebrates and plants. In addition to sorting out the evolutionary history of lineage-specific gene duplications, the comparisons also revealed the invariant amino acids essential for complexing the heme-group that is solely responsible for the conserved oxygen binding function of the globin family. The concepts of comparative sequence analysis remain unchanged and mechanisms of genome evolution and identification of functionally important domains are of prime interest, but the data set has grown to include enough nongenic sequence that we can now derive insights into gene regulation.

Today, comparative genomics is increasing in focus because researchers have free access to the genome sequences of human and a growing list of other organisms. A simple comparison of the human and mouse genomes suggests that at least 5% of human sequence is under selective constraint, and the majority of these highly conserved sequences do not correspond to known genes. There is accumulating experimental evidence showing that many of these conserved sequences participate in the transcriptional regulation of genes, implying that evolutionary conservation is a powerful predictor of function throughout the genome. As a science, biology has finally entered the "information age" of genomics. Modern biomedical investigators have an overwhelming amount of sequence at their disposal, but the databases, servers, and techniques described in this chapter are the first step in utilizing this abundance of data to explore biological questions great and small.

Acknowledgments

We thank Francis Poulin and Marcelo Nobrega for critical reading of the manuscript. This work was supported in part by the NIH-NHLBI Programs for Genomic Application, grant HL66681, NIH grant HL071954A through the U.S. Department of Energy under contract no. DE-AC03-76SF00098, and a Multidisciplinary Training Grant in Genomics at the University of California, Berkeley (NHGRI T32HG000047-03).

References

1. International Human Genome Sequencing Consortium. (2001) Initial sequencing and analysis of the human genome. *Nature* **409,** 860–921.
2. Pennisi, E. (2003) Human genome. Reaching their goal early, sequencing labs celebrate. *Science* **300,** 409.
3. Frazer, K. A., Elnitski, L., Church, D. M., Dubchak, I., and Hardison, R. C. (2003) Cross-species sequence comparisons: a review of methods and available resources. *Genome Res.* **13,** 1–12.
4. Hardison, R. C. (2003) Comparative genomics. *PLoS Biol.* **1,** 156–160.
5. Pennacchio, L. A. and Rubin, E. M. (2003) Comparative genomic tools and databases: providing insights into the human genome. *J. Clin. Invest.* **111,** 1099–1106.
6. Batzoglou, S., Pachter, L., Mesirov, J. P., Berger, B., and Lander, E. S. (2000) Human and mouse gene structure: comparative analysis and application to exon prediction. *Genome Res.* **10,** 950–958.
7. Chen, R., Bouck, J. B., Weinstock, G. M., and Gibbs, R. A. (2001) Comparing vertebrate whole-genome shotgun reads to the human genome. *Genome Res.* **11,** 1807–1816.
8. Waterston, R. H. and International Mouse Genome Sequencing Consortium (2002) Initial sequencing and comparative analysis of the mouse genome. *Nature* **420,** 520–562.
9. Oeltjen, J. C., Malley, T. M., Muzny, D. M., Miller, W., Gibbs, R. A., and Belmont, J. W. (1997) Large-scale comparative sequence analysis of the human and murine Bruton's tyrosine kinase loci reveals conserved regulatory domains. *Genome Res.* **7,** 315–329.
10. Loots, G. G., Locksley, R. M., Blankespoor, C. M., et al. (2000) Identification of a coordinate regulator of interleukins 4, 13, and 5 by cross-species sequence comparisons. *Science* **288,** 136–140.
11. Gottgens, B., Barton, L. M., Chapman, M. A., et al. (2002) Transcriptional regulation of the stem cell leukemia gene (SCL)—comparative analysis of five vertebrate SCL loci. *Genome Res.* **12,** 749–759.
12. Dermitzakis, E. T., Reymond, A., Scamuffa, N., et al. (2003) Evolutionary discrimination of mammalian conserved non-genic sequences (CNGs). *Science* **302,** 1033–1035.

13. Genome sequencing prioritization list of NIH/National Human Genome Research Institute (http://www.genome.gov/10002154).
14. Genome assembly and annotation process (http://www.ncbi.nlm.nih.gov/books/bv.fcgi?rid=handbook.chapter.ch14).
15. Kent, W. J., Sugnet, C. W., Furey, T. S., et al. (2002) The Human Genome Browser at UCSC. *Genome Res.* **12,** 996–1006.
16. UC Santa Cruz Genome Browser (http://genome.ucsc.edu).
17. Mayor, C., Brudno, M., Schwartz, J. R., et al. (2000) VISTA: visualizing global DNA sequence alignments of arbitrary length. *Bioinformatics* **16,** 1046–1047.
18. VISTA Genome Browser (http://pipeline.lbl.gov/).
19. ECR Browser (http://ecrbrowser.dcode.org/).
20. Schwartz, S., Kent, W. J., Smit, A., et al. (2003) Human-mouse alignments with BLASTZ. *Genome Res.* **13,** 103–107.
21. Brudno, M., Do, C. B., Cooper, G. M., et al., and NISC Comparative Sequencing Program. (2003) LAGAN and Multi-LAGAN: efficient tools for large-scale multiple alignment of genomic DNA. *Genome Res.* **13,** 721–731.
22. Bray, N., Dubchak, I., and Pachter, L. (2003) AVID: A global alignment program. *Genome Res.* **13,** 97–102.
23. Pollard, D. A., Bergman, C. M., Stoye, J., Celniker, S. E., and Eisen, M. B. (2004) Benchmarking tools for the alignment of functional noncoding DNA. *BMC Bioinformatics* **5,** 6.
24. Kent, W. J. (2002) BLAT—the BLAST-like alignment tool. *Genome Res.* **12,** 656–664.
25. NCBI BLAST website (http://www.ncbi.nlm.nih.gov/BLAST/).
26. Search for sequences in the NCBI unassembled trace archive (http://www.ncbi.nlm.nih.gov/BLAST/tracemb.shtml).
27. BAC library resources (http://bacpac.chori.org/).
28. DOE/Joint Genome Institute's Community Sequencing Program (http://www.jgi.doe.gov/CSP/index.html).
29. NIH/NHGRI Genome Sequencing Program (http://www.genome.gov/1000 1691).
30. Boffelli, D., McAuliffe, J., Ovcharenko, D., et al. (2003) Phylogenetic shadowing of primate sequences to find functional regions of the human genome. *Science* **299,** 1391–1394.
31. Loots, G. G., Ovcharenko, I., Pachter, L., Dubchak, I., and Rubin, E. M. (2002) rVista for comparative sequence-based discovery of functional transcription factor binding sites. *Genome Res.* **12,** 832–839.
32. Schwartz, S., Zhang, Z., Frazer, K. A., et al. (2000) PipMaker—a web server for aligning two genomic DNA sequences. *Genome Res.* **10,** 577–586.
33. PipMaker and MultiPipMaker (http://pipmaker.bx.psu.edu/pipmaker/).
34. Schwartz, S., Elnitski, L., Li, M., et al., and NISC Comparative Sequencing Program. (2003) MultiPipMaker and supporting tools: alignments and

analysis of multiple genomic DNA sequences. *Nucleic Acids Res.* **31,** 3518–3524.
35. zPicture (http://zpicture.dcode.org).
36. RankVISTA (http://genome.lbl.gov/vista/rankVISTA.shtml).
37. Bray, N. and Pachter, L. (2003) MAVID multiple alignment server. *Nucleic Acids Res.* **31,** 3525–3526.
38. eShadow (http://eshadow.dcode.org/).
39. Ovcharenko, I., Boffelli, D., and Loots, G. G. (2004) eShadow: a tool for comparing closely related sequences. *Genome Res.* **14,** 1191–1198.
40. Jung, J., Zheng, M., Goldfarb, M., and Zaret, K. S. (1999) Initiation of mammalian liver development from endoderm by fibroblast growth factors. *Science* **284,** 1998–2003.
41. Fukuchi-Shimogori, T. and Grove, E. A. (2001) Neocortex patterning by the secreted signaling molecule FGF8. *Science* **294,** 1071–1074.
42. Storm, E. E., Rubenstein, J. L. R., and Martin, G. R. (2003) Dosage of Fgf8 determines whether cell survival is positively or negatively regulated in the developing forebrain. *Proc. Nat. Acad. Sci. USA* **100,** 1757–1762.
43. Crossley, P. H., Minowada, G., MacArthur, C. A., and Martin, G. R. (1996) Roles for FGF8 in the induction, initiation, and maintenance of chick limb development. *Cell* **84,** 127–136.
44. Lewandoski, M., Sun, X., and Martin, G. R. (2000) Fgf8 signalling from the AER is essential for normal limb development. *Nat. Genet.* **26,** 460–463.
45. Sun, X., Mariani, F. V., and Martin, G. R. (2002) Functions of FGF signalling from the apical ectodermal ridge in limb development. *Nature* **418,** 501–508.
46. Tanaka, A., Kamiakito, T., Takayashiki, N., Sakurai, S., and Saito, K. (2002) Fibroblast growth factor 8 expression in breast carcinoma: associations with androgen receptor and prostate-specific antigen expressions. *Virchows Arch.* **441,** 380–384.
47. multiVISTA or mVISTA server (http://genome.lbl.gov/vista/mvista/submit.shtml).
48. Gemel, J., Jacobsen, C., and MacArthur, C. A. (1999) Fibroblast growth factor-8 expression is regulated by intronic engrailed and Pbx1-binding sites. *J. Biol. Chem.* **274,** 6020–6026.
49. Brondani, V., Klimkait, T., Egly, J. M., and Hamy, F. (2002) Promoter of FGF8 reveals a unique regulation by unliganded RARalpha. *J. Mol. Biol.* **319,** 715–728.
50. Gnanapragasam, V. J., Robson, C. N., Neal, D. E., and Leung, H. Y. (2002) Regulation of FGF8 expression by the androgen receptor in human prostate cancer. *Oncogene* **21,** 5069–5080.
51. Walsh, A., Ito, Y., and Breslow, J. L. (1989) High levels of human apolipoprotein A-I in transgenic mice result in increased plasma levels of small high density

lipoprotein (HDL) particles comparable to human HDL3. *J. Biol. Chem.* **264,** 6488–6494.
52. Collins, F. S., Green, E. D., Guttmacher, A. E., and Guyer, M. S. (2003) A vision for the future of genomics research. *Nature* **422,** 835–847.
53. Dickerson, R. E. and Geis, I. (1983) *Hemoglobin: Structure, Function, Evolution, and Pathology.* Benjamin/Cummings.

14

Developing Computational Resources in Cardiac Gene Expression

Michael B. Bober and Raimond Winslow

Summary

The development of biotechnology and the completion of the Human Genome Project have led to the rapid emergence of enormous amounts of genomic information. Computational resources are needed to enable researchers and clinicians to obtain quick, up-to-date information in targeted areas. The creation of this type of subject-specific knowledge base requires expertise in diverse fields. This chapter is meant to provide a detailed overview and blueprint for how to construct a cardiac-specific knowledge base like the Human Cardiac Gene Expression Knowledge Base (CaGE) *(1)*. We have provided an overview of some of the ways to capture, organize, and store the necessary data using Perl scripts. Each group or individual interested in developing a cardiac or gene expression knowledge base will have unique needs. We hope that this information will serve as a guide for the development of new computational tools.

Key Words: Gene expression; knowledge base; database; genomics; cardiac; computational genomics.

1. Introduction

The development of biotechnology and the completion of the Human Genome Project have led to the rapid emergence of enormous amounts of genomic information. The data that exist now, and that will only continue to grow, demand that subject-specific reference tools be created. By definition, the imposition of structure or organization onto data creates knowledge. In this way a database is turned into a knowledge base when it becomes organized by some underlying principal. The creation of this type of a subject-specific knowledge base requires expertise in diverse fields. Computer scientists, while having software development skills, often do not have the broader view of the specific subject matter needed to design the software. Research scientists and clinicians, while having the broader view, do not typically possess the programming skills necessary to create a knowledge base. Resources are needed to enable researchers and clinicians to obtain quick, up-to-date information in targeted areas. Of particular interest here is the development of computational resources to assist in the study of cardiac gene expression. With this in mind we developed and implemented a web-searchable knowledge base of gene expression in human cardiac tissue. The Human Cardiac Gene Expression Knowledge Base (CaGE) is a tool designed as a reference database for physicians and scientists interested in gene expression in human cardiac tissue (*1*). In this chapter we give an overview of some of the methods that were used to create this knowledge base.

2. Materials

2.1. NCBI Databases

The major source of freely available data is the resources at the National Center for Biotechnology Information (NCBI; http://www.ncbi.nlm.nih.gov/). This set of resources is continually changing, and a major update has recently been completed. Since the completion of CaGE (*1*), the LocusLink Database (*2,3*) has been replaced by Entrez Gene (*4*). There are many important databases in addition to Entrez Gene that are housed on the NCBI's computers, including Gen Bank (*5*), UniGene (*6,7*), OMIM (*8*), and RefSeq (*2,3*). We will not specifically discuss the details of these databases here, other than to say that all the information contained within them can be downloaded and integrated into the individual end user's software needs. We will discuss Entrez Gene and UniGene, as these data sources are necessary and sufficient to construct a local cardiac-specific database.

2.1.1. Entrez Gene

Entrez Gene is NCBI's database for gene-specific information (www.ncbi.nlm.nih.gov/entrez/query.fcgi?db=gene). This resource is similar to LocusLink, in that it provides a unique integer identifier for each gene listed. The categories of information found for each gene are nomenclature, gene structure and sequence, genomic position, citations, functional annotation, homology, expression, and related information. Nomenclature includes the official gene symbol and name as well as aliases. The name of the protein or proteins coded for by the gene is also included. Gene structure includes the reference sequence and annotation from RefSeq. Genomic position includes chromosomal location. The citations category includes both RefSeq curation information and Gene References into Function or GeneRIF, which is a type of literature review. Functional annotation includes Gene Ontology (GO) (*9,10*) terms and Online Mendelian Inheritance in Man (OMIM) data. The homology category contains information regarding homology of both the gene and the protein. The expression category contains among other information the UniGene cluster information. The related information category includes data that may include links to clinical laboratories testing for mutations in the given gene or databases of described mutations in the gene. All the data in the Entrez Gene database is freely downloadable from the web at ftp://ftp.ncbi.nlm.nih.gov/gene/. The data are available in several different formats and are updated frequently. It is possible to download only data subsets.

2.1.2. UniGene

Another of the important NCBI databases is UniGene (http://www.ncbi.nlm.nih.gov/UniGene/). UniGene is a largely automated analytical tool used to create an organized view of the transcriptome. UniGene was first designed and implemented in the mid-1990s, well before the completion of the Human Genome Project. At that time many small cDNAs were being sequenced and deposited into GenBank, the NCBI's repository for raw sequence data. These small cDNAs often represented no more than 70% of the actual genetic information being transcribed. These fragments were termed expressed sequence tags (ESTs). The fundamental concept behind UniGene was that the ESTs that arose from the same gene should be overlapping fragments that could be assembled computationally into clusters. These clusters were thought to represent distinct genes. As of the time of this writing, the most recent UniGene Build (#181) for *Homo sapiens* includes 5,080,380 sequences grouped into 52,924 clusters. For the purposes of this chapter, there are two major uses for UniGene. The first use is to allow each of the over 5 million sequences to be related to a unique cluster and therefore a specific gene. The second use is that for each of the UniGene clusters there is a field termed EXPRESS, which contains a list of all the tissue types from which the ESTs in the cluster have originated.

2.2. Other Resources

The Internet has many different resources that provide evidence different from or not included in the NCBI data. Furthermore, the individual labs interested in creating their own databases will no doubt have internal data they wish to house in the database. When creating CaGE, we used cardiac gene expression data from multiple sources aside from the NCBI. These included microarray expression data generated in our lab and expression data from the now inactive Toronto Cardiac Genes web site *(11)* and also from the Body Maps web site. Body Maps is a Japanese-based web site that contains tissue-specific gene expression data (http://bodymap.ims.u-tokyo.ac.jp/) *(12,13)*.

3. Methods

The primary necessity in creating a computational resource will of course be the hardware. The type needed will depend on the desired goal. On the minimalist side, a database could be constructed on a desktop computer with web access and the most basic database software. A lab-wide system could be constructed on a small local network with a central computer having web access and more sophisticated database software that could be simultaneously accessed from several of the workstations using either an html interface or forms created in the database software. On an even larger scale, a large Internet server housing a professional-grade database software package could be used as the central computer and allow for multiple worldwide simultaneous users via a web interface. There is little doubt that hardware poses a significant expense, and the costs may prove defining in which type of system to construct.

Several different operating system options, relational database software packages, and programming languages can be employed to build a cardiac database. We chose to create CaGE initially on an IBM web-server (RS/6000) with a UNIX operating system (AIX 4.3). We used the IBM DB2 relational database and programmed in Perl. The most up-to-date Perl compiler for Windows and UNIX can be freely downloaded from the perl.com website (http://www.perl.com/download.csp#stable). There are obviously many different permutations of software, databases, and programming languages, and we will attempt to present general information so readers may use this information to create a system based on their needs and equipment. Specifically, when examples are provided, we will base them on CaGE.

3.1. Data Flow

The flow of data from various sources and their organization into the local database is probably the single most important step in building a successful software tool. It is critical for the development of the system to understand what ultimately the software is meant to do in order to structure properly how the data are collected, parsed, or decomposed and then reorganized. This is illustrated in **Fig. 1**. The desired files from the various web data sources must first be downloaded onto the local hardware.

3.1.1. Downloading Data

Many of the files from NCBI are located on their servers in a compressed format. These files have a .gz or .z file extension and require unzipping software to be installed on the local machine in order for the data within the files to be manipulated. An example of a Perl script used by CaGE to download an NCBI UniGene file is found just below. In order for Perl to access and down-

Computational Resources in Cardiac Gene Expression

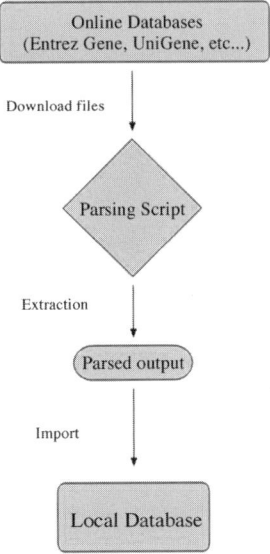

Fig. 1. Data flow diagram.

load web files, the LWP::Simple module must be included in the initial Perl installation. Lines that begin with the character "#" are comment lines.

```
#! /perl/bin/perl
#This script downloads and unzips the file HS.data.Z from #the NCBI.
#module loading
use LWP::Simple;
#define what directory we're using.
$dir="/data/CAGE/downloads";

Getstore ("ftp://ncbi.nlm.nih.gov/repository/UniGene/Hs.data.Z", "$dir/Hs.data.Z");
'/usr/local/bin/gunzip -f -q $dir/HS.data.Z';
```

This type of simple script can be used to download files from the web. Sites like the BodyMaps, however, do not have files available for download, rather, they have web pages with the data wanted listed on them. The following is an example of a Perl script we used to create a file with the some of the data available at BodyMaps. This specific script downloads the information on the genes expressed within atrial tissue. It takes the table present on the webpage row by row and temporarily stores the first five columns in a hash. A file is then created that has a single column of the accession numbers representing the genes expressed in atrial tissue.

```perl
#! /usr/bin/perl
```

##Script Description: stores html file from BodyMap site as a string. Parses the
string for output, then outputs the parsed data to a file named
##"BMoutputAtri.txt".

```perl
use LWP::Simple;

$outputFilename="/data/CAGE/downloads/BMoutputAtri.txt";

$source= get("http://bodymap.ims.u-tokyo.ac.jp/human/doquery_composition.php?tissue_id=37");
$source=~s/ //isg;
$source=~s/^.*?<\/table>//is;
$source=~s/<\/table>.*?$//is;
print "parsing BodyMap atrial webpage...\n";
$counter=0;

##parse out the fields of the webpage and store in hash
while ($source=~/<tr><td.*?><.*?>(.*?)<.*?><\/td>.*?<td.*?>(?:<A.*?>)?(.*?)(?:<\/A.*?>)?<\/td>.*?<td.*?>(.*?)<\/td>.*?<td.*?>(.*?)<\/td>.*?<td.*?>(.*?)<\/td>.*?<\/tr>/gi)
{
      $hash{$counter}{"GS"}=$1;
      $hash{$counter}{"primer"}=$2;
      $hash{$counter}{"unigene"}=$3;
      $hash{$counter}{"ACC"}=$4;
      $hash{$counter}{"name"}=$5;
      $counter++;
}
##output the hash to a file
open (FILE, ">$outputFilename");
for each $key (keys (%hash))
{
      if (($hash{$key}{ACC} ne "") && (!defined ($Check{$hash{$key}{ACC}})))
      {
             print FILE "$hash{$key}{ACC}\n";
             $Check{$hash{$key}{ACC}}++;
      }
}
print "BodyMap atrial parsing complete\n";
print "\n";

close (FILE);
```

Computational Resources in Cardiac Gene Expression

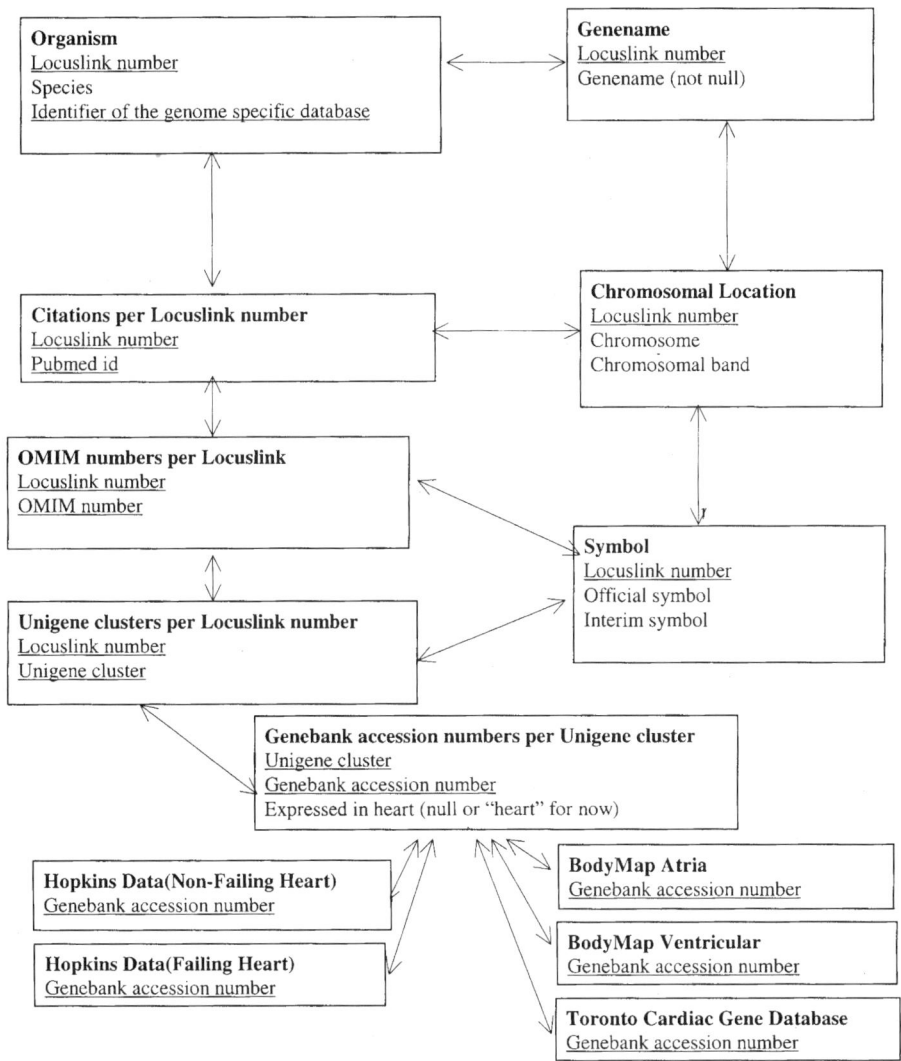

Fig. 2. Sample CaGE database schema.

3.1.2. Data Organization

Once all the desired data are downloaded to the local hard drive, they must then be organized so they may be imported into the database as tables. The most efficient way to organize the tables is very much dependent on the way the database will be used. For the execution of queries on the database, it is best that not all data be stored in a single large table. Computational efficiency can be achieved if smaller tables are used. To design the database in the best way, a schema should be created to list the fields in each table and to define the relationships between the various tables and fields.

Figure 2 shows one of the early schema that we used in developing CaGE. The boxes represent tables, with the bold face text being the name of the table. The fields are listed beneath the table name. The underlined fields represent the key fields. Although this schema worked perfectly to store our data, it was felt that several of our most common queries took too long to execute, and we adapted our schema to accelerate these queries.

Once the schema are determined, scripts can be written to format the existing downloaded files into the tabular format, which will later be imported into the database. This is a sample script that parses the HS.data file previously downloaded. It creates a pipe-delimited, three-column file. In each row there is a unique GenBank accession number and its associated UniGene cluster; information on whether that cluster is known to be expressed in cardiac tissue "heart" is present in the third column.

```perl
#! /usr/bin/perl
## This script takes the Hs.data file, opens it, parses it, and reorganizes the data into file
## named HSoutput.txt

$dir="/data/CAGE/downloads";

use LWP::Simple;

$filename="$dir/HSoutput\.txt";
open (FILE,"$dir/Hs.data");
while (defined( $line = <FILE> ))
{
        if ($line!~/^\/\//)
                {
                        if ($line =~/^ID/)
                                {
                                        ($HS_ID)=$line=~/^ID\s*(\S.*)$/;
                                }
                        if ($line =~/^EXPRESS/)
                                {
                                        ($express)=$line=~/EXPRESS\s*(\S.*)$/;
                                        $HS_Data_Hash{$HS_ID}{"express"}=$express;
                                }
                        if ($line =~/^SCOUNT/)
                                {
                                        ($scount)=$line=~/SCOUNT\s*(\S.*)$/;
                                        $HS_Data_Hash{$HS_ID}{"scount"}=$scount;
                                }
                        if ($line =~/^SEQUENCE/)
                                {
```

```perl
                                    ($Acc)=$line=~/\bACC=(\S.*?)\;.*/;
                                    $seq_append="sequence".$scount;
                                    $HS_Data_Hash{$HS_ID}{$seq_append}=$Acc;
                                    $scount—;
                            }
                }
        }
}
close (FILE);

##output to file
open (OUT, ">$filename ");
foreach $key_to_hash (keys (%HS_Data_Hash))
{
        ##only write "heart" if it is in the expression field else just leave it blank
        if ($HS_Data_Hash{$key_to_hash}{express}=~/heart/i)
                {$heart="heart"}
        else {$heart="";}

        for ($i=$HS_Data_Hash{$key_to_hash}{scount};$i>0;$i—)
        {
                $seq_append2="sequence".$i;
                print OUT
"$HS_Data_Hash{$key_to_hash}{$seq_append2}|$key_to_hash|$heart\n";
        }
close OUT;
```

This process should be repeated for all the newly downloaded data until each table to be placed into the database is represented by a pipe-delimited text file.

3.1.3. Populating the Database

Once all the data are organized into pipe-delimited text files, they are ready to be placed into the database. Depending on the database software being used, there may be choices for automatically populating the database. Some software has a user interface that can schedule the creation and loading of tables on a regular basis. If this feature is present, it may be used. If it is not present, then once again Perl can be used. There are Perl modules for every major relational database. For the IBM DB2 software we used, the module was named DB2::DBI.

3.1.4. Automating the Process

Given the rapid growth of genetic information, some of the NCBI data are updated daily. Therefore, to have the most up-to-date information populating the database, it is necessary to execute most, if not all, of the scripts daily.

There are once again different methods to accomplish this. Since the scripts are executable files, the operating system the computer is using will determine the choice. On a Windows-based system, there is an "automated tasks" function that can execute pre-specified files at a given time. Unix-based systems have a similar feature termed a "chron job." One advantage to using a Perl database interface (DBI) is that following the downloading and parsing of the data, populating the database can be accomplished in a similar fashion. To populate CaGE, approximately 170 MB of relevant data was downloaded and parsed every Sunday through Thursday, in the early morning hours.

3.1.5. Using the Database

As we discussed previously, there are many different ways in which the database could be used, and each group will almost certainly have different needs. Most database software has built-in features that allow users to query and visualize data in defined ways. For stand-alone machines and lab networks, these tools are probably adequate. They are beyond the scope of this chapter, as they each have their own idiosyncrasies. For those groups wishing to create a web-based search functionality and possessing the server and software necessary to do so, this should not prove difficult.

4. Conclusion

Computational resources are needed to enable researchers and clinicians to obtain quick, up-to-date information in targeted areas. The creation of this type of subject-specific knowledge base requires expertise in diverse fields. Computer scientists, while having software development skills, often do not have a broader view of the specific subject matter needed to design the software. Research scientists and clinicians, while having the broader view, do not typically possess the programming skills necessary to create a knowledge base. This chapter aimed to provide a detailed overview and blueprint for how to construct a knowledge base like CaGE *(1)*. We have discussed various web-based resources that have cardiac gene expression data including the NCBI *(2–10)* and BodyMaps *(12,13)*. We have provided an overview of some of the ways to capture, organize, and store these data using Perl scripts. It is likely that each group or individual interested in developing a cardiac or gene expression knowledge base will have unique needs. We hope that this information will serve as a guide for the development of new computational tools.

References

1. Bober, M. B., Wiehe, K., Yung, C., et al. (2002) CaGE: Cardiac Gene Expression Knowledgebase. *Bioinformatics* **18,** 1013–1014.
2. Pruitt, K. D. and Maglott, D. R. (2001) RefSeq and LocusLink: NCBI gene-centered resources. *Nucleic Acids Res.* **29,** 137–140.

3. Pruitt, K. D., Katz, K. S., Sicotte, H., and Maglott, D. R. (2000) Introducing RefSeq and LocusLink: curated human genome resources at the NCBI. *Trends Genet.* **16,** 44–47.
4. Maglott, D., Ostell, J., Pruitt, K. D., and Tatusova, T. (2005) Entrez Gene: gene-centered information at NCBI. *Nucleic Acids Res.* **33,** D54–58.
5. Benson, D. A., Karsch-Mizrachi, I., Lipman, D. J., Ostell, J., Rapp, B. A., and Wheeler, D. L. (2000) GenBank. *Nucleic Acids Res.* **28,** 15–18.
6. Schuler, G. D., et al. (1996) A gene map of the human Genome. *Science* **274,** 540–546.
7. Pontius, J. U., Wagner, L., and Schuler, G. D. (2003) UniGene: a unified view of the transcriptome, in *The NCBI Handbook*. National Center for Biotechnology Information, Bethesda, MD.
8. Maglott, D., Amberger, J. S., and Hamosh, A. (2002) Online Mendelian Inheritance in Man (OMIM): a directory of human genes and genetic disorders, in *The NCBI Handbook*. National Center for Biotechnology Information, Bethesda, MD.
9. Camon, E., Magrane, M., Barrell, D., et al. (2004) The Gene Ontology Annotation (GOA) database: sharing knowledge in Uniprot with Gene Ontology. *Nucleic Acids Res.* **32,** D262–D266.
10. Gene Ontology Consortium. (2004) The Gene Ontology (GO) database and informatics resource. *Nucleic Acids Res.* **32,** D258–D261.
11. Hwang, D. M., Dempsey, A. A., Wang, R. X., et al. (1997) A genome-based resource for molecular cardiovascular medicine: toward a compendium of cardiovascular genes. *Circulation* **96,** 4146–4203.
12. Okubo, K., Hori, N., Matoba, R., Niiyama, T., and Matsubara, K. (1991) A novel system for large-scale sequencing of cDNA by PCR amplification. *DNA Seq.* **2,** 137–144.
13. Okubo, K., Hori, N., Matoba, R., et al. (1992) Large scale cDNA sequencing for analysis of quantitative and qualitative aspects of gene expression. *Nat. Genet.* **2,** 173–179.

V

CARDIAC SINGLE NUCLEOTIDE POLYMORPHISMS

15

In Silico Analysis of SNPs and Other High-Throughput Data

Neema Jamshidi,* Thuy D. Vo,* and Bernhard O. Palsson

Summary

The availability and accessibility of high-throughput and biological legacy data have allowed mathematical analyses of genome-scale metabolic networks and models. Model formulation is centered on the conservation principles of mass and charge. Thermodynamic information is generally incorporated by means of reaction reversibility. If further experimental data are available, such as kinetic parameters, models describing system evolution over time can be developed. The type of data available largely determines the type of model (and subsequently the type of analysis) that can be performed. Different modeling approaches offer different advantages. Detailed kinetic models can make specific predictions about network functional states given knowledge about the enzyme parameter variations resulting from single-nucleotide polymorphisms (SNPs). They also require a large amount of experimental data, which is rarely available. On the other hand, although current formulations using the constraint-based optimization framework do not offer information about metabolite concentrations or time-dependent changes, it is a remarkably flexible modeling framework and permits the integration of a large amount of very different data types.

Key Words: Systems biology; high-throughput data; single-nucleotide polymorphism; flux balance analysis; linear optimization; red blood cell; kinetic; genotype; phenotype; constraint-based models.

1. Introduction

The elucidation of the human genome sequence has created the potential for great advances in the medical and biological sciences. Recent research identifying single-nucleotide polymorphisms (SNPs) as the genetic root of many diseases also holds much promise for the future of medicine. Discoveries

---

(*) The first two authors contributed equally to this work.

From: *Methods in Molecular Biology, vol. 366: Cardiac Gene Expression: Methods and Protocols*
Edited by: J. Zhang and G. Rokosh © Humana Press Inc., Totowa, NJ

relating SNPs to complex diseases such as hemochromatosis and hemolytic anemias have not only shed light on the etiology of these diseases but have also been shown to play significant roles in the person-to-person variations in drug metabolism *(1,2,24,25)*.

Many organizations are collaborating to identify and characterize these variations. The SNP Consortium *(3)* and the HapMap Project have successfully mapped 1.43 millions SNPs in the human genome. Such work has helped bring the genotype-to-phenotype relationship into focus, with greater appreciation for nuances in phenotypes. An SNP may not necessarily result in a truncated protein or a totally dysfunctional protein, but rather a protein with altered or limited catalytic activity. Therefore, instead of a clear black and white picture, in which a person either has or does not have a fully functional ability to carry out a particular enzymatic reaction, there appear multiple shades of gray in the spectrum of metabolic capabilities.

As more detailed biological information is discovered and computational capabilities improved, so has our ability to describe biological systems quantitatively and to analyze them using engineering techniques and principles. With the exponentially growing volume of data from high-throughput experiments, it is important to organize this information into a framework that allows systematic studies. The development of *in silico* biological models based on physical principles thus becomes a necessary and integral aspect of building the genotype-to-phenotype bridge in the 21st century.

A well-constructed *in silico* model can be a valuable tool for identifying emergent properties and gaining further insight into normal and pathological states of the biological system of interest. These models are typically built in an iterative fashion such that results from each model can help direct new experiments, which, if verified, will lead to a more accurate model. The remaining sections of this chapter will describe a modeling approach that integrates high-throughput and sequence variation data with established biochemical knowledge to quantitatively analyze a biological system.

2. Materials

2.1. Data Acquisition

The first step in constructing a metabolic model is to identify metabolites and biochemical reactions contributing to the metabolism of the system of interest. These components can be collected from two main sources: accumulated experimental data from primary literature and high-throughput, *-omics*, data. The informational databases referred to here are not meant to be comprehensive, but are simply examples for possible data resources. Those that the authors have found particularly useful are mentioned in this chapter.

2.1.1. Biochemical Data

Direct biochemical information is usually the most reliable source. It usually includes the stoichiometry and directionality of the reaction. Collections of biochemical data on an organism's metabolism are often found in review articles, biochemistry textbooks, and volumes that are focused on the biology of single organisms *(26)*. When inconsistencies were found among these sources, one should refer to the primary literature in which the reaction is characterized.

2.1.2. High-Throughput Data

2.1.2.1. GENOMICS

A large number of organisms, particularly bacteria, have publicly available genomic sequences with an initial genome annotation. Genome annotation usually includes both experimentally verified functions of known proteins and putative functions for open reading frames (ORFs) annotated with *in silico* methods. *In silico* methods (implemented in programs such as BLAST and FASTA) rely on sequence similarity between closely related organisms and can assign putative function to 40 to 70% of identified ORFs on a freshly sequenced microbial genome. However, gene annotation based on homology methods is hypothetical and is subjected to revision until the gene has been experimentally studied. Genomic data can be obtained from databases such as the National Center for Biotechnology Information (NCBI) Entrez (http://www.ncbi.nlm.nih.gov/entrez), TIGR (www.tigr.org), the Kyoto Encyclopedia of Genes and Genomes (KEGG, http://www.genome.jp/kegg) or directly from the group that sequenced the organism. Approaches to the automation of network reconstruction from annotated sequences are also under development *(4)*.

2.1.2.2. SNP-OMICS

Given the great interest in SNPs from scientists in academia to pharmaceutical companies, extensive resources and concerted efforts have been put forth to identify and study SNPs. The SNP Consortium and the International HapMap Project (http://www.hapmap.org/cgi-perl/gbrowse/gbrowse/hapmap) have discovered and characterized nearly 2 million SNPs in the human genome and made the data freely available to the public early on. The efforts of the SNP Consortium have been joined with the International HapMap Project, which now maintains and updates the databases. The web site provides the capability of searching for specific SNPs, allele frequency data, coding SNPs, or proximity to a particular gene.

The NCBI and the National Human Genome Resource Institute have also created an SNP database, dbSNP (http://www.ncbi.nlm.nih.gov/SNP), which accepts submissions of new SNPs from the public *(5,6)*. In contrast to the International HapMap Project, the dbSNP web site is not confined to the human genome. A wide range of search fields is available, including specific SNPs, organisms, chromosomes, genes, alleles, and base positions.

In general, to ensure the quality of sequence variants retrieved from these databases, it is necessary to collect and compare data from multiple sources *(7)*. Furthermore, to incorporate causal SNPs into a computational model, quantitative information about the change in the resulting protein activity is needed. These alterations often lead to changes in concentration of active enzymes, catalytic activities, and binding constants.

2.1.2.3. Proteomics

The proteome of a biological system (an organelle, a cell, or an organism) is characterized by its protein content, posttranslational modification states, localization, and abundance. Therefore proteomic data are dependent on the environmental and internal conditions of the system. These data are typically more difficult to obtain than genomic data, but subcellular proteomes and proteomes of small bacteria are being identified *(8,9)*. Proteomic data are particularly important for studying systems such as the mitochondria, in which most enzymes are not encoded in the organelle's genome, or for differentiated cells, in which only a subset genes are expressed at the same time. These data can be obtained from protein databases such as Swiss Prot (http://au.expasy.org/sprot), the Nuclear Protein Database (NPD; http://npd.hgu.mrc.ac.uk), or organelle/cell/organism-specific databases, or directly from the primary literature.

2.2. Modeling Software

Owing to the nonlinear and stiff nature of most mass balance differential equations describing the kinetics of biological networks, it is advisable to use one of the widely available software packages for numerical integration. Some commonly used programs include Matlab® (The MathWorks, Inc.), Mathematica® (Wolfram Research, Inc.), Mathcad® (MathSoft, Inc.), Maple® (Maplesoft, Inc.), Berkeley Madonna™, and Gepasi. In addition, various commercial software packages are available to solve linear programming problems. Smaller programs can be run on Microsoft Excel, Matlab, or Mathematica; larger ones may require the use of more specialized software such as LINDO (LindoSystems, Inc.), MOSEK (MOSEK ApS), TOMLAB (The MathWorks, Inc.), and GAMS (GAMS Development Corporation). Palsson et al. have also developed a collection of publicly available programs, including FBA (http://systemsbiology.ucsd.edu), which was specially designed to solve flux balance analysis problems. This program can handle matrices of dimension up to 100×100.

The choice of a particular program is largely determined by the numerical difficulties that may be encountered in a given model and the comfort level of the modelers in using a particular tool.

3. Methods
3.1. Constraint-Based Modeling Approach

With the increasing availability of high-throughput data, there is a growing need for integrating and reconciling these heterogeneous data sets to increase its consistency and reliability *(10,11)*. A constraint-based model can serve as a model-centric database that provides quantitative predictions. The constraint-based modeling approach involves the application of a series of constraints arising from the consideration of stoichiometry, thermodynamics, enzymatic capacities, and regulatory and kinetic constraints when they are available. This method involves identifying the molecular composition of the biological network and defining constraints on their interactions. The key components of a metabolic model are metabolites and enzymes. Regulated or kinetic metabolic models will also include regulatory rules and regulator molecules. The analysis assumes that under any given environmental conditions, the organism will reach a homeostatic state that satisfies all physiochemical constraints. The general method can be summarized as follows:

1. Define the molecular composition and interactions in the network.
2. Construct the stoichiometric matrix to represent the network.
3. Identify constraints on molecular components and links in the network.
4. Define rate expressions if a dynamic model is desired, or an objective function for flux balance analysis.
5. Analysis:
   a. Employ the model to predict verifiable results.
   b. Incorporate additional data (e.g., SNP) and apply the model to predict outcomes under new conditions.

3.2. Model Formulation

A metabolic network, no matter how extensive and complex, must always adhere to the fundamental laws of physics and chemistry. A metabolic reaction can be viewed as a conversion of substrates into products by the action of an enzyme, whose activity can be described by a rate law. Therefore the change in concentration of a metabolite over time is equal to the sum of all of fluxes that lead to the production of the particular metabolite minus the sum of fluxes that consume the metabolite, or

$$\frac{d\mathbf{X}}{dt} = S \cdot \mathbf{v} \tag{1}$$

where **X** is the metabolite vector (length m), **v** is the flux vector (length n), and S is the $m \times n$ stoichiometric matrix consisting of the appropriate coefficients for all the reactions participating in the network. In general, v_i is a function of metabolite concentrations,

$$v_i = f(x_1, x_2, \ldots, x_m) \tag{2}$$

S is structured such that every column in the matrix corresponds to a reaction and every row to a metabolite. An element S_{ij} represents the coefficient of metabolite i in reaction j. By convention, the coefficient of a metabolite is positive, if it is the product of the forward reaction, and negative otherwise *(27)*.

More comprehensive models will account for other physical effects, such as osmotic forces and electroneutrality. Formulation of these principles is simple, but solving the resulting equations is often more difficult. Models that account for osmotic forces require incorporation of the volume. Metabolite concentrations can be re-expressed in terms of

$$M = V\mathbf{X} \tag{3}$$

where V is the volume of the cell and M is a vector of the total mass of metabolites, both of which are time dependent. Differentiating Eq. 3 with respect to time and rearranging the terms *(28)* yields

$$\frac{d\mathbf{X}}{dt} = S \cdot \mathbf{v} - \frac{\mathbf{X}}{V}\frac{dV}{dt} \tag{4}$$

Balancing the external and internal osmotic pressures requires

$$\Pi_e = \Pi_i \tag{5}$$

where Π_e and Π_i are the external and internal osmotic pressures, respectively. Given the osmotic coefficients for each metabolite, Φ_i, one can calculate the osmotic pressures for the internal and external metabolites. Specifically,

$$\Pi_e = RT \sum_{i=1}^{m} \Phi_{ei} C_{ei} \tag{6}$$

where R is the ideal gas constant, T is the temperature, and C_{ei} is the concentration of species i in the external environment.

Conservation of charge within each reaction can also be satisfied. This requires defining the pH of the environment and calculating the charge of the metabolites based on their respective pK_a values. For example, at a pH of 7.2, which reflects the cytosolic environment of most cells, succinate and lysine are represented as $C_4H_4O_4^{2-}$ and $C_6H_{15}N_2O_2^{1+}$, respectively. This representation will allow one to account for the presence of free proton or hydroxide ions, which may otherwise be overlooked. The principle of electroneutrality states that the sum of the charges inside and outside the cell must be zero.

$$\sum_{j=1}^{m} z_j C_j^e = 0 \tag{7}$$

$$\sum_{j=1}^{m} z_j C_j^i = 0 \tag{8}$$

Equations 7 and **8** describe a system of m metabolites, where z denotes the charge of the metabolite j and C_j^i represents the concentration of the metabolite j inside the cell. The superscripts e and i correspond to the external and internal environments, respectively.

The thermodynamics of the reactions must be accounted for in the models. The Gibbs free energy is used to characterize directions of reactions. Furthermore, in dynamic models, the kinetic expressions describing the enzyme fluxes are related to the enzymatic catalytic rate, which is also a function of the Gibbs free energy. Directionality of a flux, v_i, can be expressed by

$$\alpha \leq v_i \leq \beta \tag{9}$$

where α and β are the lower and upper limits, respectively. Maximum flux values (β) can be estimated based on enzymatic capacity limitations or maximal measured uptake rates for transport reactions. The lower limit (α) is zero for irreversible reactions.

In dynamic models, Michaelis-Menten kinetics are generally assumed to be valid. A simple expression for an unregulated enzyme flux has the form

$$v = V_{max} \frac{X}{K_m + X} \tag{10}$$

where V_{max} is the maximum enzymatic rate, K_m is the Michaelis constant, and X is the concentration of the substrate metabolite *(12)*.

Depending on data availability and computational capability, one can choose which type of analysis to use, dynamic or flux balanced, based on the availability of the appropriate information. The FBA method requires only the stoichiometric matrix, S, directionality of network reactions, and some knowledge of the maximum fluxes, whereas dynamic analysis requires knowledge of reaction mechanisms and kinetic parameters for enzymes in the system.

3.3. Kinetic Model Analysis

Once a kinetic-based description of the model has been developed, the set of ordinary differential equations can be solved to calculate the concentrations of metabolites over time. Given the highly nonlinear nature of biological systems, the wide range of time scales over which reactions occur, and the numerical complications that may result, it is particularly advisable to utilize one of the commercial software packages suggested in the Materials section (**Subheading 2.**).

In general, we are interested in the concentrations of network components and reaction rates once the system reaches a steady state. It is useful to have a good approximation of the anticipated steady state, so that appropriate initial conditions are chosen. These differential equations can be numerically integrated over a finite time interval to produce concentration profiles of metabolites at discrete time periods. The individual fluxes can be calculated given the concentration profiles; thus a complete kinetic description of the cell is achieved.

When one works with a new system for the first time, standard practice is to begin investigating the kinetics via dynamic phase plane diagrams of the system. Since changes in the metabolite dynamics depend not only on time, but also on other metabolites, phase plane diagrams are useful to visualize the regulation and transient changes in metabolite concentrations, as well as enzyme activities, over a prescribed time interval. Phase plane diagrams are plots of two metabolite concentrations, or reaction fluxes, parameterized with respect to time *(19)*. Therefore, the trajectories in the diagram move chronologically.

For a given set of initial conditions, a dynamically stable system will have only one orbit (the trajectories of the system from time equals zero to the last time point specified) that approaches fixed points. A complete orbit of a system for two metabolites or fluxes is the *path* that is plotted over time from the initial condition to the fixed point. During the course of the orbit, the path may assume a series of different trajectories, which are characterized by the shape of the curve (i.e., linear, parabolic, and so on) and the slope (i.e., positive, negative, zero). Changes in trajectories of the curves are indications of the regulatory effects on metabolites. The qualitative characteristics of phase planes can provide insight into the inner workings of the system. In particular, one should take note of the slopes of the different trajectories of phase planes. There are a plethora of texts and resources that discuss phase plane diagrams in greater detail *(13)*.

With regards to **Eq. (1)**, as the cell reaches a steady state, changes in metabolite concentrations, $d\mathbf{X}/dt$, approach zero. Practically speaking, however, most cellular systems approach the steady state within the order of 100 h (much shorter for bacteria). If one were only interested in the changes that occur in the steady state (which is reasonable in the analysis of biological systems; see discussion of FBA for justifications), then **Eq. (1)** can be simplified to

$$\mathbf{S} \cdot \mathbf{v} \, (x_1, x_2, \ldots, x_m) = 0 \tag{11}$$

This eliminates the need to integrate the equations since the time derivatives equal zero. The **v** vector in **Eq. (11)** is composed of the kinetic rate laws of the enzymes. Although the time dependence has been eliminated, the interdependence of the metabolites still exists. Thus the simplification from **Eq. (1)** to **Eq. (11)** involves moving from a nonlinear set of ordinary differential equations to a nonlinear set of algebraic equations. Solving for the roots of the above equations results in the steady-state concentrations of the metabolites, from which the fluxes can be calculated, as discussed above.

3.3.1. Assessment Using SNP Data

Normal physiological and pathological states can be simulated by altering the appropriate parameters. For example, changing the initial conditions of the metabolites can represent the changes in the availability of the metabolites. One can also introduce new fluxes or alter existing ones in order to simulate changes in the demands on a cell. For the human red blood cell, this involved introducing an energy load flux, which "consumes" ATP and an oxidative load flux, which decreases the redox potential of the cell. Similarly, one can quantitatively simulate the effects of causal SNP mutations on the activity of the resultant enzyme. For example, given an enzyme described by Michaelis-Menten kinetics, if the altered V_{max} and Michaelis constants are known, then the rate expression can be modified accordingly.

We illustrate the present method using pyruvate kinase (PK) deficiency in red blood cells as an example. PK is a key glycolytic regulatory enzyme, and it accounts for 90% of the enzyme deficiencies found in red blood cell glycolysis. Approximately 400 SNP mutations have been associated with PK deficiency *(14,15)*. The rate expression for PK formulated by Holzhutter et al. accounts for the inhibitory effect of ATP as well as the activation of PEP and FDP *(16)*,

$$v_{PK} = \frac{V_{max}^{PK}}{N_{PK}} \left(\frac{[MgADP]}{K_{ADP}^{PK} + [MgADP]} \right) \left(\frac{[PEP]}{K_{PEP}^{PK} + [PEP]} \right) \quad (12)$$

$$N_{PK} = 1 + L \frac{(1 + [ATP]/K_{ATP}^{PK})^4}{(1 + [PEP]/K_{PEP}^{PK})^4 (1 + [FDP]/K_{FDP}^{PK})^4} \quad (13)$$

Data from the literature are available for kinetic parameters of people with normal and deficient PK activities *(14)*. The altered V_{max} and Michaelis binding constants (K_{PEP}) were entered into the rate expressions representing the altered maximum efficiency and altered substrate-binding properties. Since different laboratories often report different values for the same parameter, one may need to scale the new values with respect to the "normal" value as measured by each respective laboratory group in order to keep subsequent simulations internally consistent.

Another benefit of kinetic models is that they can be used to quantitatively describe changes in multiple parameters, such as the binding constants. Conclusions can then be drawn about the effect of the SNP by comparing the SNP cell with the wild-type cell and available clinical and experimental data *(17)*. The greater the availability of quantitative experimental data, the more rigorously the model can be tested and the more reliable its predictions will be. Further *in silico* investigations of systemic consequences of causal SNPs can be pursued via dimensionless biologically meaningful parameters *(18)*.

3.4. Flux Balance Analysis (FBA)

Although kinetic models can provide detailed information about a biological system, they general require and are sensitive to a large amount of quantitative data and experiments. Even when such data are available, systems with more than tens of differential equations approach computation infeasibility. In addition, metabolism usually involves fast reactions and high turnover of substances such that most processes generally reach steady state at an observable time scale. In such cases, FBA, which is much more scalable to large networks, can be applied to study the steady state of a system. Further discussion on the classical kinetic and FBA modeling approach can be found in a recent review *(29)*.

As mentioned earlier, mass balance equations can be written for each metabolite by taking the dot product **S** and a flux vector **v**. The vector **v** contains flux values for all reactions in the network. Applying the steady-state assumption, the time derivatives of metabolite concentrations are zero, and **Eq. (1)** can be simplified to **Eq. (11)**, $\mathbf{S} \cdot \mathbf{v} = 0$. It follows that in order for a flux vector **v** to satisfy this relationship, the rate of production must equal the rate of consumption for each metabolite. This mass balancing equation forms the key constraint in the flux balanced analysis. The flux distribution, **v**, in this equation can be solved by finding the null space of the matrix **S**. The null space of a matrix contains all the vectors whose product with that matrix equals zero. Constraints on the directionality of each reaction can be applied in the form of $\alpha \leq v_i \leq \beta$ as described in **Subheading 3.2.**

Generally these constraints are not sufficient to define a unique flux distribution in network; therefore, multiple mathematically valid vectors **v** can be found. FBA searches within the allowable solution space to identify solutions of interest. Linear optimization is often used to find a particular solution in the allowable steady-state flux space that maximizes a chosen objective function, such as cellular growth. A linear programming (LP) problem can be formulated as follows:

$$\text{Maximize} \quad Z = \mathbf{f}^T v$$
$$\text{Subjected to} \quad \mathbf{S} \cdot \mathbf{v} = 0 \qquad (14)$$
$$\alpha \leq v_i \leq \beta \text{ for all reactions } i$$

3.4.1. Assessment of Enzymopathies and Single-Nucleotide Polymorphism Effects

Outcomes of knockout experiments and enzymopathic conditions can be quickly predicted by constraining the flux of the corresponding reaction to either zero or some suboptimal values and solving the LP problem with new constraints. Therefore, given the SNPs and clinical data on the activities of the enzyme of interest, one can readily predict its phenotypic outcome using this method. We illustrate the analysis method also using PK as a case study below.

The stoichiometric matrix of the red blood cell and constraints on its enzymes are formed using the stoichiometry of metabolites in network reactions and constraints on each reaction rates, respectively. The objective function representing the energy load can be written as the maximization of ATP hydrolysis

$$ATP + H_2O \rightarrow ADP + P_i + H^+$$

The red blood cell's energy production capability can be evaluated by solving a linear programming problem given the above stoichiometric matrix and the ATP hydrolysis as the objective function. This value represents the maximal energy supply a wild-type cell can afford under demanding situations. This number can be used for later comparisons with mutants.

For each SNP variant with kinetic data for PK, one can translate its V_{max} value into an upper bound constraint on the corresponding enzyme and resolve the LP problem. The value of the objective function in the resulting problem represents the maximum ATP load this particular SNP variant can withstand. Comparisons with the maximal ATP production in the previous calculation and with the regular ATP maintenance load allow one to quantitatively evaluate the systemic effects each SNP has on the entire cell.

In summary, this chapter describes how one can integrate high-throughput data such as genomics and proteomics and sequence variation data into a biochemical and physical structure framework, within which systemic properties of a system can be studied. Moreover, as additional data become available, they can be readily incorporated to update the model and improve its predictions.

Appendix

For clarification purposes, an example of a five-flux, five-metabolite system is illustrated below.

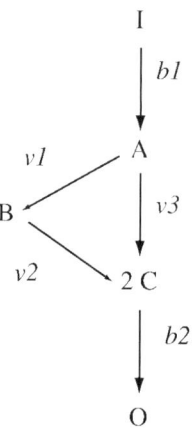

Kinetic Analysis

From the schematic illustration of the network, the time derivative for each metabolite can be written as follows:

$$\frac{dA}{dt} = b_1 - v_1 - v_3$$

$$\frac{dB}{dt} = v_1 - v_2 \qquad \text{(A1–A3)}$$

$$\frac{dC}{dt} = 2v_2 + 2v_3 - 2b_2$$

Information about the mechanism of each reaction is necessary to construct a kinetic model. To keep the example simple, we assume that the first two rate laws are linear with first-order kinetics: v_2 obeys simple Michaelis-Menten kinetics, whereas v_1 obeys similar kinetics, but with inhibition by a metabolite. The last rate law is assumed to follow the Bi-Bi reaction mechanism. Applying these assumptions, we can express the reaction rates in terms of metabolite concentrations as follows:

$$v_1 = \frac{k_{v1}\,[A]}{1 + [B]/K_B^{v1}} \qquad \text{(A4)}$$

Table 1
Parameter Value for "Wild-Type" Condition

Parameter	Value
k_{v1} (1/h)	500
K_B^{v1} (mM)	0.1
V_{max}^{v2} (mM/h)	8
K_B^{v2} (mM)	0.05
V_{max}^{v3} (mM/h)	5
K_A^{v3} (mM)	0.9
K_C^{v3} (mM)	0.5
k_{v1} (1/h)	0.9
k_{b2} (1/h)	10
I (mM)	10

$$v_2 = V_{max}^2 \frac{[B]}{K_B^{v2} + [B]} \qquad (A5)$$

$$v_{v3} = V_{max}^3 \left(\frac{[A]}{K_A^{v3} + [A]}\right)\left(\frac{[C]}{K_C^{v3} + [C]}\right) \qquad (A6)$$

$$b_1 = k_{b1}[I] \qquad (A7)$$

$$b_2 = k_{b2}[C] \qquad (A8)$$

In the above equations, constants abbreviated with K_X^v denote the Michaelis-Menten constants. These parameters are defined to be the substrate concentrations at which the corresponding reaction rates are at half their maximal value (V_{max}). The linear, first-order dependence of v_1 on metabolite A is represented by the numerator of **Eq. A4**. The denominator of this equation accounts for the inhibitory effect of metabolite B on v_1. The expression for v_2 follows the classic Michaelis-Menten kinetics based on assumption. The last rate expression, v_3, is described in terms of the ordered Bi-Bi reaction mechanism—low concentrations of metabolites A or C severely limit their activity. Expressions for b_1 and b_2 both follow first-order kinetics. **Table 1** specifies the values for the parameters in **Eqs. A4** to **A8**.

Equations A4 to A8 were substituted into **Eqs. A1** to **A3**. Next, **Eqs. A1** to **A3** were solved with Mathematica using the ordinary differential equation (ODE) solver for the phase planes and the polynomial equation root finder for the definitive, stable steady-state solutions. Initial concentrations of 10, 0.2, and 3 were used for metabolites A, B, and C; the ODEs were integrated for $t =$

Table 2
Calculated Steady-State Metabolite Concentrations and Fluxes for Wild Type and SNPs

	Wild type	SNP
A (mM)	0.489	3.783
B (mM)	3.004	23.595
C (mM)	0.9	0.9
b_1 (mM/h)	9.0	9.0
v_1 (mM/h)	7.869	7.983
v_2 (mM/h)	7.869	7.983
v_3 (mM/h)	1.131	1.017
b_2 (mM/h)	9.0	9.0

0 to 500 h. Phase plane diagrams for metabolites and fluxes are produced as described in **Subheading 3.3.** Steady-state metabolite concentrations and flux values are shown in **Table 2**.

To simulate an altered variant (SNP), V_{max}^{v3} was decreased from 5 to 2, and K_A^{v3} was increased from 0.9 to 1.0. Phase planes, as well as steady-state metabolite concentrations and fluxes, are shown in **Fig. 1**.

Note that since this is a very simple model with fixed inputs and outputs, dramatic variations are unlikely to be observed. Metabolite phase planes show that the regulatory trends are not changed (they have similar slopes), although the magnitude of the regulatory effects are affected. The wild-type and SNP variant phase planes show some differences in the competing motions as the system moves from the same initial condition to the steady state. The flux-phase planes nicely demonstrate the independent time scales on which they operated. Note that metabolite concentration differences at the steady state are much greater than the flux changes. The increased steady state concentrations of metabolites A and B simply reflect a "bottleneck" effect resulting from the decreased V_{max}^{v3}. Knowledge of metabolite concentrations and their changes under different conditions can be very useful when one is making experimental and clinical correlations *(17,19)*.

Flux Balance Analysis

If one is not interested in the immediate dynamic changes in the system and focuses only on the steady-state condition, the model does not require assumptions about the kinetics of each reaction nor extensive information for param-

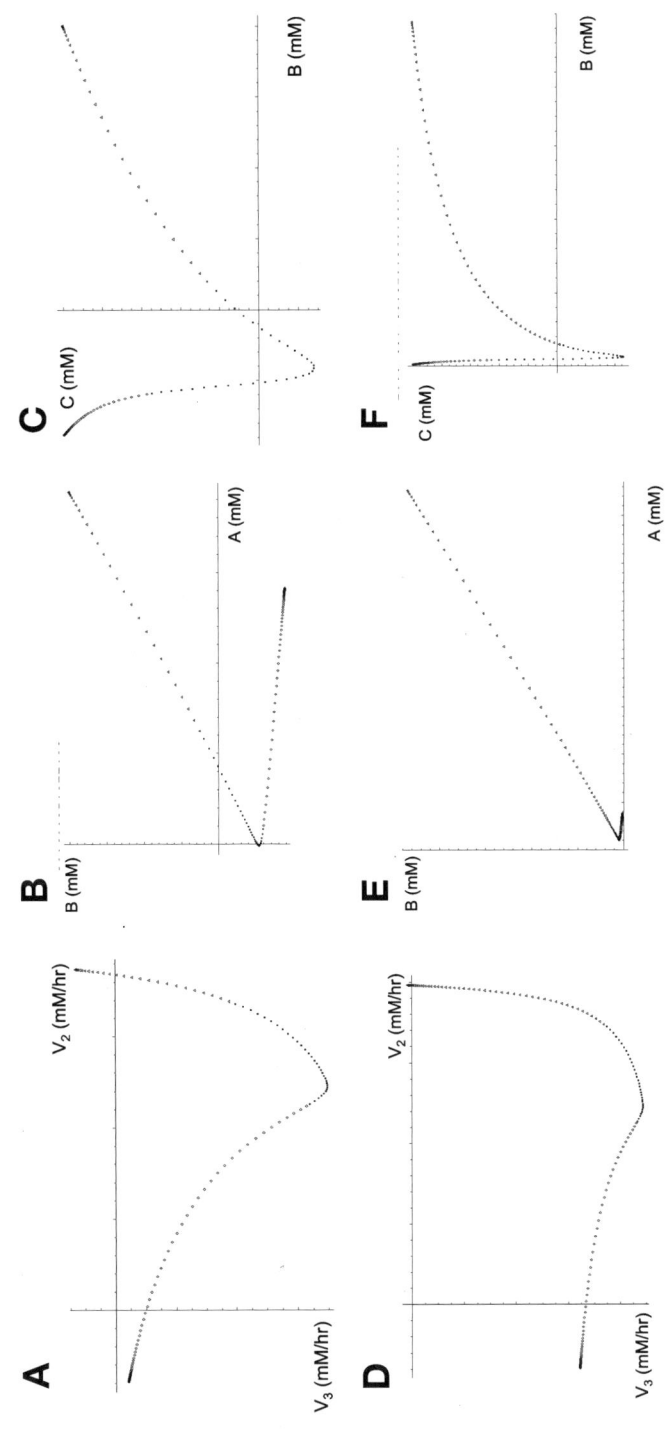

Fig. 1. Diamonds, 0 to 1 min; triangles, 1 min to 1 h; boxes, 1 to 500 h. Flux and metabolite phase planes for the wild-types (A–C) and SNP variants (E,F).

eter values. Consequently, the stoichiometric matrix can be written with all linear coefficients:

$$\frac{d\mathbf{X}}{dt} = \begin{bmatrix} -1 & 0 & -1 & 1 & 0 \\ 1 & -1 & 0 & 0 & 0 \\ 0 & 2 & 2 & 0 & -1 \end{bmatrix} \begin{bmatrix} v_1 \\ v_2 \\ v_3 \\ b_1 \\ b_2 \end{bmatrix} \quad \text{(B1)}$$

where the vector $\mathbf{X}$ contains the concentrations of metabolites A, B, and C, respectively. The coefficients in the matrix from row 1 to row 3 correspond to **Eqs. A1** to **A3**.

Under the steady-state assumption, the variations in metabolite concentrations, $d\mathbf{X}/dt$, can be set to zero. One can decide which steady-state flux distributions are of interest by selecting an appropriate objective function. The corresponding linear programming problem can be formulated as follows:

$$\text{Maximize} \quad Z = \mathbf{f}^T \mathbf{v} \quad \text{(B2)}$$

$$\text{Subjected to} \quad \mathbf{S} \cdot \mathbf{v} = 0 \quad \text{(B3)}$$

$$\alpha \leq v_i \leq \beta \text{ for all reactions i} \quad \text{(B4)}$$

Equation B2 represents the objective function. One can maximize or minimize for the output of a particular metabolite or combinations of metabolite as in the case of biomass. For example, to maximize for the production of metabolite C or the output flux b_1

$$\mathbf{f} = b_1$$

$$\mathbf{f}^T \mathbf{v} = [0\ 0\ 0\ 0\ 1] \cdot [v_1\ v_2\ v_3\ b_1\ b_2]^T$$

To maximize for the production of intermediate B

$$\mathbf{f} = v_1$$

$$\mathbf{f}^T \mathbf{v} = [1\ 0\ 0\ 0\ 0] \cdot [v_1\ v_2\ v_3\ b_1\ b_2]^T$$

Equations B2 and **B3** represent the constraints on the fluxes in the network. The values α and β specify in the lower bound (LB) and upper bound (UB) for each reaction flux. As all reactions in this network are irreversible, the LB vector can be set to zero. In general, reversible reactions can be decoupled into two irreversible reactions. To determine values for the UB vector, we can incorporate the V_{max} values given in **Table 2** or choose an arbitrary large number (for example, 100 in this case). Applying this information, we derive

$$\text{LB} = [0\ 0\ 0\ 0\ 0]$$

$$\text{UB\_wild type} = [100\ 8\ 5\ 10\ 100]$$

$$\text{UB\_SNP} = [100\ 8\ 2\ 10\ 100]$$

We now have all information necessary to solve the linear programming problem defined by **Eqs. B2** to **B4**. Using function **linprog( )** Matlab, we can solve for a steady-state flux distribution that maximizes for the output of C under wild-type and altered condition (SNP). The function linprog( ) in Matlab accepts six arguments and returns two values, x and Fval, which are the solution (in our case, the flux vector v) and the value of the objective function. Specifically,

[x,Fval] = LINPROG(f,A,b,Aeq,beq,LB,UB)

solves for the linear programming problem

```
min f'*x    subject to: A*x ≤ b
                        Aeq*x = beq
                        LB ≤ x ≤ UB
```

As we do not have any inequality constraints other than bounds for the flux vector, we set A to the identity matrix I, so that A*x ≤ b is equivalent to x ≤ UB.

```
>>S =
    [ -1   0  -1   1   0
       1  -1   0   0   0
       0   2   2   0  -1 ]
>>f = [0 0 0 0 1]'
>>b = [0 0 0]'
>>I = eye(5)
>>LB = [0 0 0 0 0]'
>>UB_wildtype = [100 8 5 10 100]'
>>UB_SNP = [100 8 2 10 100]'
>>[v,Z] = linprog(-f,I,UB,S,b,LB,UB)
Optimization terminated successfully.
v =
    7.6870
    7.6870
    2.3130
   10.0000
   20.0000
Z =
  -20.0000
>>[v_SNP,Z_SNP] = linprog(-f,I,UB,S,b,LB,UB_SNP)
Optimization terminated successfully.
v_SNP =
    8.0000
    8.0000
    2.0000
   10.0000
   20.0000
Z_SNP =
  -20.0000
```

In comparing results from the kinetic and FBA methods, we should note that

1. The formulation of a kinetic model is much more mathematically complex and requires the use of many parameters compared with the FBA method. The benefit of the increased complexity is the ability to describe changes, such as SNPs, with greater fidelity to nature. Additionally, one can analyze metabolic regulatory effects between metabolites and fluxes. Furthermore, kinetic models allow for the calculation of metabolite concentrations, which can offer great insight into biological mechanisms and pathophysiology *(17)*.
2. The FBA method is much more scalable and computationally feasible compared with the kinetic model. Application of this method has been successfully employed to study genome-scale networks of organisms such as *Escherichia coli (20)* and *Saccharomyces cerevisiae (21–23)*.

References

1. McCarthy, J. J. and Hilfiker, R. (2000) The use of single-nucleotide polymorphism maps in pharmacogenomics. *Nat. Biotechnol.* **18,** 505–508.
2. Kwok, P. Y. and Gu, Z. (1999) Single nucleotide polymorphism libraries: why and how are we building them? *Mol. Med. Today* **5,** 538–543.
3. Thorisson, G. A. and Stein, L. D. (2003) The SNP Consortium website: past, present and future. *Nucleic Acids Res.* **31,** 124–127.
4. Karp, P. D., Paley, S., and Romero, P. (2002) The Pathway Tools software. *Bioinformatics* **18 Suppl 1,** S225–232.
5. Sherry, S. T., Ward, M. H., Kholodov, M., et al. (2001) dbSNP: the NCBI database of genetic variation. *Nucleic Acids Res.* **29,** 308–311.
6. Smigielski, E. M., Sirotkin, K., Ward, M., and Sherry, S. T. (2000) dbSNP: a database of single nucleotide polymorphisms. *Nucleic Acids Res.* **28,** 352–355.
7. Reich, D., Gabriel, S., and Altshuler, D. (2003) Quality and completeness of SNP databases. *Nat. Genet.* **33,** 457–458.
8. Taylor, S. W., Fahy, E., Zhang, B., et al. (2003) Characterization of the human heart mitochondrial proteome. *Nat. Biotechnol.* **21,** 281–286.
9. Jiang, X. S., Zhou, H., Zhang, L., et al. (2004) A high-throughput approach for subcellular proteome: Identification of rat liver proteins using subcellular fractionation coupled with two-dimensional liquid chromatography tandem mass spectrometry and bioinformatic analysis. *Mol. Cell. Proteomics* **3,** 441–455.
10. Ge, H., Walhout, A. J., and Vidal, M. (2003) Integrating 'omic' information: a bridge between genomics and systems biology. *Trends Genet.* **19,** 551–560.
11. Covert, M. W., Schilling, C. H., Famili, I., et al. (2001) Metabolic modeling of microbial strains in silico. *Trends Biochem. Sci.* **26,** 179–186.
12. Dixon, M. and Webb, E. (1979) *Enzymes*, Longman Group Limited.
13. Hubbard, J. H. and West, B. H. (1991) *Differential Equations: A Dynamical Systems Approach*, Springer-Verlag, New York.
14. Zanella, A. and Bianchi, P. (2000) Red cell pyruvate kinase deficiency: from genetics to clinical manifestations. *Baillieres Best Pract. Res. Clin. Haematol.* **13,** 57–81.

15. Bianchi, P. and Zanella, A. (2000) Hematologically important mutations: red cell pyruvate kinase (third update). *Blood Cells Mol. Dis.* **26**, 47–53.
16. Holzhutter, H. G., Jacobasch, G., and Bisdorff, A. (1985) Mathematical modelling of metabolic pathways affected by an enzyme deficiency. A mathematical model of glycolysis in normal and pyruvate-kinase-deficient red blood cells. *Eur. J. Biochem.* **149**, 101–111.
17. Jamshidi, N., Wiback, S. J., and Palsson, B. B. (2002) In silico model-driven assessment of the effects of single nucleotide polymorphisms (SNPs) on human red blood cell metabolism. *Genome Res.* **12**, 1687–1692.
18. Reich, J. and Selkov, E. (1981) *Energy Metabolism of the Cell: A Theoretical Treatise.* Academic Press, New York.
19. Kauffman, K., Pajerowski, J., Jamshidi, N., Bo, B. P., and Edwards, J. (2002) Description and analysis of metabolic connectivity and dynamics in the human red blood cell. *Biophys. J.* **83**, 646–662.
20. Reed, J. L., Vo, T. D., Schilling, C. H., and Palsson, B. O. (2003) An expanded genome-scale model of *Escherichia coli* K-12 (iJR904 GSM/GPR). *Genome Biol.* **4**, R54.51–R54.12.
21. Forster, J., Famili, I., Fu, P. C., Palsson, B. O., and Nielsen, J. (2003) Genome-scale reconstruction of the *Saccharomyces cerevisiae* metabolic network. *Genome Res.* **13**, 244–253.
22. Forster, J., Famili, I., Palsson, B. O., and Nielsen, J. (2003) Large-scale evaluation of in silico gene knockouts in *Saccharomyces cerevisiae*. *Omics* **7**, 193–202.
23. Famili, I., Forster, J., Nielsen, J., and Palsson, B. O. (2003) *Saccharomyces cerevisiae* phenotypes can be predicted by using constraint-based analysis of a genome-scale reconstructed metabolic network. *Proc. Natl. Acad. Sci. USA* **100**, 13134–13139.
24. Lee, P. L. and Barton, J. C. (2006) Hemochromatosis and severe iron overload associated with compound heterozygosity for TFR2 R455Q and two novel mutations TFR2 R396X and G792R. *Acta Haematol.* **115**, 102–105.
25. Wada, M. (2006) Single nucleotide polymorphisms in ABCC2 and ABCB1 genes and their clinical impact in physiology and drug response. *Cancer Lett.* **234**, 40–50.
26. Neidhardt, F. C. and Curtiss, R. (1996) *Escherichia coli* and *Salmonella*: cellular and molecular biology. Washington, DC, ASM Press.
27. Reed, J. L., Famili, I., Thiele, I., and Palsson, B. O. (2006) Towards multidimensional genome annotation. *Nat. Rev. Genet.* **7**, 130–141.
28. Joshi, A. and Palsson, B. O. (1989) Metabolic dynamics in the human red cell. Part I—A comprehensive kinetic model. *J. Theor. Biol.* **141**, 515–528.
29. Vo, T. D. and Palsson, B. (2006) Building the power house: Recent advances in mitochondrial studies through proteomics and systems biology. *Am. J. Physiol. Cell Physiol.*

16

Discovery and Identification of Sequence Polymorphisms and Mutations with MALDI-TOF MS

Dirk van den Boom and Mathias Ehrich

Summary

Matrix-assisted laser desorption ionization (MALDI) time-of-flight (TOF) mass spectrometry (MS) has become a widely used technology for the detection of nucleic acids. In this chapter we introduce its use for the discovery of novel sequence polymorphisms and the identification of known DNA changes. We first provide a brief overview about MALDI-TOF MS analysis of nucleic acids. We then elucidate the concept of base-specific cleavage and its use for the discovery of sequence polymorphisms. We also introduce the use of primer extension assays for the classification of known genomic alterations. Finally, we provide a detailed protocol for the implementation of both methods for practical use in a high-throughput setting.

Key Words: MALDI-TOF MS; genotyping; SNP discovery; DNA mutations; DNA sequence polymorphisms; high throughput; mutation detection; MassEXTEND; MassCLEAVE.

1. Introduction

1.1. Overview

The invention of the first powerful DNA sequencing methods like Sanger sequencing *(1)* has allowed us to study the nucleotide sequence of entire genomes. Today several eukaryotic and prokaryotic organisms have been fully sequenced, and an almost complete sequence of the human genome is available. Knowledge of a genome sequence creates the opportunity to study genotype-phenotype correlations systematically and to elucidate the genetic pathways that contribute to the physiology of an organism. One major focus of current international research projects is to investigate and catalog genetic variation. The most abundant genetic markers, e.g., microsatellites and single-nucleotide polymorphisms (SNPs), have found widespread use in academic as well as commercial areas.

A variety of methods have been developed in the last two decades to investigate genetic variation. These include methods for *de novo* discovery of polymorphisms and mutations as well as the large-scale genotyping of known polymorphisms/mutations in various sample populations (as required in disease association studies and pharmacogenomics studies). Among the methods developed, mass spectrometry (MS) has left a mark as a versatile nucleic acid analysis technology that offers the highest analytical accuracy and throughput.

This contribution focuses on the use of matrix-assisted laser desorption ionization (MALDI) time-of-flight (TOF) MS for high-throughput discovery of sequence variants and genotyping.

1.2. Principle of MALDI-TOF MS

MALDI-TOF MS was developed in the late 1980s by Karas and Hillenkamp *(2)*. Both, MALDI-TOF MS and electrospray ionization (ESI) MS *(3)*, for the first time enabled the mass spectrometric analysis of large biomolecules such as proteins and nucleic acids and are nowadays cornerstones of routine proteomics and genomics research.

With MALDI-TOF MS, the analysis of proteins and nucleic acids is accomplished by embedding the analyte in a crystalline structure of organic molecules, referred to as a matrix. This matrix later serves as the "launching" material for mass spectrometric analysis. The matrix-analyte co-crystal is volatilized with laser bursts of 337 nm wavelength. The matrix type and laser wavelength are aligned such that during this volatilization the matrix molecules absorb the laser energy. During the last decade the matrix of choice for nucleic acid analysis has been 3-hydroxy-picolinic acid. Introduction of energy into the crystal structure leads to a microexplosion, which generates a particle cloud. Analyte molecules are desorbed into the gas phase along with the matrix molecules. Because the matrix absorbs the energy, analyte molecules remain intact and can be analyzed as intact molecules. The volatilization process is accompanied by gas-phase proton transfer reactions, which generate analyte and matrix ions. An electric field of approximately 20 kV is used to accelerate the ions to nearly uniform kinetic energy. The ions then travel through a field-free drift region (usually 1 m length) and separate by their mass-to-charge ratio. The ions finally reach a detector, and their impact allows the measurement of their TOF. This TOF is directly proportional to the mass-to-charge ratio. Because the process generates predominantly singly charged ions, measured mass signals directly represent the analyte mass. The molecular mass of the analyte is an intrinsic molecular property, which allows direct and highly accurate characterization of the underlying nucleic acid reaction product

1.3. Assay Formats for MALDI-TOF MS DNA Sequence Analysis

1.3.1. Discovery of Sequence Polymorphisms and Mutations

Two main features drove the hope in the early phases of the Human Genome Project that MALDI-TOF MS could be among the new technologies revolutionizing DNA sequence analysis and genome sequencing: the speed of signal acquisition and the inherent accuracy of MS. In these original concepts, MALDI-TOF MS was proposed as a separator and detector for Sanger sequencing ladders (4). It was thought that MS could replace the cumbersome and time-consuming gel or capillary electrophoresis. In this concept, the mass difference between adjacent mass signals representing the nested dideoxy-terminated sequencing products provides the base-sequence readout.

Despite tremendous progress during the last decade, MALDI-TOF MS analysis of Sanger sequencing reaction could never be established as a standard method. To a large degree, this was caused by the very short read length. Sequencing reads up to 100 bp have been reported, but the average sequence length on a routine basis settled in the range of 25 bases (5–7). This "short" read length originates from some of the fundamental limitations in MALDI-TOF MS of nucleic acids. The desorption/ionization efficiency decreases with increasing length of the nucleic acid products. Longer DNA molecules fragment much more strongly in MALDI, and thus the sensitivity decreases nearly exponentially.

In recent years, new biochemical concepts have been applied to circumvent some of the existing issues in measuring larger DNA products with MALDI-TOF MS. They use base-specific cleavage of amplification products as a means to analyze the amplified sequence for potential sequence changes (8–12). Complete base-specific cleavage generates a set of short oligonucleotides from the amplification product, in which the distance between the cleavage sites determines the length of the oligonucleotides. Except for microsatellite repeats, the products thus most often fall into a mass range preferable for current MALDI-TOF MS. This alleviates the major issues limiting the read length and allows sequence determination by MS.

Conceptually, these approaches resemble earlier methods for RNA sequencing and also methods for protein identification by peptide mapping.

The identification of sequence changes using base-specific cleavage requires a different analysis approach than Sanger sequencing. In its current format, base-specific cleavage is used as a resequencing method and is not applied to *de novo* sequencing. Thus, we assume that a reference sequence is available for each target region. This reference sequence can be used to generate *in silico* cleavage and mass signal patterns. If a sample carries a sequence change in the amplified target region, then this sequence change will have an impact on the cleavage pattern: the sequence change can introduce a new cleavage site leading to two shorter cleavage products; it can remove a cleavage site, leading to a longer cleavage product; if no cleavage-base is affected, it can shift the mass of an existing cleavage product higher or lower. Hence, a comparison of experimental mass signal patterns with *in silico* patterns can be employed to identify additional and missing mass signals, which can then be used to interpret the sequence change discovered. This last interpretation step is based on compositional analysis of the identified mass signals. Each cleavage product can consist of a permutation of the three noncleavage nucleotides and a single nucleotide residue of the cleavage nucleotide. It is thus fairly straightforward to determine for each additional mass signal which combination of A, C, G, and T calculate to the measured molecular mass.

Once the composition is identified, an algorithm based on the reference sequence can determine which sequence change can account for the observed cleavage product. Usually this requires the integration of multiple base-specific cleavage reactions. For further reading, refer to the original articles describing the concept *(12–14)*.

Several methods have been developed to obtain base-specific cleavage *(8–10,15,16)*. Among them are enzymatic as well as chemical means of cleaving DNA and RNA. The most prominent methods use the following process. The target region is amplified by conventional polymerase chain reaction (PCR) using primers tagged with a T7 promoter sequence. This allows subsequent generation of a single-stranded RNA transcript from either the forward or reverse direction. The RNA transcript can then be cleaved to completion with a base-specific RNAse, such as RNAse T1, which yields, for example, G-specific cleavage *(10,11,17)*. A variation of this concept uses a mutant T7 RNA polymerase capable of incorporating dNTPs *(12)*. The selective incorporation of a dNTP in the RNA transcript allows the use of less specific RNAses. RNAse A, for example, cleaves at every C and U residue of a transcript. This would degrade the transcript into too small cleavage products, and most of the information would be lost. However, if rCTP is fully replaced by dCTP during the transcription, RNAse A is rendered specific for U (T) cleavage. Similarly, the use of dTTP during transcription renders RNAse A C-specific. The combination of C- and U-specific cleavage from forward and reverse directions in four separate cleavage reactions allows cleavage after virtually all four bases. This provides the most comprehensive scan of a target region for potential polymorphisms or mutations.

The process has been applied successfully to target regions ranging in size from 150 bp up to 1 kB. The detection rates depend on amplicon length and sequence context. In a 500-bp amplicon, on average 95% of all single-base sequence changes can be detected when solely heterozygous sequence changes are considered, and about 85% of these can be mapped unambiguously to a nucleotide position in the amplicon. For haploid organisms or for detection of homozygous sequence changes, the detection rates increase to 99% of all possible single-base changes. (Here the combination of additional and missing mass signals can be used.)

Figure 1 depicts a schema of the current SEQUENOM®, Inc. base-specific cleavage process, called MassCLEAVE™, as implemented on the MassARRAY® system. The process is homogeneous and does not require any intermediate purification steps. The reagents for each reaction step are simply added to the well of a microtiter plate. After completion of the cleavage reaction, ion-exchange resin is added to each well in order to remove salts from the phosphate backbone of the nucleic acid cleavage products. These are then transferred onto a miniaturized chip array and analyzed by MALDI-TOF MS. Spectra interpretation and identification of polymorphisms are performed by algorithms as described in **ref. *14***.

1.3.2. High-Throughput Genotyping of Known Polymorphisms and Mutations

Several methods have been developed in the past decade to analyze known SNPs and mutations by MALDI-TOF MS. The most prominent method, the MassEXTEND assay, relies on post-PCR primer extension to generate allele-specific reaction products.

As part of the process, the target region harboring the SNP or mutation is first amplified by PCR. Usually these PCR products are relatively small and range between 70 and 100 bp in length. This allows efficient amplification even from highly degraded nucleic acid samples such as paraffin-embedded tissue as a source of genomic DNA or RNA.

Following PCR, a primer is annealed immediately adjacent to the nucleotide position of interest. This primer is extended using a DNA polymerase and a defined mixture of deoxy- and dideoxynucleotides generating allele-specifically terminated extension products. The allele-specific extension products differ in length and nucleotide composition and can hence be unambiguously identified in a subsequent MALDI-TOF MS analysis. **Figure 2** depicts a schematic overview of the MassEXTEND assay. Other approaches for MALDI-TOF MS-based genotyping include the GOOD assays *(18–20)* and the Invader assay *(21)*.

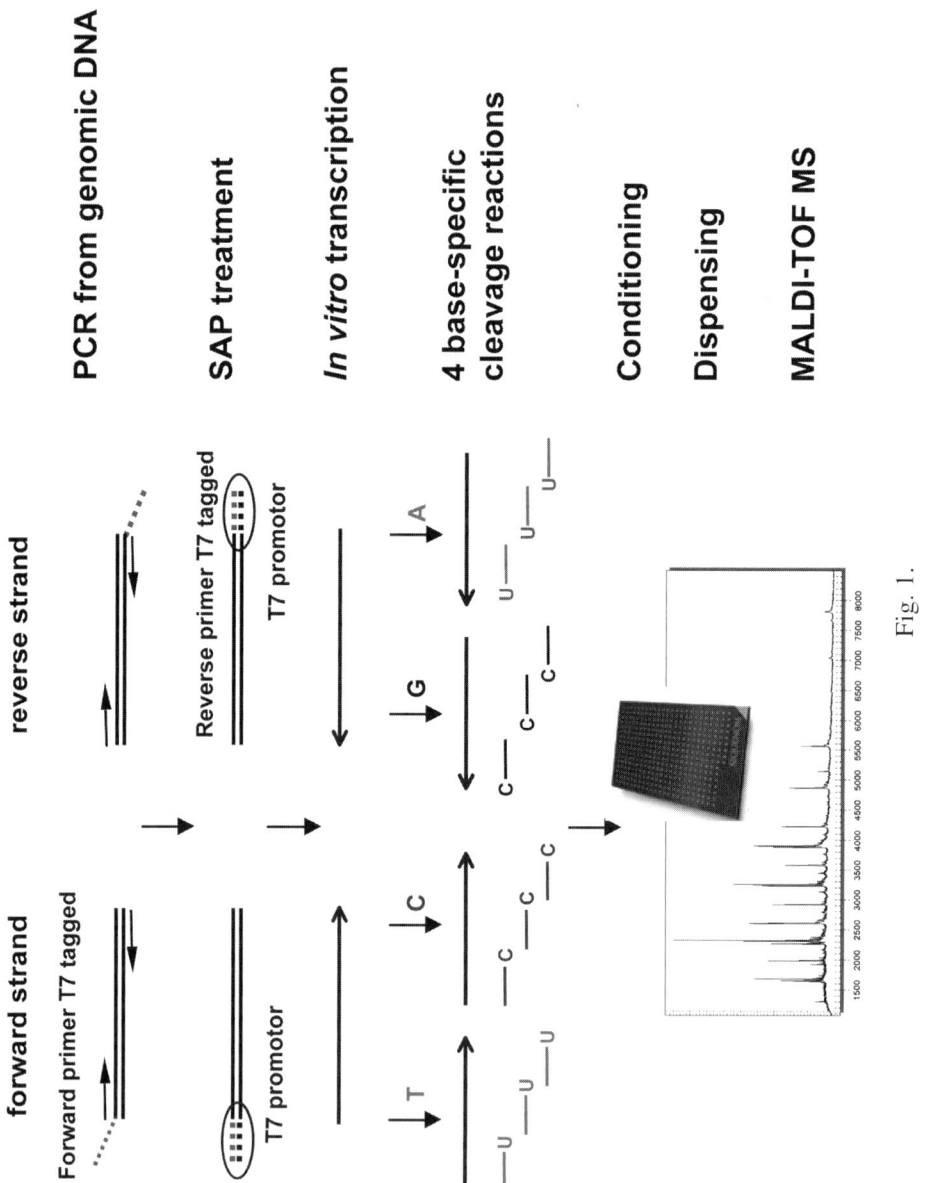

Fig. 1.

Two main aspects render the MassEXTEND assay the method of choice for higher throughput genotyping. First, it is simple to automate, because the assay can be performed in a homogeneous format: for each subsequent reaction step, the reagent cocktail is simply added to each reaction vessel. Following PCR, unincorporated nucleotides are degraded by addition of shrimp alkaline phosphatase. Then the primer extension reaction cocktail is added. After cycling, the products are conditioned for mass spectrometric analysis by simple addition of ammonium ion-loaded cation-exchange beads. After the conditioning, the reaction products are transferred onto a miniaturized SpectroCHIP® array for automated MALDI-TOF MS readout.

Second, the assay can be multiplexed to interrogate multiple SNPs or mutations in a single reaction. For multiplexing, the following aspects should be considered: PCR primers should be selected for optimal length (20mer), GC content (50%), and melting temperature (T_m 60°C). Primers should be mapped against the human genome (if applicable) to avoid amplification from multiple sites. Furthermore, primers generating hairpins or cross-hybridization should be avoided. MassEXTEND primers should be selected under similar considerations. Additionally, primer lengths should be chosen such that each primer and extension product has a unique molecular mass. Products should be mass-spaced by at least 50 Daltons to allow unambiguous identification under high-throughput conditions even when salt adducts, such as sodium adducts (+22 Daltons) are present. Automated assay design tools for design of optimal uniplexed and multiplexed MassEXTEND reactions are commercially available through SEQUENOM. (For further reading, see **refs. 22** and **23**).

With a sufficient set of SNPs or mutations, up to 15-plexed MassEXTEND reactions can be performed with generic conditions. Further assay optimization may allow for even higher plexing levels. A 384-element chip array can nowadays be analyzed on the MassARRAY system within 1 h. Thus, at 15-fold multiplexing, around 100 genotypes can be measured per minute (or 5760 genotypes/h). With this throughput and reaction costs of around 10 cents per genotype, large-scale analysis of SNPs and mutations becomes feasible.

Fig. 1. *(previous page)* Schematic overview of the MassCLEAVE™ process for discovery of sequence polymorphisms. The target region is amplified by PCR using PCR primers tagged with a T7 promoter sequence. After PCR, unincorporated nucleotides are inactivated by shrimp alkaline phosphatase (SAP) treatment. The PCR product is transcribed into an RNA transcript in four separate reactions, two from the forward direction and two from the reverse direction. Each transcription reaction is performed with a special nucleotide mix, which leads to either C- or U-specific cleavage upon addition of RNAse A. The cleavage products are conditioned by addition of ion-exchange resin and then transferred onto a SpectroCHIP array for automated MassARRAY system analysis.

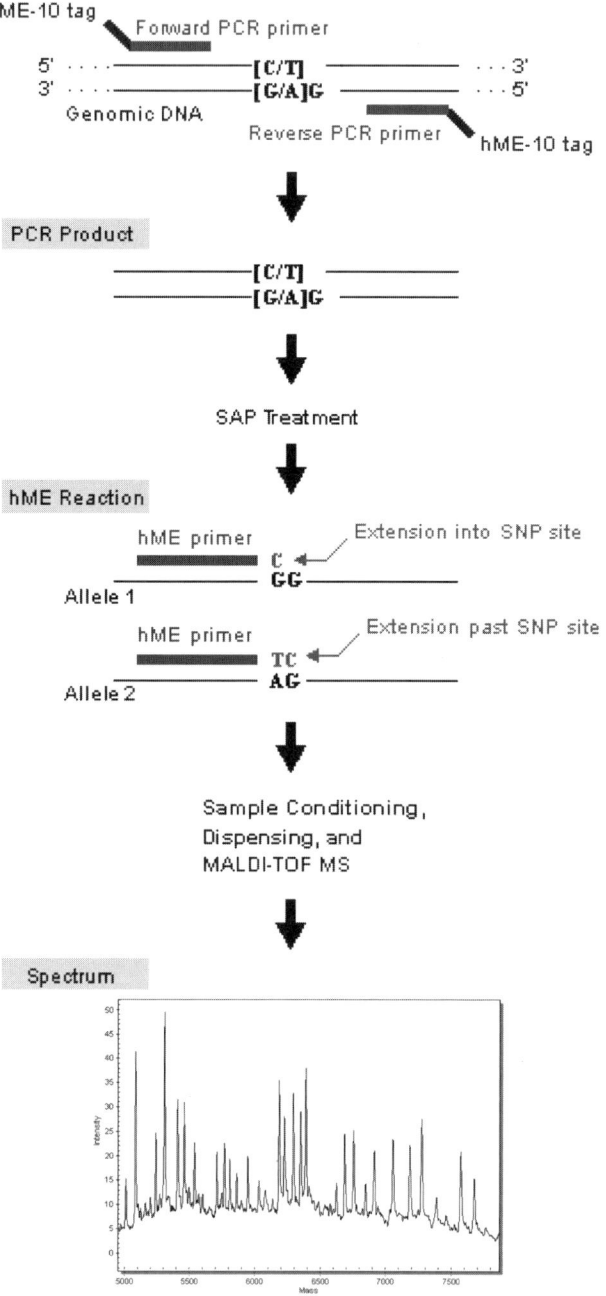

Fig. 2.

2. Materials

This section gives the materials required to perform base-specific cleavage reactions for the discovery of sequence polymorphisms by MALDI-TOF MS. For details on the materials required to perform primer extension-based MALDI-TOF MS high-throughput genotyping, refer to Storm et al. *(23)*.

2.1. Materials Common to PCR and hMC

The following instruments and components are used to process PCR and hMC reactions:

1. Multimek 96 Automated 96-channel pipetor (Beckmann Coulter, Fullerton, CA; also available through SEQUENOM as MassARRAY Liquid Handler). Used with 20-µL tips, also from Beckman Coulter (cat. no. 717254).
2. Thermal Cycler: either GeneAmp® PCR System 9700 (Applied Biosystems, Foster City, CA); or PTC-225 DNA Engine Tetrad Cycler (MJ Research, Watertown, MA).
3. Rotator capable of holding microplates (e.g., Fisher Scientific, Pittsburgh, PA, model 346).
4. MassARRAY Nanodispenser (pintool instrument for nanoliter dispensing onto SpectroCHIP arrays), available through Sequenom, Inc.
5. MALDI-TOF MS instruments (3K or 7K system available through SEQUENOM) in combination with MassARRAY Typer RT and Discovery RT software.

2.2. PCR-Specific Materials

1. 384-Well microplates (Marsh Biomedical Products, Rochester, NY, cat. no. TF-0384).
2. Autoclaved nanopure water (18.2 MΩ resistance).
3. Forward and reverse PCR primers: including promoter tags, the average length is 45 bases. Primers should be ordered desalted, resuspended in water, and stored at –20°C.
4. Ultrapure dNTP set (Amersham Pharmacia Biotech, Piscataway, NJ). Store at –20°C.
5. HotStarTaq DNA polymerase, $MgCl_2$, and buffer (Qiagen, Valencia, CA). Store at –20°C.
6. Genomic DNA (5 ng/µL). Store at 4°C.

Fig. 2. *(previous page)* Schematic overview of the MassEXTEND process for high-throughput genotyping of polymorphisms and mutations. The target region is amplified by PCR. Following PCR, unincorporated nucleotides are removed by addition of shrimp alkaline phosphatase (SAP). In a subsequent reaction, a primer is annealed immediately adjacent to the polymorphic site and extended with a selected mix of d/ddNTPs and a DNA polymerase. The extension reaction generates allele-specifically terminated products, which are conditioned by ion-exchange resin, transferred onto a SpectroCHIP array, and automatically analyzed by MassARRAY (in the depicted case a mass spectrum of a multiplexed MassEXTEND reaction).

2.3. HMC-Specific Materials

1. Autoclaved nanopure water (18.2 MΩ resistance).
2. Shrimp alkaline phosphatase (SEQUENOM). Store at –20°C.
3. T7 R&DNA polymerase (Epicentre or SEQUENOM).
4. T-cleavage mix, C-cleavage mix (SEQUENOM). Store at –20°C.
5. RNAse A (SEQUENOM). Store at –20°C.
6. CLEAN resin for sample desalting prior to mass spectrometric analysis (SEQUENOM). Stored at room temperature (RT).
7. 384-Element silicon chip (SpectroCHIP, SEQUENOM).

3. Methods

The MassCLEAVE assay for discovery of sequence variations consists of the following main steps. The target region of interest is first amplified using PCR and PCR primer carrying a promoter tag. The PCR product is then treated with shrimp alkaline phosphatase (SAP) to remove unincorporated nucleotides. Subsequently the PCR product is transcribed into a single-stranded RNA transcript (from either the forward or reverse reaction). The RNA transcript is then cleaved base-specifically with RNAse A. The cleavage products are conditioned for MS by addition of ion-exchange resin. The products are then transferred onto a miniaturized chip array and analyzed by MALDI-TOF MS. Finally, the information encoded in the four mass signal patterns is deconvoluted with advanced software algorithms, which provide a list of identified single-base sequence changes *(14)*. A flowchart is depicted in **Fig. 1**. The following sections provide experimental details of each process step (*see* also **Notes 1** and **2**).

3.1. Assay Setup

The current assay format relies on the use of a single type of RNA polymerase (T7 RNA polymerase). As depicted in **Fig. 1**, to perform four base-specific cleavages, the method currently requires two PCR reactions. For each of the PCR reactions, a different pair of primers is required. The two pairs of primers share the same gene-specific sequence but differ in the direction of the attached T7 promoter sequence. The first primer set contains the T7 promoter tag on the 5' end of the forward primer. In the second set the T7 promoter tag is attached to the 5' end of the reverse primer. Thus, one PCR set allows for generation of forward RNA transcript; the second PCR set allows for generation of reverse RNA transcript. For each set there is a 10-bp tag attached to the remaining primer that minimizes differences in length of the forward and the reverse primer.

Figure 3 illustrates the building blocks of primers designed to allow for forward and reverse transcription after PCR. PCR products that will subsequently be used in forward transcription reactions require a T7-tagged, gene-specific primer with an 8-bp insert as a forward primer and a gene-specific reverse primer with a 10-bp tag.

SNP Discovery by MALDI-TOF MS

1. PCR products that will subsequently be used as templates for reverse transcription reactions require a gene-specific, T7-tagged reverse PCR primer with an 8-bp insert and a gene-specific forward primer with a 10-bp tag. This allows for forward and reverse transcription reactions following PCR.
   T7 promoter tag: 5'-CAGTAATACGACTCACTATAGGG-3'.
2. The 8-bp insert is placed between the T7-promoter sequence and the gene-specific portion of the primer, which prevents excessive abortive cycling during transcription start. It also generates a constant 5'-cleavage product during base-specific cleavage by RNase A.
   8-bp tag: 5'-AGAAGGCT-3'.
3. The 10-bp tag attached to gene-specific primers provides for more uniform amplification during PCR.
   10-bp tag for opposite primer: 5'-AGGAAGAGAG-3'.
4. In general, gene-specific primers can be designed using conventional primer design software (e.g., http://www.basic.nwu.edu/biotools/Primer3.html).
5. After the gene-specific parts are designed, add the T7-promoter tag, including an 8-bp insert and the 10mer tag, to the gene-specific part of the forward and reverse primer.
6. The PCR amplicons should be 300 to 700 bp, and the recommended melting temperature is 62°C.
7. Avoid T-stretches in amplicons, if possible. Place T-stretches as close as possible to the 3' end of the amplicon, and place A-stretches as close as possible to the 5' end of the amplicon (*see* **Note 3**).

3.2. Assay Protocol

3.2.1. PCR Amplification

Amplify 1 μL DNA (5 ng) in a 5 μL PCR volume using a 384-microtiter format. (Use 1.00 μL of 5 ng/μL DNA.)

1. Prepare a PCR cocktail as described in the following table:

Reagent	Final conc.	Volume for single reaction (μL)
Nanopure water	NA	1.44
HotStar Taq PCR buffer containing 15 mM MgCl$_2$ (10X)	1X	0.5
Fresh dNTPs (25 mM each)[a]	200 μM	0.04
Hot Star *Taq* (5 U/μL; Qiagen)	0.1 U/reaction	0.02
Forward PCR primer (100 μM)[b]	200 nM	1
Reverse PCR primer (100 μM)[c]	200 nM	1
Genomic DNA (5 ng/μL)	5 ng/rxn	1
Total volume		5

[a]Maximum of 5 freeze/thaws.
[b]Containing T7 promoter tag and 8-bp insert on the 5' end.
[c]Containing 10mer tag.

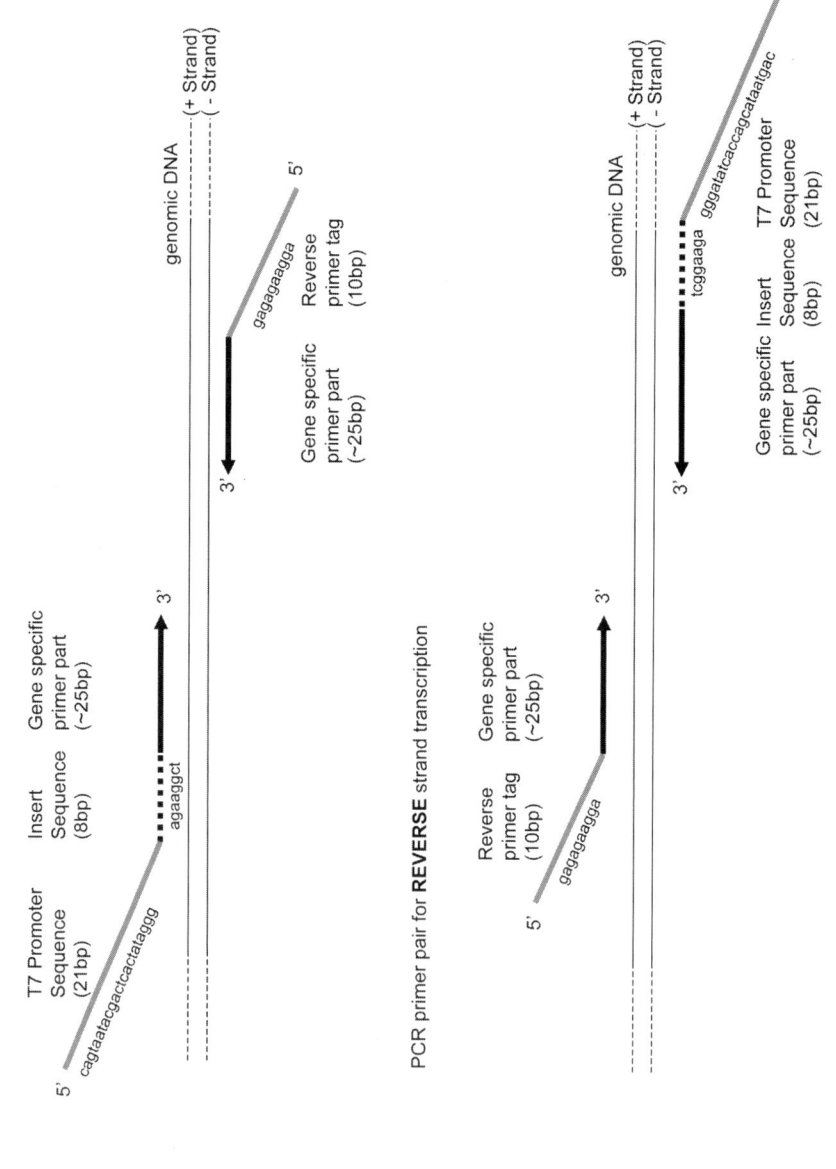

Fig. 3.

2. Seal the plates and cycle in a standard thermal cycler as follows:

Cycle	Condition
1	95°C for 15 min
45	94°C for 20 s
45	62°C for 30 s
45	72°C for 1 min
1	72°C for 3 min
1	4°C on hold

3. Dephosphorylation. Add 2 µL (0.3 U) of SAP enzyme to each 5 µL PCR to dephosphorylate unincorporated dNTPs from the PCR. Incubate the plates for 20 min at 37°C. Then incubate at 85°C for 5 min.

Reagent	Volume for single reaction (µL)
Nanopure H$_2$O	1.7
SAP 1U/µL	0.3
Total volume	2

Cycle	Condition
1	37°C for 20 min
1	85°C for 5 min
1	4°C on hold

Fig. 3. *(previous page)* Primer design by MassCLEAVE. Schematic description of the two primer pairs needed to allow all four-base-specific cleavage reactions. The first set of primers carries the T7 promoter tag (CAGTAATACGAC TCACTATAGGG) on the 5' end of the forward primer, enabling the binding of the T7 R&DNA polymerase to the amplification product. The T7 promoter tag is followed by an 8-bp insert (AGAAGGCT) that prevents the incorporation of deoxynucleotides during the initial and fragile phase of transcription, in which the polymerase has not yet formed the stable elongation complex. This reduces abortive cycling and increases the transcription yield. Finally, the forward primer carries the gene-specific portion at the 3' end. The reverse primer is tagged with a 10mer tag (GAGAGAAGGA) on the 5' end, which helps to remove nonspecific elongation products from the analyzed mass range. The second primer set contains identical gene-specific portions, but the tags are switched to the opposite primer. Now the reverse primer carries the T7 tag and 8-bp insert, whereas the forward primer is tagged with the artificial 10mer sequence.

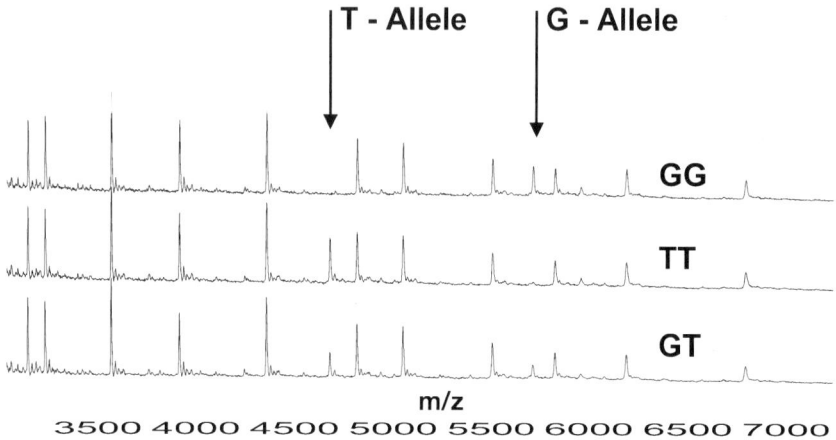

Fig. 4. Exemplified results for MassCLEAVE. Section of mass spectra from the T-specific forward cleavage reaction of a 451-bp amplicon from chromosome 22 with a G/T polymorphism at position 341. (Forward PCR primer: CAGTAATAC GACTCACTATAGGGAGAAGGCT GATGTACACCACTCCCTGCC; reverse PCR primer: GAGAGAAGGAG GTGACTTGCAGAGGAGGAC). The mass signal at 5712.5 Daltons represents a cleavage product (5'-...GAGGGGGAAGCAA**G**GGT-3') derived from the wildtype sequence, and the mass signal at 4676.9 Daltons is based on a cleavage product (5'-...GAGGGGGAAGCAA**T**-3') that results from a mutation G to T at position 341. The depicted mass spectra demonstrate the shift in signal intensities according to the underlying genotype. The mass spectrum at the top is derived from a wildtype GG sample and shows only the signal at 5712 Daltons; the spectrum in the middle is derived from a homozygote TT sample and shows only the signal at 4677 Daltons; the spectrum at the bottom is derived from a heterozygous GT sample and correspondingly displays both mass signals.

3.2.2. Post-PCR Processing

Each PCR will be split into two transcription and cleavage reactions: a T-specific cleavage reaction and a C-specific cleavage reaction. For each cleavage reaction a separate transcription cocktail has to be prepared.

3.2.2.1. IN VITRO TRANSCRIPTION

For each transcription reaction, 2 µL of transcription cocktail and 2 µL per liter of PCR/SAP sample are used. For a standard setup, prepare one transcription cocktail per plate. Per well of a microtiter plate, add 2 µL of transcription cocktail and 2 µL of PCR/SAP sample into a new, uncycled microtiter plate. Incubate the plates at 37°C for 2 h.

1. Prepare transcription mix as described in the following table.
2. Dispense 2 µL transcription cocktail into each well of a 384-well plate.
3. Step add 2 µL PCR/SAP mix to each well.
4. Seal plates and incubate for 2 h at 37°C.

T-cleavage reaction[a]	Final conc.	Volume for single reaction (µL)	Volume for a 384-well microtiter plate (µL)
Nanopure H$_2$O	NA	0.38	215.8
T7 Polymerase buffer (5X)	1X	0.80	454.4
T-cleavage mix	NA	0.22	125.0
DTT (100 mM)	5 mM	0.20	113.6
T7 R&DNA polymerase	20 U/reaction	0.40	227.2
Total volume		2.00	1136.0

[a]Add the reagents in the order in which they appear in the table. Includes 48% overhang.

C-cleavage reaction[a]	Final conc.	Volume for single reaction (µL)	Volume for a 384-well microtiter plate (µL)
Nanopure H$_2$O	NA	0.38	215.8
T7 Polymerase buffer (5X)	1X	0.80	454.4
C-cleavage mix	NA	0.22	125.0
DTT (100 mM)	5 mM	0.20	113.6
T7 R&DNA polymerase	20 U/reaction	0.40	227.2
Total volume		2.00	1136.0

[a]Add the reagents in the order in which they appear in the table. Includes 48% overhang.

3.2.2.2. RNase A Cleavage

After transcription cocktails (C-cleavage and T-cleavage) and PCR/SAP mix have been incubated, the next step is addition of RNase A for base-specific cleavage. RNAse A cleaves at every C and U. With the use of the respective cleavage mix containing either deoxyTTP or deoxyCTP, RNAse A is forced to cleave either at every U (T-cleavage mix) or every C (C-cleavage mix). The transcription mixes are labeled to correspond to the cleavage specificity they generate.

1. Prepare RNAse A cocktail as described in the following table.
2. Add 2.5 µL RNase A cocktail to each reaction (T-cleavage and C-cleavage).
3. Seal plates and incubate at 37°C for 1 h.

Reagent	Final conc.	Volume for single reaction (µL)	Volume for a 384-well microtiter plate (µL)	Volume for two 384-well microtiter plates (µL)
Nanopure H$_2$O	NA	2.45	1391.6	2783.2
RNAse A	0.08 mg/mL	0.05	28.4	56.6
Total volume		2.5	1420.0	2840.0

3.2.3. Sample Cleanup

After the hMC reaction has been processed (see **Subheading 3.2.2.2.**), the next step is to condition the base-specific cleavage reaction products with Clean Resin. This conditioning step is important to optimize mass spectrometric analysis.

1. Add 20 µL of ddH$_2$O to each sample within the 384-well plate.
2. Add 6 mg of Clean Resin to each well using the resin plate.
3. Rotate for 10 min and spin down for 5 min at 3200*g*.

Note: To avoid pH shifts and undesired side products, always add ddH$_2$O first before adding the Clean Resin.

3.2.4. Sample Transfer on Chip Array

For automated, high-performance analysis of nucleic acid mixtures by MALDI-TOF MS, the sample needs to be transferred from the microtiter plate format to a chip array. The SpectroCHIP array (SEQUENOM) provides miniaturized prefabricated arrays of matrix sample spots on a silicon chip. The miniaturization improves homogeneity of the sample and leads to increased performance in qualitative and quantitative analysis. Although an ample volume of sample is generated during the MassCLEAVE process, only 15 nL are required for the subsequent analysis. The corresponding volume of analyte is transferred onto the chip array. This is best performed using a piezoelectric pump-based dispensing system able to transfer low sample volumes or a pintool system as provided by SEQUENOM (Nanodispenser device). The robotic system transfers analyte from 384-well microtiter plates onto 384-SpectroCHIP arrays in 9 min. The chip array can be used immediately for MALDI-TOF MS analysis.

3.2.5. MALDI-TOF MS Analysis

Analysis of chip-transferred samples proceeds in a linear, delayed-extraction TOF mass spectrometer. Mass spectra are acquired in positive ion mode. (All positively charged molecular ions are accelerated.) The chips are introduced into the ion source, and high-vacuum conditions are applied. Image processing aligns the laser position automatically to the chip element raster for fully automated scanning of each chip position. Each matrix crystal is addressed individually and irradiated with a 337-nm laser pulse of 1 ns duration. The time-resolved mass spectrum is then translated into a mass spectrum by comparison with known calibrants. Usually 15 single laser shots are accumulated and averaged into a single spectrum. This average spectrum is then further processed and analyzed using dedicated software (MassARRAY SNP Discovery, SEQUENOM) that performs baseline correction, peak identification, and quality assessments. Spectrum data quality is analyzed in real time during data acquisition. In case of insufficient mass spectrum quality, the software will automatically reacquire new data points from the same chip position before it finally moves to the next chip position.

Reliable detection of sequence polymorphisms requires acquisition of all four-cleavage spectra. Once all data are acquired, they are saved to an Oracle database. The data can be assessed using client analysis software. The software compiles all four-cleavage spectra for one sample and generates a so-called composite spectrum. The composite spectrum allows calculation of differences compared with the reference sequence and, thus, discovery and identification of sequence polymorphisms *(14)*. A typical result for a MassCLEAVE-based discovery of a reference polymorphism is depicted in Fig. 4.

4. Notes

1. *Validating SNPs.* When MassARRAY SNP Discovery software is used for analysis, all detected sequence polymorphisms are listed automatically. However, sometimes the software does not find enough evidence to make an unambiguous call. Here it allows the user to interact and validate the correctness of a call. When you validate the correctness of a detected SNP, it is recommended to follow general guidelines:
    a. Explore the affected signals for each reaction and determine which signals represent the wildtype sequence (later referred to as "wildtype" signals) and which are caused by the mutated sequence (later referred to as "mutant" signals).
    b. Select a signal pair of wildtype and mutant signals from a reaction in which the sequence polymorphism does not affect the cleavage base. Also, each signal within this pair should be standing alone and not overlap with other signals that are independent of the polymorphic event. It is recommended that all signals be located in the optimal mass range between 2000 and 6500 Daltons.
    c. The selected "wildtype" and "mutant" signals build a corresponding signal pair. This signal pair can be compared across multiple samples. When "real" sequence polymorphisms are present, a shift in signal intensities can be observed. A decrease in intensity for one signal goes along with an increase of intensity for the corresponding signal. Hence homozygote wildtype samples will not show any "mutant" signals and vice versa. A heterozygote sample

will show both "wildtype" and "mutant" samples but with half the intensity of the surrounding signals.
    d. (Optional.) When the sample size is large enough and the SNP frequency is appropriate, the presence of Homozygote wildtype, Homozygote Mutant, and Heterozygote samples according to the Hardy Weinberg equilibrium will aid in the confirmation of a "real" SNP event.
2. *Spectra Quality*.
    a. T-rich fragments. Additionally, peak intensities of fragments with 50% or more Ts (for example, TTATGT or TTATT) may also be reduced. This is caused by weaker desorption/ionization in MALDI-TOF MS analysis.
    b. Unexplained signals. If a "mutant" signal appears in every sample for a given reaction (it may also appear in the negative control), the signal may be an artificial signal, which does not support a valid SNP. Double-check any unexplained signals with results of the negative control.
    c. Weak spectra. Low yield or weak spectra may indicate a problem during transcription or PCR. It is recommended that you perform gel checks. You may also want to verify the PCR primer design. Adjust PCR conditions to increase spectra yield.
    d. Failed matching. If your spectrum looks good but you get Failed Matching on the **Spectra Quality** tab, make sure you have the correct combination of reference sequence and cleavage reaction assigned to the SpectroCHIP.
3. *PCR Design*. During PCR design consider that T-stretches (multiple Ts in a row) may result in premature termination of the transcription. Subsequently, all fragments following the T-rich fragment might be missed. Whenever possible, try to exclude T-stretches or position multiple T-stretches toward the 3' end and multiple A-stretches toward the 5' end of the amplification region

    In rare cases the addition of sequence tags to the 5' and 3' end of the gene-specific primer part may cause unfavorable secondary structures leading to formation of Primer-Dimers, Crossdimers, or Hairpins. It is recommended that you check all primer pairs after the addition of necessary tags with appropriate software.

References

1. Sanger, F., Nicklen, S., and Coulson, A. R. (1977) DNA sequencing with chain-terminating inhibitors. *Proc. Natl. Acad. Sci. USA* **74,** 5463–5467.
2. Karas, M. and Hillenkamp, F. (1988) Laser desorption ionization of proteins with molecular masses exceeding 10000 Daltons. *Anal. Chem.* **60,** 2299–2301.
3. Fenn, J. B., Mann, M., Meng, C. K., Wong, S. F., and Whitehouse, C. M. (1989) Electrospray ionization for mass spectrometry of large biomolecules. *Science* **246,** 246.
4. Smith, L. M. (1993) The future of DNA sequencing. *Science* **262,** 530–532.
5. Koster, H., Tang, K., Fu, D. J., et al. (1996) A strategy for rapid and efficient DNA sequencing by mass spectrometry. *Nat. Biotechnol.* **14,** 1123–1128.

6. Nordhoff, E., Luebbert, C., Thiele, G., Heiser, V., and Lehrach, H. (2000) Rapid determination of short DNA sequences by the use of MALDI-MS. *Nucleic Acids Res.* **28,** E86.
7. Taranenko, N. I., Allman, S. L., Golovlev, V. V., Taranenko, N. V., Isola, N. R., and Chen, C. H. (1998) Sequencing DNA using mass spectrometry for ladder detection. *Nucleic Acids Res.* **26,** 2488–2490.
8. Hahner, S., Ludemann, H. C., Kirpekar, F., et al. (1997) Matrix-assisted laser desorption/ionization mass spectrometry (MALDI) of endonuclease digests of RNA. *Nucleic Acids Res.* **25,** 1957–1964.
9. Elso, C., Toohey, B., Reid, G. E., Poetter, K., Simpson, R. J., and Foote, S. J. (2002) Mutation detection using mass spectrometric separation of tiny oligonucleotide fragments. *Genome Res.* **12,** 1428–1433.
10. Hartmer, R., Storm, N., Boecker, S., et al. (2003) RNase T1 mediated base-specific cleavage and MALDI-TOF MS for high-throughput comparative sequence analysis. *Nucleic Acids Res.* **31,** e47.
11. Krebs, S., Medugorac, I., Seichter, D., and Forster, M. (2003) RNaseCut: a MALDI mass spectrometry-based method for SNP discovery. *Nucleic Acids Res.* **31,** e37.
12. Stanssens, P., Zabeau, M., Meersseman, G., et al. (2004) High-throughput MALDI-TOF discovery of genomic sequence polymorphisms. *Genome Res.* **14,** 126–133.
13. Pomerantz, S. C., Kowalak, J. A., and McCloskey, J. A. (1993) Determination of oligonucleotide composition from mass spectrometrically measured molecular weight. *J. Am. Soc. Mass Spectrom.* **4,** 204–209.
14. Bocker, S. (2003) SNP and mutation discovery using base-specific cleavage and MALDI-TOF mass spectrometry. *Bioinformatics* **19 Suppl 1,** I44–I53.
15. von Wintzingerode, F., Bocker, S., Schlotelburg, C., et al. (2002) Base-specific fragmentation of amplified 16S rRNA genes analyzed by mass spectrometry: a tool for rapid bacterial identification. *Proc. Natl. Acad. Sci. USA* **99,** 7039–7044.
16. Shchepinov, M. S., Denissenko, M. F., Smylie, K. J., et al. (2001) Matrix-induced fragmentation of P3'-N5' phosphoramidate-containing DNA: high-throughput MALDI-TOF analysis of genomic sequence polymorphisms. *Nucleic Acids Res.* **29,** 3864–3872.
17. Hahner, S., Schneider, A., Ingendoh, A., and Mosner, J. (2000) Analysis of short tandem repeat polymorphisms by electrospray ion trap mass spectrometry. *Nucleic Acids Res.* **28,** 18.
18. Sauer, S., Lechner, D., Berlin, K., et al. (2000) A novel procedure for efficient genotyping of single nucleotide polymorphisms. *Nucleic Acids Res.* **28,** E13.
19. Sauer, S., Gelfand, D. H., Boussicault, F., Bauer, K., Reichert, F., and Gut, I. G. (2002) Facile method for automated genotyping of single nucleotide polymorphisms by mass spectrometry. *Nucleic Acids Res.* **30,** e22.
20. Sauer, S. and Gut, I. G. (2003) Extension of the GOOD assay for genotyping single nucleotide polymorphisms by matrix-assisted laser desorption/ionization mass spectrometry. *Rapid Commun. Mass Spectrom.* **17,** 1265–1272.

21. Griffin, T. J., Hall, J. G., Prudent, J. R., and Smith, L. M. (1999) Direct genetic analysis by matrix-assisted laser desorption/ionization mass spectrometry. *Proc. Natl. Acad. Sci. USA* **96,** 6301–6306.
22. Tang, K., Oeth, P., Kammerer, S., et al. (2004) Mining disease susceptibility genes through SNP analyses and expression profiling using MALDI-TOF mass spectrometry. *J. Proteome Res.* **3,** 218–227.
23. Storm, N., Darnhofer-Patel, B., van den Boom, D., and Rodi, C. P. (2003) MALDI-TOF mass spectrometry-based SNP genotyping. *Methods Mol. Biol.* **212,** 241–262.

VI

GENE OVEREXPRESSION AND TARGETING IN THE MYOCARDIUM

17

Conditional Targeting

Inducible Deletion by Cre Recombinase

Kelly R. O'Neal and Ramtin Agah

Summary

Tissue specific gene knockouts using Cre recombinase can have broad applicability in murine disease models of cardiovascular disease. The Cre system has been shown to have broad experimental versatility for both temporal and spatial control of gene deletion. By and large this is achieved by first generating mice with an inducible tissue specific promoter for expression of Cre. These mice can then be crossed with a second line of mice where the gene of interest in 'knocked in' flanked by Cre recognition sequences Lox-P sites. The double transgenic lines are then induced, through administration of an exogenous agent, to allow tissue specific, i.e. cardiac, knockout of the gene of interest at the desired time. An experimental protocol delineating this technique is described in the chapter.

Key Words: Cre Recombinase; conditional knockout; murine models.

1. Introduction

Murine gene knockouts have made significant contributions to our understanding of the biological function of many genes. However, many gene knockouts are embryonically lethal and, therefore, do not permit a definitive loss of function study in an adult context *(1–4)*. This limitation of traditional knockouts led to a search for alternative techniques that would allow the consequences of gene inactivation to be studied in a regulatable or tissue-restricted manner. Inducible systems in gene targeting have been developed in response to this need. The numerous inducible systems are covered fully in a subsequent chapter. This chapter will focus on a method of inducible deletion using Cre recombinase.

Cre is a P1 bacteriophage recombinase commonly used in conditional targeting. It was first confirmed as a viable tool for use in manipulating the mouse genome by Lakso et al. *(5)*. They found that it was possible to define the

A
Mechanism of Cre Deletion

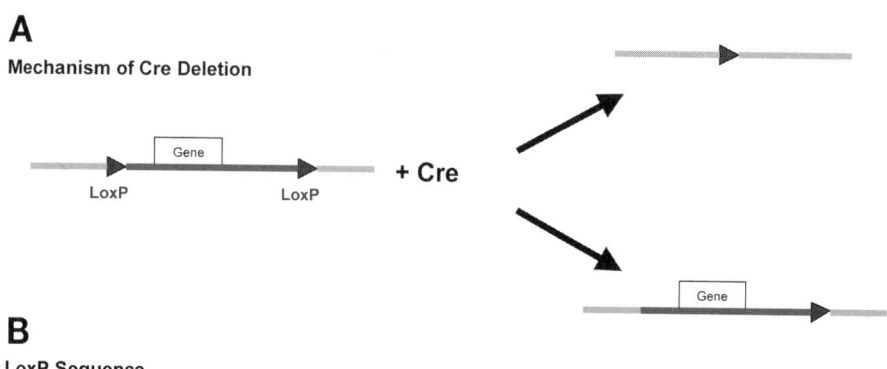

B
LoxP Sequence

5' ATAACTTCGTATAATGTATGCTATACGAAGTTAT 3'

Fig. 1. (**A**) Mechanism of Cre deletion. Cre recognizes the LoxP sites and excises the gene of interest through recombination. (**B**) LoxP sequence.

specific site of mutation by flanking the gene of interest with Cre recognition sequences, LoxP sites. A gene flanked by LoxP sites is said to be "floxed." These LoxP sites are palindromic sequences that Cre utilizes to excise the genetic code between these sites (**Fig. 1**).

FLP is another recombinase often used in conjunction with Cre. Similar to Cre, it also recognizes specific palindromic sequences known as FRT sites *(6,7)*. It is often used in the intermediate steps of transgenesis to remove a selection marker such as the neomycin resistance gene (neo^r) from the targeted allele *(8)*. This is useful because neo^r has been shown to decrease nontargeted gene expression over time if left in the transgenic genome *(9)*.

A fundamental characteristic of the Cre/LoxP system is that it has experimental versatility for both temporal and spatial control of gene deletion. By and large, this is accomplished by controlling the site and timing of the expression of Cre recombinase. Spatial control is accomplished with the use of tissue-specific promoters. The first cardiac-specific Cre-mediated deletion was generated in vivo utilizing an α-myosin heavy chain (αMHC) promoter *(10)*. Temporal control is maintained in Cre systems by adding a ligand binding domain in the promoter site to confer the ability to control Cre in a temporal fashion. Examples of such ligands include interferon *(11)*, progesterone *(12)*, tetracyline *(13)*, and tamoxifen *(14–16)*. Tamoxifen-inducible Cre (MerCreMer) is one of the more common models of inducible Cre-mediated gene deletion in use today *(8,15–19)*. A useful model in cardiovascular research that utilizes tamoxifen-inducible Cre is the αMHC/MerCreMer mouse generated by Sohal et al. *(16)*.

Conditional Targeting

The purpose of this chapter is to describe a general protocol for inducing cardiac-specific gene deletions using the Cre/loxP system. As previous chapters focused on the generation of a cardiac-specific promoter in an inducible system, i.e., as needed for a Cre-expressing mouse, this chapter will focus on the method for creating a gene flanked by loxP sites, a "floxed" transgene. The mouse with the floxed gene can then be crossed with the mouse line that carries the cardiac-specific Cre expression to produce cardiac tissue-restricted gene deletion. For simplicity, we will use a tamoxifen-inducible Cre mouse in our examples.

2. Materials

2.1. Cells/DNA

1. C57BL6 genomic DNA (*see* **Note 1**).
2. Embryonic stem (ES) cells/media (provided by a transgenic lab).
3. Cloning plasmid(s) (e.g., pMC-lox Neo lox, Specialty Media, Phillipsburg, NJ).
4. Competent *E. coli* (e.g., DH5α, Invitrogen).

2.2. Mice

1. FLPe deleter mouse (human β-actin FLPe deleter strain, stock no. 003800, The Jackson Laboratory).
2. Inducible, cardiac-expressing Cre mouse.

2.3. Reagents

1. Standard polymerase chain reaction (PCR) reagents/primers.
2. Standard Southern blotting reagents/probe.
3. Standard Western blotting reagents.
4. Anti-Cre antibody (cat. no. BIOT-106L, Covance, Berkeley, CA).
5. Restriction enzymes.
6. Standard reagents for gel electrophoresis.
7. Tamoxifen (Sigma; if using a tamoxifen-inducible Cre mouse).

3. Methods

The methods described below outline (1) designing a LoxP/Frt targeting vector with a PCR/Southern blot screening strategy, (2) generating a floxed/Frt transgenic mouse, (3) removing the neo resistance gene, (4) introducing the inducible, cardiac-specific Cre construct into floxed transgenic, (5) inducing Cre-mediated gene deletion, (6) analyzing the phenotype, and (7) the timeline for generating an inducible Cre transgenic mouse.

Important: Although we will describe the general steps in generating a floxed construct, this entire process, from the planning stages to the phenotype analysis, must be performed in coordination with a transgenic mouse lab.

3.1. Design of the LoxP/Frt Vector

3.1.1. General Design

The initial step in designing a targeting vector is to obtain information about your gene of interest including location, intron/exon sequences, functional domains, and restriction sites. C57BL/6J sequencing data are available from a variety of Internet databases: Ensembl, EntrezGene, Mouse Genome Informatics, and NCBI Mouse Genome Resource. After obtaining sequence information, decide which exon(s) to remove. The best strategy is usually to remove an earlier exon(s) because this results in a frameshift for the remaining exons and produces many stop codons, which lead to a functional gene deletion.

Numerous commercial plasmids designed for Cre-targeted cloning are available. For example, one that contains a neomycin selection marker bound by LoxP sites is pMC-lox Neo lox (Specialty Media, Phillipsburg, NJ). However, one plasmid will not work for every targeted gene deletion. The selection of a plasmid backbone will depend on multiple factors, including gene structure, restriction sites, selection markers, screening strategy, and individual cloning preferences. Multiple plasmids are often utilized to create the targeting vector. For cloning methods, refer to a standard molecular biology text *(20)*.

The general strategy in this protocol utilizes both Flp and Cre recombinase to create an inducible deletion. The strategy is illustrated in **Fig. 2**. The neomycin resistance gene (Neo) serves as a positive selection marker, and the HSV-tk gene (HSV-tk) serves as a negative selection marker. Because Neo has been shown to decrease protein expression in transgenic constructs *(9)*, Flp serves as a secondary recombinase to remove the selection cassette (*see* **Note 2**).

3.1.2. Building the Targeting Vector

Build the targeting vector with standard molecular biology techniques *(20)*. The subtleties of vector design and screening will vary according to the structure of the gene of interest, the plasmid(s) used for the backbone, the chosen screening strategy, and the preferences of the participating transgenic lab. However, there are several recommendations in this FLP/CRE strategy:

1. Insert the positive selection marker, the neo resistance gene, into an intron within the target deletion site.
2. Position the negative selection marker (HSV-tk) outside your targeted homologous recombination site at one end of your construct.
3. Flank the neo with Frt sites at the other end of your construct.
4. Flank the target deletion sequence and the Frt-flanked neo gene with LoxP sites.
5. Use restriction enzymes that cut outside your construct insertion sites in order to screen for recombination (*see* **Note 3**).

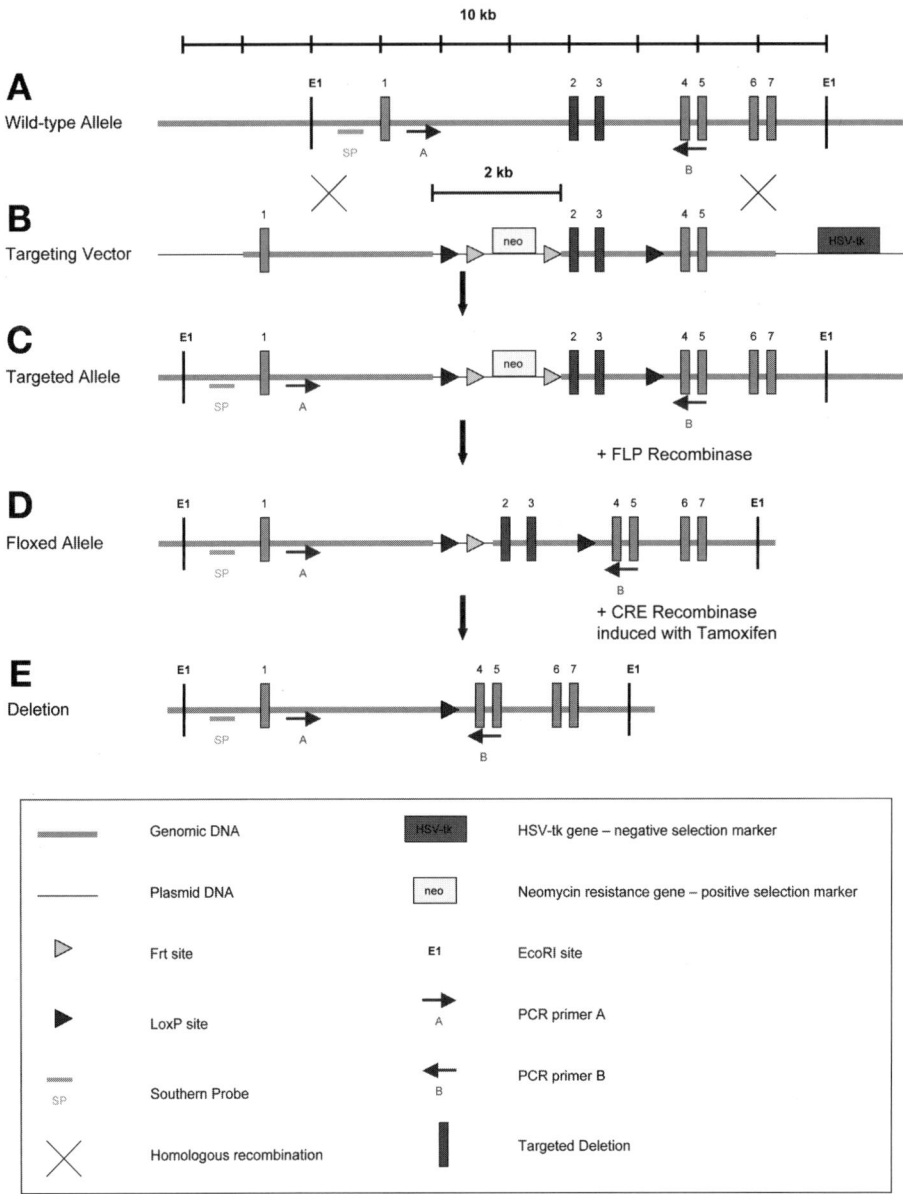

Fig. 2. Example of inducible deletion using Cre recombinase. FLP is used as a secondary recombinase to remove the neomycin selection marker.

Consult a standard molecular biology text for cloning techniques *(20)*. Confirm the accuracy of the construct with sequencing (*see* **Note 4**).

3.1.3. Purifying and Linearizing the DNA Construct and Sending to the Transgenic Lab for Electroporation

Particular transgenic labs have preferences for DNA purification prior to electroporation. Some mouse cores will even request that they purify the DNA themselves. However, a general protocol is as follows:

1. Grow a large bacterial culture (>250 mL) transformed with your targeting vector.
2. Isolate the DNA (Maxiprep Kit, Qiagen).
3. Linearize the construct, and excise the targeting vector from the bacterial plasmid with restriction enzymes according to standard molecular biology methods *(20)*.
4. Confirm digestion by running on an agarose gel.
5. Purify the construct (e.g., with the Gel Extraction Kit, Qiagen), and quantify the DNA.
6. Suspend the DNA in TE at a concentration of 1 µg/µL.
7. Send at least 100 µg of DNA.
8. Contact the transgenic core for shipping requirements and send the construct for electroporation.

3.2. Knocking in the Gene of Interest for Generation of the Appropriate Transgenic Mice

3.2.1. Screening ES Cell DNA for Homologous Recombination

The isolated DNA from individual ES cell colonies will be sent back to you for analysis with the PCR and Southern strategies designed during vector construction. Perform PCR and Southern blotting with standard molecular biology methods *(20)*. Screen to determine which wild-type alleles have undergone homologous recombination with your targeting vector, i.e., which colonies contain your targeted allele.

A screening strategy for identifying homologous recombinants by DNA analysis is fundamental to the design of your targeting vector. The genotyping protocol should be well established prior to sending your vector for electroporation into ES cells. Most transgenic labs require proof of a working screening assay prior to submission of the construct DNA.

PCR is often used as an initial screen, but most strategies also incorporate Southern blotting for confirmation of recombination. The advantage of PCR is that it is easier to perform but is more susceptible to DNA contamination and provides less information. Southern blotting provides additional information about structure, integrity, copy number of inserted vector segments, and mosaicism.

PCR primers as well as the Southern probe should be positioned to detect size changes in your allele after a recombination event. Design the primers and

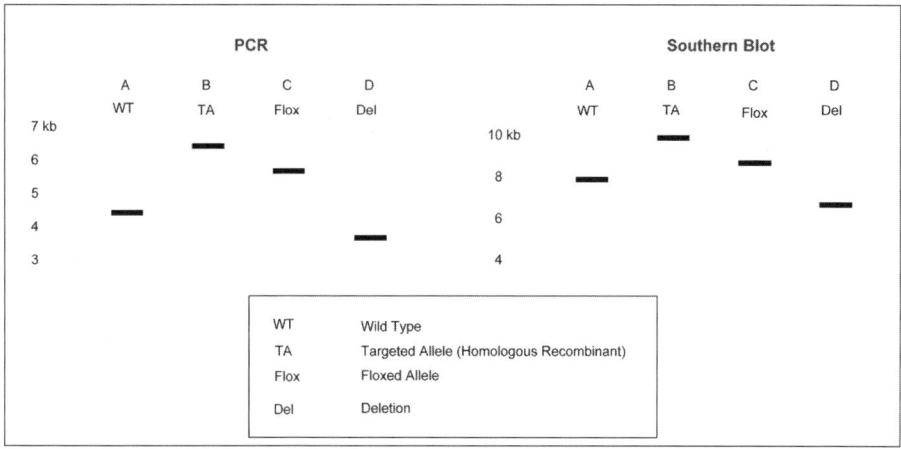

Fig. 3. Expected PCR and Southern blotting results (from theoretical example in **Fig. 2**).

probe using standard molecular biology methods *(20)*. Multiple strategies exist, but a simple rule is that the Southern probe as well as one of the PCR primers should be positioned outside the flanking regions. This is illustrated in **Fig. 2**. The expected PCR products and Southern blot from the example are depicted in **Fig. 3**.

3.2.2. Generation of Transgenic Mouse with the Targeted (Floxed/Frt) Allele

After identification of the targeted homologous recombinants, colonies will then be further cultured in the transgenic lab and injected into a blastocyst to produce the transgenic mouse. This process will result in the birth of a transgenic mouse containing your floxed/Frt allele. Screen the offspring by DNA analysis using the PCR/Southern blotting strategy previously developed (*see* **Subheading 3.2.**) Hemizygotes can be distinguished from homozygotes with a flanking Southern probe that reveals two products in hemizygous mice versus a single band in homozygotes *(21)*. PCR using two sets of primers can be designed to reveal one distinct band for homozygotes, another distinct band for wild-type mice, and two bands for hemizygotes *(22)*.

3.3. Removal of the neo Resistance Gene

Because of the potential toxicity of the neomycin resistance gene, FLP recombinase is utilized to remove this selection marker. This step is efficiently executed by breeding the floxed/FRT transgenic mouse with an FLP transgenic mouse. In general terms, this mouse is said to serve the role of a FLP "deleter."

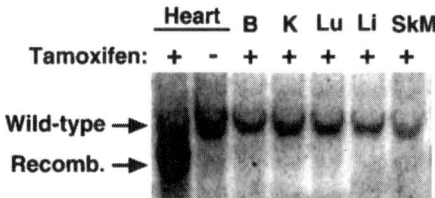

Fig. 4. Southern blot of cardiac-specific gene deletion pre/post Cre induction with tamoxifen. (From **ref. 16**.)

A specific mouse commonly used for this purpose is the FLPe mouse originally generated by Susan Dymecki *(8,23–25)*. This mouse is now commercially available through The Jackson Laboratory (human β-actin FLPe deleter strain). Screen the offspring of this cross for evidence of FRT-mediated recombination and the absence of the neo resistance gene with the PCR/Southern blotting strategy previously developed (*see* **Subheading 3.2.**). PCR and Southern blotting should be performed according to standard molecular methods *(20)*. These mice will contain the floxed allele without the neo insertion (**Fig. 2C**).

3.4. Introduction of the Inducible, Cardiac-Specific Cre Construct

The floxed allele in these mice is expressed in its endogenous fashion. To introduce cardiac specificity into your system, breed the floxed transgenic mouse with a Cre transgenic mouse, for example, a tamoxifen-inducible, αMHC Cre mouse (αMHC/MerCreMer) *(8,16)*. Genotype as described in **Subheadings 3.2.** and **3.3.** Confirm Cre transgene expression with Western blotting of cardiac tissue from offspring using a Cre-specific antibody (Covance, Berkeley, CA) *(16)*.

3.5. Induction of Cre Recombination to Produce Gene Deletion

Induce Cre with daily injections of tamoxifen (20 mg/kg i.p. for 4 to 6 d; dissolved in 60% ethanol at a concentration of 5 mg/mL). This protocol has been shown to induce >70% recombination in the heart *(16)*.

3.6. Analysis of the Phenotype

Confirm cardiac-specific gene deletion with PCR and Southern blotting pre/post Cre induction using DNA from multiple tissues including the heart *(10,11,15,16)*. The Southern blot example in **Fig. 4** illustrates tamoxifen-mediated Cre recombination only in the heart *(16)*.

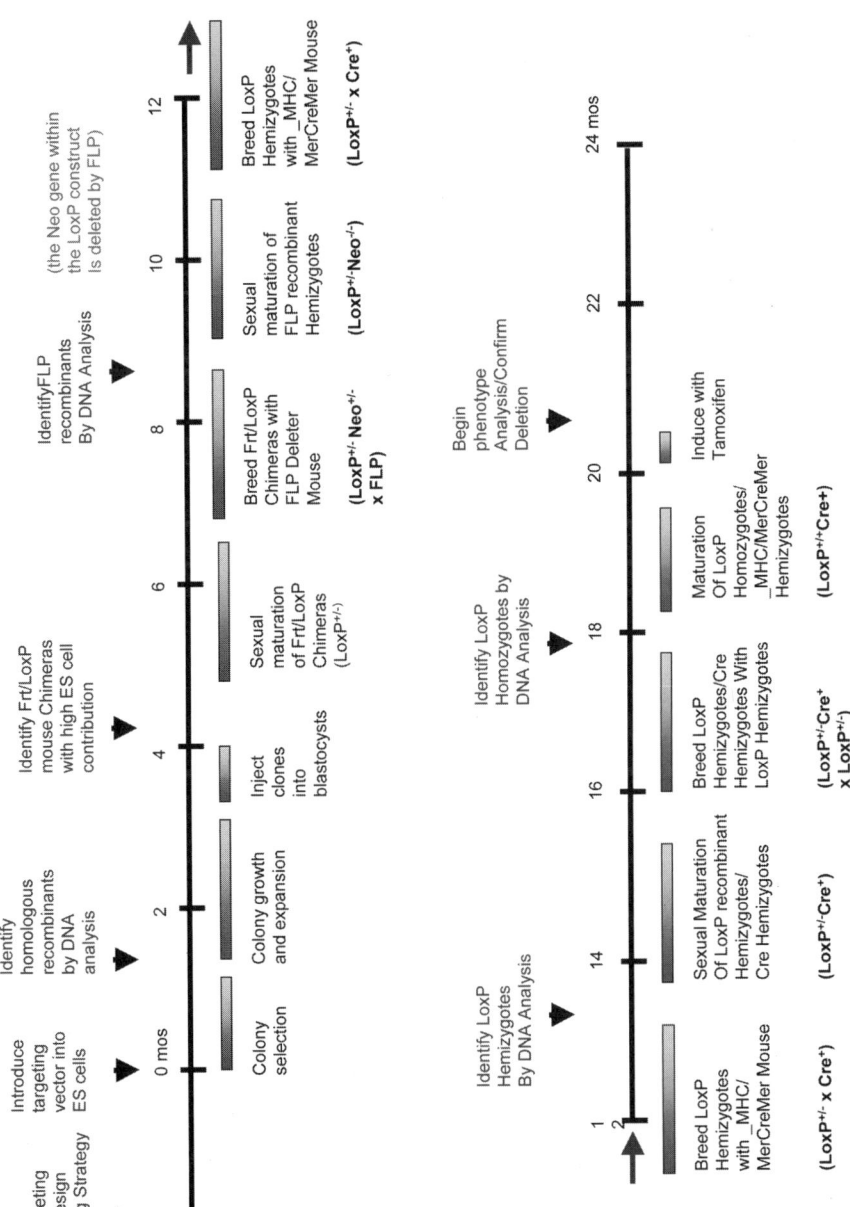

Fig. 5. Cre transgenic timeline.

3.7. Overview/Timeline for Generating an Inducible Cre Transgenic Mouse

The generation of a cardiac-specific, inducible Cre transgenic mouse is a lengthy process requiring upward of 2 yr. Numerous crossing strategies exist, but one example of a general strategy we recommend is displayed in the timeline in **Fig. 5**. We recommend initially crossing the hemizygous Frt/LoxP chimeras (LoxP$^{+/-}$Neo$^+$) with the FLP deleter mouse to delete the neo selection gene. Then the Neo$^{-/-}$ recombinant hemizygotes (LoxP$^{+/-}$) are crossed with the αMHC/MerCreMer mouse (Cre$^+$). The resulting LoxP hemizygotes/αMHC/MerCreMer are back-crossed with the LoxP hemizygotes (LoxP$^{+/-}$). Homozygous offspring will be end points (LoxP$^{+/+}$Cre$^+$). Induction of Cre in these endpoint mice will generate the desired cardiac deletion. Then begin phenotype analysis.

4. Notes

1. Transgenesis is also routinely performed in mice from a 129Sv background. The choice of strain will depend on the final strain on which the mutation will be maintained as well as the preferences of the participating transgenic lab.
2. Prior to the discovery of FLP recombinase, the classic strategy for removing the neo selection marker was to use three LoxP sites, two of which bound the selection marker. Cre recombination followed by negative selection results in two products: a conventional knockout and a conditional deletion *(26)*. The FLP/CRE design is generally considered a less cumbersome approach.

3. Some researchers will insert an additional restriction site within the floxed construct in order to distinguish the homologous recombinants further from the random insertions with screening *(8,27)*.
4. Functional in vitro assays serve as alternative methods to confirm the accuracy of the construct. Assess for the functionality of the Cre/LoxP construct in cell culture transfected with the construct along with an adenovirus-Cre (AdCre) vector by confirming changes in protein expression pre/post Cre transfection *(10,28)*. A similar assay with transfection of AdFLP confirms the inducibility of the FLP/Frt segment of the construct *(29)*.

Acknowledgments

We would like to thank Dr. Guy Zimmerman for his generous support and encouragement. We would also like to acknowledge Mark Cody for his dedication to our research. R.A. is supported by a VA Career Development award.

References

1. Liu, J. L., Grinberg, A., Westphal, H., et al. (1998) Insulin-like growth factor-I affects perinatal lethality and postnatal development in a gene dosage-dependent manner: manipulation using the Cre/loxP system in transgenic mice. *Mol. Endocrinol.* **12**, 1452–1462.
2. Haigh, J. J., Gerber, H. P., Ferrara, N., and Wagner, N. F. (2000) Conditional inactivation of VEGF-A in areas of collagen2a1 expression results in embryonic lethality in the heterozygous state. *Development* **127**, 1445–1453.
3. Farese, R. V., Jr., Ruland, S. L., Flynn, L. M., Stokowski, R. P., and Young, S. G. (1995) Knockout of the mouse apolipoprotein B gene results in embryonic lethality in homozygotes and protection against diet-induced hypercholesterolemia in heterozygotes. *Proc. Natl. Acad. Sci. USA* **92**, 1774–1778.
4. Bonyadi, M., Rusholme, S. A., Cousins, F. M., et al. (1997) Mapping of a major genetic modifier of embryonic lethality in TGF beta 1 knockout mice. *Nat. Genet.* **15**, 207–211.
5. Lakso, M., Sauer, B., Mosinger, B., Jr., et al. (1992) Targeted oncogene activation by site-specific recombination in transgenic mice. *Proc. Natl. Acad. Sci. USA* **89**, 6232–6236.
6. Seibler, J., Schubeler, D., Fiering, S., Groudine, M., and Bode, J. (1998) DNA cassette exchange in ES cells mediated by Flp recombinase: an efficient strategy for repeated modification of tagged loci by marker-free constructs. *Biochemistry* **37**, 6229–6234.
7. Sternberg, N. and Hamilton, D. (1981) Bacteriophage P1 site-specific recombination. I. Recombination between loxP sites. *J. Mol. Biol.* **150**, 467–486.
8. Kostetskii, I., Li, J., Xiong, Y., et al. (2005) Induced deletion of the N-cadherin gene in the heart leads to dissolution of the intercalated disc structure. *Circ. Res.* **96**, 346–354.
9. Meyers, E. N., Lewandoski, M., and Martin, G. R. (1998) An Fgf8 mutant allelic series generated by Cre- and Flp-mediated recombination. *Nat. Genet.* **18**, 136–141.
10. Agah, R., Kirshenbaum, L. A., Abdellatif, M., et al. (1997) Adenoviral delivery of E2F-1 directs cell cycle reentry and p53-independent apoptosis in postmitotic adult myocardium in vivo. *J. Clin. Invest.* **100**, 2722–2728.
11. Kuhn, R., Schwenk, F., Aguet, M., and Rajewsky, K. (1995) Inducible gene targeting in mice. *Science* **269**, 1427–1429.
12. Minamino, T., Gaussin, V., DeMayo, F. J., Schneider, M. D., et al. (2001) Inducible gene targeting in postnatal myocardium by cardiac-specific expression of a hormone-activated Cre fusion protein. *Circ. Res.* **88**, 587–592.
13. Bujard, H. (1999) Controlling genes with tetracyclines. *J. Gene. Med.* **1**, 372–374.
14. Metzger, D., Clifford, J., Chiba, H., and Chambon, P. (1995) Conditional site-specific recombination in mammalian cells using a ligand-dependent chimeric Cre recombinase. *Proc. Natl. Acad. Sci. USA* **92**, 6991–6995.
15. Li, M., Indra, A. K., Warot, X., et al. (2000) Skin abnormalities generated by temporally controlled RXRalpha mutations in mouse epidermis. *Nature* **407**, 633–636.

16. Sohal, D. S., Nghiem, M., Crackower, M. A., et al. (2001) Temporally regulated and tissue-specific gene manipulations in the adult and embryonic heart using a tamoxifen-inducible Cre protein. *Circ. Res.* **89,** 20–25.
17. Petrich, B. G., Molkentin, J. D., and Wang, Y. (2003) Temporal activation of c-Jun N-terminal kinase in adult transgenic heart via cre-loxP-mediated DNA recombination. *FASEB J.* **17,** 749–751.
18. Eckardt, D., Theis, M., Degen, J., et al. (2004) Functional role of connexin43 gap junction channels in adult mouse heart assessed by inducible gene deletion. *J. Mol. Cell Cardiol.* **36,** 101–110.
19. van Rijen, H. V., Eckardt, D., Degen, J., et al. (2004) Slow conduction and enhanced anisotropy increase the propensity for ventricular tachyarrhythmias in adult mice with induced deletion of connexin43. *Circulation* **109,** 1048–1055.
20. Ausubel, F. M. (2001) *Current Protocols in Molecular Biology.* Wiley Inter-Science, New York.
21. Jaenisch, R., Harbers, K., Schnieke, A., et al. (1983) Germline integration of moloney murine leukemia virus at the Mov13 locus leads to recessive lethal mutation and early embryonic death. *Cell* **32,** 209–216.
22. Radice, G., Lee, J. J., and Costantini, F. (1991) H beta 58, an insertional mutation affecting early postimplantation development of the mouse embryo. *Development* **111,** 801–811.
23. Buchholz, F., Angrand, P. O., and Stewart, A. F. (1998) Improved properties of FLP recombinase evolved by cycling mutagenesis. *Nat. Biotechnol.* **16,** 657–662.
24. Schaft, J., et al. (2001) Efficient FLP recombination in mouse ES cells and oocytes. *Genesis* **31,** 6–10.
25. Rodriguez, C. I., Buchholz, F., Galloway, J., et al. (2000) High-efficiency deleter mice show that FLPe is an alternative to Cre-loxP. *Nat. Genet.* **25,** 139–140.
26. Gu, H., Marth, J. D., Orban, P. C., Mossmann, H., and Rajewsky, K. (1994) Deletion of a DNA polymerase beta gene segment in T cells using cell type-specific gene targeting. *Science* **265,** 103–106.
27. Chan, T. A., Hermeking, H., Lengauer, C., Kinzler, K. W., and Vogelstein, B. (1999) 14-3-3Sigma is required to prevent mitotic catastrophe after DNA damage. *Nature* **401,** 616–620.
28. Akagi, K., Sandig, V., Vooijs, M., et al. (1997) Cre-mediated somatic site-specific recombination in mice. *Nucleic Acids Res.* **25,** 1766–1773.
29. Nakano, M., Odaka, K., Ishimura, M., et al. (2001) Efficient gene activation in cultured mammalian cells mediated by FLP recombinase-expressing recombinant adenovirus. *Nucleic Acids Res.* **29,** E40.

18

Cardiomyocyte Preparation, Culture, and Gene Transfer

Alexander H. Maass and Massimo Buvoli

Summary

Neonatal rat ventricular myocytes (NRVMs) cultured in vitro have been used as a model system for easily recreating and studying several cardiac molecular conditions, such as hypertrophy, oxygen deprivation, and gene expression. However, low efficiency of gene transfer has often represented one of the major limitations of this technique. In this chapter we describe in detail how to isolate NRVMs from neonatal rat heart and the optimal conditions for their long-term culture. Different cardiomyocyte transfection methodologies, based on viral or viral/chemical delivery carriers, are also discussed.

Key Words: Neonatal rat ventricular myocytes; NRVM; cell culture; transfection adenovirus; gene transfer; polyethylenimine; PEI; augmentation.

1. Introduction

Over the past years, cultured neonatal rat ventricular myocytes (NRVMs) have been widely employed as a model to study cardiac physio/pathology. NRVMs are postmitotic terminally differentiated cells, which do not divide, express both cardiac myosin isoforms (α/β-MyHC), beat spontaneously, and respond to different pharmacological stimuli.

The first neonatal rat cardiac myocyte culture model was developed and characterized in the Simpson laboratory to study the pathways and mechanisms that underlie the development of ventricular hypertrophy *(1–4)*. The ability to culture NRVMs for an extended period has allowed the study of other complex pathways, such as cardiomyocyte response to hypoxia *(5,6)*, cardiomyocyte apoptosis *(7,8)*, or cardiac gene expression *(9–11)*. A few cell lines (AT1 and HL-1) derived from atrial tumor cells have also been established *(12,13)*. However, the intrinsic limitations associated with these cells have restricted their application. Thus, primary cardiomyocytes remain the system of choice for most in vitro heart studies. The low percentage of cardiomyocytes that can be transfected by conventional gene transfer methods represents the main limitation of this technique *(14)*. Nevertheless, the use of adenoviral vectors has overcome this technical barrier. In fact, even at low multiplicity of infection (moi), recombinant adenoviruses allow highly efficient delivery of transgenes to most cells *(15)*. A simpler but less efficient technique that employs a mixture of adenovirus-polylysine-plasmid DNA has also been developed and successfully tested *(16)*. Moreover, we have extended this last study by using the different cation polyethylenimine, which appears to give more reproducible results.

This chapter describes a procedure to isolate, culture, and transfect efficiently primary cardiomyocytes from newborn rats.

2. Materials

2.1. Cell Culture Reagents

1. Bovine serum albumin (BSA) (Sigma, cat. no. A 4919), 100X stock solution (100 mg/mL): dissolve in water, filter, aliquot, and store at 4°C.
2. Bromodeoxyuridine (BrdU; Sigma, cat. no. B 5002), 100X stock solution (10 mM): dissolve in water, filter, aliquot, and store at –20°C.
3. Calf serum (CS; Hyclone, cat. no. 30073.03): aliquot (10 or 50 mL) and store at –20°C.
4. DNase (Sigma, cat. no. DN25), 2 mg/mL stock solution: dissolve in 0.15 M NaCl, filter, aliquot, and store at –20°C.
5. Gelatin (Sigma, cat. no. G 9391), 0.1% solution: dissolve in water and sterilize by autoclaving. Make fresh before use.
6. HEPES (Sigma, cat. no. H 4034), 50X stock solution (1 M): dissolve in water, adjust to pH 7.4 with NaOH, filter, and store at 4°C.
7. Insulin (Sigma, cat. no. I 1882), 100X stock solution (1 mg/mL): dissolve in 10^{-3} M HCl, filter, aliquot, and store at 4°C.
8. P/B_{12}, 500X stock solution: make a 1:1 solution with PCN-G (50,000 U/mL) and vitamin B_{12} (2 mg/mL). Filter, aliquot, and store at –20°C.
9. Penicillin G sodium salt (PCN-G; Sigma, cat. no. P 3032), 1000X stock solution (50,000 U/mL): dissolve in water, filter, aliquot, and store at –20°C.
10. Streptomycin (SM; Sigma, cat. no. S 9137), 1000X stock solution (50 µg/mL): dissolve in water, filter, aliquot, and store at –20°C (*see* **Note 1**).
11. Transferrin (Sigma, cat. no. T 1147), 100X stock solution (1 mg/mL): dissolve in water, filter, aliquot, and store at 4°C.
12. Trypan blue (Gibco, cat. no. 15250-061).
13. Trypsin (BD, cat. no. 215210), 1:250: working solution (2–3 mg/mL). On the day of preparation, dissolve in calcium and bicarbonate-free HEPES-buffered Hanks' (CBFHH) (*see* **Note 2**).
14. Vitamin B12 (Sigma, cat. no. V 6629), 1000X stock solution (2 mg/mL): dissolve in water, filter, aliquot and store at –20°C.

2.2. Cell Culture Solutions

1. CBFHH: 137 mM NaCl, 5.36 mM KCL, 0.81 mM MgSO$_4$, 5.55 mM dextrose, 0.44 mM KH$_2$PO$_4$, 0.34 mM Na$_2$HPO$_4$, 20 mM HEPES, pH 7.4. Filter and store at 4°C. Before use add SM (1X final) and PCN-G (1X final).
2. CBFHH + trypsin (2–3 mg/mL): just before use, mix 200 mL of CBFHH with 400 to 600 mg of trypsin, and warm to 37°C. When trypsin is completely dissolved, refilter and reequilibrate at 37°C in a water bath (*see* **Note 3**).
3. MEM with Hanks' balanced solution (Gibco, cat. no. 11575-032).
4. MEM H: add PB12 (1X final) and HEPES (1X final) per 500 mL MEM. Make fresh for use (*see* **Note 4**).
5. MEM 5: MEM H containing 5% CS.
6. MEM 5B: MEM 5 containing 1X BrdU.
7. MEM TI-BSA: MEM H containing transferrin (1X final), insulin (1X final), and BSA (1X final).
8. Phosphate-buffered saline (PBS), pH 7.3: 137 mM NaCl, 2.68 mM KCl, 0.4 mM MgSO$_4$, 0.75 mM CaCl$_2$, 1.45 mM KH$_2$PO$_4$, 8 mM Na$_2$HPO$_4$. Filter and store at 4°C.

2.3. Gene Transfer Reagents and Solutions

1. Adenovirus vectors containing the gene of interest. Stocks expanded according to standard procedures.
2. Replication-deficient adenovirus (e.g., dl327) (*see* **Note 5**).
3. Plasmid DNA of interest (*see* **Note 6**).
4. HEPES-buffered saline (HBS): 20 mM HEPES, 150 mM NaCl, pH 7.0. Filter and store at 4°C.
5. Polyethylenimine (PEI) solution (Aldrich, cat. no. 408727) average MW 25 kDa: prepare a 10 mM solution calculated toward the nonbranched monomer (9 mg of a 50% [w/v] solution in 10 mL water). Adjust to pH 7.0 with HCl. Filter and store at 4°C.

2.4. Equipment

1. Dissecting tools: medium and small scissors, forceps, and microdissecting curved forceps. Autoclave to sterilize.
2. Sterile gauze pads.
3. Three 60-mm and one 100-mm tissue culture dishes.
4. 50-mL Sterile glass beaker containing a micro spin bar. Autoclave to sterilize.
5. Hemocytometer.
6. 1- to 2-L sterile Nalgene bottles.
7. Cell strainer, 40-μm mesh size (BD, cat. no. 352340).
8. 50-mL sterile tubes with caps.
9. Benchtop centrifuge.
10. 37°C Water bath.
11. Sterile 1.5-mL tubes.
12. Large ice bucket.
13. 1% CO_2 tissue culture incubator.

3. Methods
3.1. Dissection

1. Place a piece of aluminum foil in a tissue culture hood. Arrange around it: the sterile surgical tools, two trays for pup cleaning, paper towels, sterile gauze pads, a beaker filled with 70% ethanol (for tool cleaning), and the sterile beaker containing a micro spin bar. Put three 60-mm and one 100-mm tissue culture plates in the ice bucket.
2. Add 10 and 4 mL of CBFHH to the tissue culture plates (100- and 60-mm, respectively), leaving one 60-mm plate empty.
3. Collect Sprague-Dawley pups and transfer them to a clean cage (*see* **Note 7**).
4. Disinfect pup neck and sternum area with 70% isopropyl alcohol in the first pup-cleaning tray. Transfer pups to the second tray to dry (up to five to six at a time).
5. When dried, immobilize each pup between the fingers of the left hand (**Fig. 1A** and **B**).
6. Quickly decapitate the animal with small scissors, and immediately open the thorax following a straight line along the sternum (**Fig. 1C**).
7. Push the heart out of the thoracic cavity by applying pressure with the fingers of the left hand (**Fig. 1D**).
8. With curved tweezers or scissors, separate the atria (**Fig. 1E** and **F**), and place the ventricle in the 100-mm plate containing 10 mL of CBFHH (*see* **Note 8**).
9. Clean surgical tools in alcohol before next dissection.
10. When all hearts have been harvested, clean the work area before the next steps of the procedure.

3.2. Dissociation

1. Transfer the hearts into the first 60-mm plate containing 4 mL of CBFHH.
2. Rinse, and if necessary, remove remaining atria and large vessels before transferring the ventricles to the second plate with 4 mL CBFHH.
3. Divide the ventricles in half and transfer to the third empty plate.
4. Cut each half into two to four pieces (about 2–3 mm per piece) (*see* **Note 9**).

3.3. Trypsin Digestion

1. Transfer minced tissue into the 50-mL sterile beaker containing the micro spin bar and 10 mL of prewarmed CBFHH + trypsin solution. Add 100 µL of DNase stock solution (*see* **Note 10**).
2. Cover the beaker with sterile aluminum foil and stir at room temperature for 15 min at approx 30 rpm.
3. Dissociate cellular debris and epi/endocardial cells by slowly pipeting the tissue pieces about 10 times through a 10-mL sterile pipet (*see* **Note 11**). After the tissue has settled for a few seconds, remove the supernatant and check under the microscope that only unwanted cells have been removed after this wash. Discard the supernatant.
4. Add 10 mL of fresh prewarmed CBFHH + trypsin and 100 µL of DNase stock solution into the beaker.

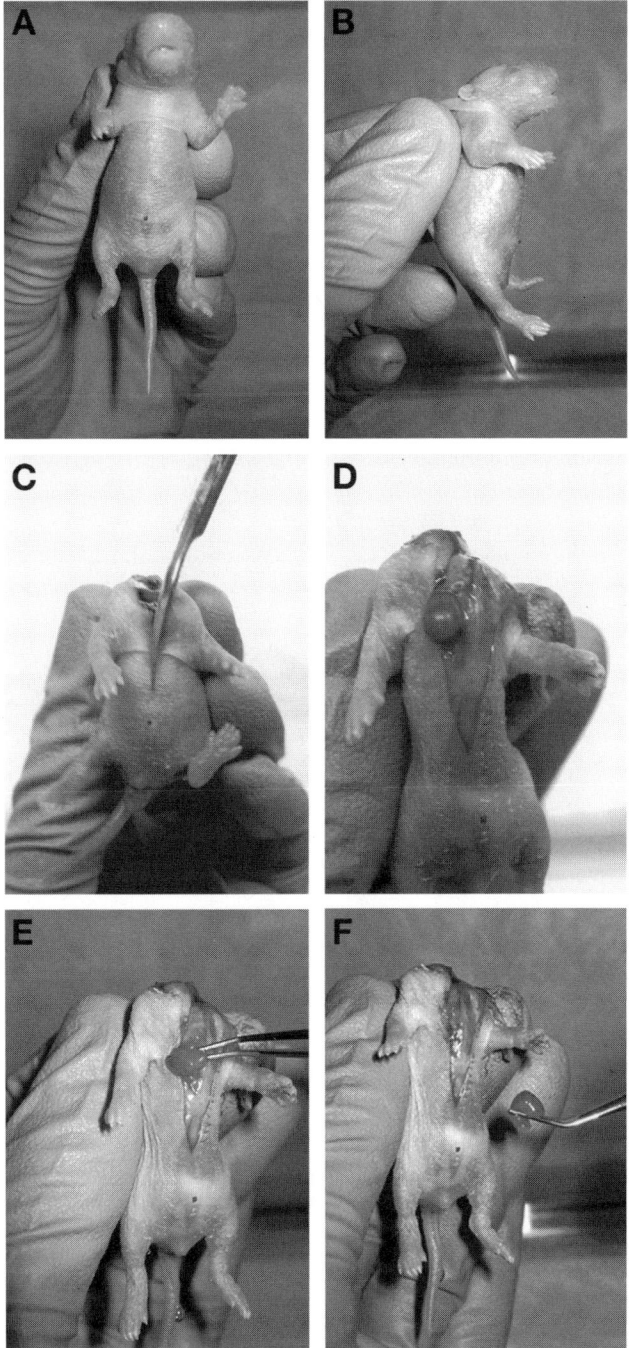

Fig. 1. Exteriorization of the heart and removal of the ventricles. (**A,B**) Immobilization of the pup. (**C**) Incision along the sternum. (**D**) Heart exteriorization. (**E,F**) Ventricle separation from atria.

5. Repeat **steps 2** and **3**.
6. Add 10 mL of fresh CBFHH + trypsin + 100 µL DNase to the tissue pieces and this time stir for 4 to 5 min only.
7. Slowly pipet the tissue pieces about 10 times through a 10-mL sterile pipet, let them settle, and then transfer the supernatant, containing the dissociated cardiomyocytes, into a 50-mL tube containing 7 mL of bovine CS. Analyze few drops of the supernatant under the microscope to verify the presence of cardiomyocytes (*see* **Note 12**).
8. Add another 10 mL of CBFHH + trypsin + 100 µL DNase to the tissue pieces and stir for an additional 4 to 5 min.
9. Repeat **step 7** and pool the supernatants (do not add more CS).
10. Collect approx 40 to 45 mL of supernatant per litter of pups by repeating **steps 6** to **9**. However, progressively decrease the volume of CBFHH + trypsin + DNase (approx 1 mL every time) added to the tissue pieces. If necessary, collect a second tube of supernatant (*see* **Note 13**).
11. Add an additional 0.5 mL of DNase stock solution to the 50-mL tube(s) and mix by inversion.
12. Pellet cells by centrifugation at 1500 rpm (approx 700 g) for 10 min in a tabletop centrifuge. Decant the supernatant, and carefully resuspend the cells in 10 mL of MEM 5 per litter of pups.

3.4. Preplating

1. To remove clumped cells and/or residual debris, filter the cells through a cell strainer (*see* **Note 14**) into a sterile 50-mL tube. Subsequently transfer the cell suspension into 100-mm tissue culture dishes, using 5 mL of cells for each 100-mm dish (two 100-mm dishes for every litter of pups). Rinse the first tube with 10 mL of MEM 5 and add 5 mL of this wash (filtered through the cell strainer) to each dish.
2. Incubate plates at 37°C for 30 min in 1% CO_2 to allow the remaining nonmyocyte cells to attach to the plates (*see* **Note 15**).
3. Remove the cardiomyocyte-enriched media from each plate and transfer them into a sterile 500-mL polypropylene bottle.
4. Add 10 mL of MEM 5 to each plate. Swirl, collect, and transfer into the polypropylene bottle.
5. Tap the empty plates against the wall of the hood several times to remove additional myocytes. Rinse each plate with 10 mL of MEM 5, and transfer into the polypropylene bottle.
6. Add 10 mL of growth media to each plate and repeat **step 4** two more times.
7. Add 10 mL of MEM 5 to each preplate and incubate at 37°C in 1% CO_2 to culture nonmyocytes (*see* **Note 16**).

3.5. Counting Cells and Plating

1. Gently mix the cardiomyocyte suspension by pipeting.
2. Aliquot 500 µL of cell suspension into two microfuge tubes.

3. Add 500 µL of Trypan blue and mix.
4. Determine the average number of viable (nonblue) and dead cells (blue) by counting the two aliquots separately in a hemocytometer
5. Dilute cells to approx 300,000/mL in MEM 5. If desired, add BrdU (*see* **Note 17**).
6. Carefully mix the cell suspension and transfer the cells into tissue culture dishes as follows: 1 mL for 12 wells, 2 mL for a 35-mm plate, 4 mL for a 60-mm plate, and 10 mL for a 100-mm plate (*see* **Note 18**).
7. Uniformly disperse the cells by gently rocking every plate three times in each direction (North-South, East-West). Do not swirl or overmix.
8. Incubate plates at 37°C in 1% CO_2.
9. After 24 h, wash the cells twice with PBS and incubate them in MEM TI-BSA.

3.6. Adenovirus Infection

1. Use an moi ranging from 2 to 100 infectious particles per cell (*see* **Note 19**).
2. Dilute the needed amount of recombinant adenovirus in MEM TI-BSA in a volume equal to or greater than 10% of the plate volume.
3. Add the adenovirus dropwise while swirling the plate. Incubate at 37°C in 1% CO_2 overnight.
4. Wash the cells once with PBS and refeed with fresh MEM TI-BSA. Incubate according to the planned experiment.

3.7. Adenovirus-Assisted Transfection of Plasmid DNA

1. In a sterile 1.5-mL tube dilute 1 µg of recombinant plasmid DNA in 30 µL HBS. (This amount of DNA is enough for a 12-well plate. Scale up for larger plates.)
2. In a different sterile 1.5-mL tube, dilute 3 µL of the 10 m*M* PEI solution (vortex before use) in 30 µL HBS. Vortex for a few seconds (medium speed) and incubate for 5 min at room temperature (*see* **Note 20**).
3. Add the PEI dilution to the DNA solution (dropwise). Vortex as above and incubate for 15 min at room temperature.
4. Add a minimum of approx 1000 adenovirus particles per cell to the DNA/PEI complex. Vortex briefly and incubate for another 10 min at room temperature (*see* **Note 21**).
5. Add the DNA/PEI/adenovirus complex directly to the cells cultured in MEM H, swirl to mix, and incubate for at least 4 h (maximum overnight).
6. Wash the cells with PBS, add fresh MEM TIBSA, and incubate according to the planned experiment.

4. Notes

1. Streptomycin is only used during the dissociation step, since it can interfere with MyHC expression and cell growth.
2. The optimal concentration of trypsin must be established for every new lot, as trypsin activity changes from one lot to another.
3. To avoid loss of activity, the trypsin solution has to be made fresh, just before the beginning of the preparation, as prolonged equilibration at 37°C results in trypsin autodigestion.

4. HEPES is added to buffer the media to avoid pH changes that could result in cell death. (NRVMs do not tolerate alkaline conditions.)
5. Each virus preparation needs to be tested for infection efficiency. We usually calculate the amount of virus required as particle number instead of plaque-forming units.
6. Our plasmids are purified using the Qiagen Plasmid Maxi Kit, since CsCl plasmid preparations have lower transfection efficiency.
7. Pups are 1 d old when sacrificed. Preparations from older animal often result in lower yield and cell viability. A good NRVM preparation generates about 5×10^6 cells/heart, with a 30% plating efficiency (corresponding to the number of cells that will attach to the culture plate).
8. Remove the heart as quickly as possible, since unnecessary delays can result in lower myocyte viability. Ten milliliters of CBFHH are adequate for about one to two litters of pups (up to approx 20 hearts).
9. The size of the pieces represents a critical parameter for cardiomyocyte yield and viability. Cell damage can occur (1) during mincing and digestion if the pieces are too small, or (2) during the slower trypsin digestion, if the pieces are too large. Mincing should be performed as quickly as possible.
10. Ten milliliters of CBFHH + trypsin solution are adequate for about one litter of pups.
11. This mechanical dissociation is a critical step. Avoid shearing the cardiomyocytes by pipeting them too fast. A "wide-tip" sterile pipet should be used. The correct pipeting speed is approx 5 s for 10 mL solution. Avoid bubbles.
12. Healthy cardiomyocytes are either rod-shaped or rounded cells with sharp membranes. The presence of cell clumps or cells with irregular membranes denotes that digestion was not complete, or that mechanical dissociation was too harsh.
13. Most of the cardiomyocytes are usually isolated in the first 10 dissociations. The digestion is complete when the tissue becomes mostly white and clumpy and the supernatant contains only damaged cells. If the digestion is not complete after 3 h, the concentration of trypsin needs to be increased for the next cardiomyocyte preparation, since longer exposure to trypsin will damage the cells.
14. The filtration is an important step because it retains unwanted cell clumps containing cardiac fibroblasts. By growing rapidly, fibroblasts can become the predominant cell population.
15. Since healthy cardiomyocytes can adhere quite rapidly too, the optimal time for this step should be experimentally determined by frequently monitoring the process under the microscope.
16. The preplates contain a predominant nonmyocyte cell population contaminated by a small number NRVMs. A few passages will remove NRVMs and leave mostly cardiac fibroblasts.
17. For general purposes, cells are seeded at low confluency (approx 20–30%) to avoid activation of spontaneous hypertrophy (occurring at approx 30–50%). BrdU inhibits growth of nonmyocyte cells. However, it should be used only if a very pure population of NRVMs is needed, since we observed negative effects on cellular integrity.

18. NRVM attachment can be enhanced by precoating the tissue culture plates with gelatin. (A 0.1% gelatin solution is added to the plates to cover the bottom completely. Plates are then incubated at room temperature for few hours, drained, and air-dried in a tissue culture hood under UV light for about 15 min.)
19. We found that levels of recombinant gene expression can be modulated by promoter choice. For example, the CMV promoter, one of the most active transcription elements generally used, can be pharmacologically induced in NRVMs (approx 100-fold). Therefore, viral promoters should be selected according to each experimental design *(17)*.
20. A PEI solution at pH 7.0 is charged with protons and forms an ionic interaction with the negatively charged DNA. The advantage of PEI over poly-L-lysine lies in the easier preparation procedure. Furthermore, the intrinsic ability of PEI to target recombinant plasmids to the nucleus leads to higher levels of heterologous transcription *(18)*. We routinely use an optimized nitrogen/phosphate ratio of PEI/DNA of about 10:1.
21. The number of active viral particles differs between each virus preparation. Usually, 20 to 50% of the cells can be transfected by using between 1000 and 10,000 viral particles per cell. Higher amounts (up to 100,000) are not toxic but do not improve the efficiency of transfection dramatically. Interestingly, "empty" adenoviruses that do not express any recombinant DNA appear to enhance plasmid transgene expression.

References

1. Simpson, P. and Savion, S. (1982) Differentiation of rat myocytes in single cell cultures with and without proliferating nonmyocardial cells. Cross-striations, ultrastructure, and chronotropic response to isoproterenol. *Circ. Res.* **50,** 101–116.
2. Simpson, P., McGrath, A., and Savion, S. (1982) Myocyte hypertrophy in neonatal rat heart cultures and its regulation by serum and by catecholamines. *Circ. Res.* **51,** 787–801.
3. Simpson, P. (1983) Norepinephrine-stimulated hypertrophy of cultured rat myocardial cells is an alpha 1-adrenergic response. *J. Clin. Invest.* **72,** 732–738.
4. Simpson, P. (1985) Stimulation of hypertrophy of cultured neonatal rat heart cells through an alpha 1-adrenergic receptor and induction of beating through an alpha 1- and beta 1-adrenergic receptor interaction. Evidence for independent regulation of growth and beating. *Circ. Res.* **56,** 884–894.
5. Musters, R. J., Post, J. A., and Verkleij, A. J. (1991) The isolated neonatal rat-cardiomyocyte used in an in vitro model for 'ischemia'. A morphological study. *Biochem. Biophys. Acta* **1091,** 270–277.
6. Iwaki, K., Chi, S. H., Dillmann, W. H., and Mestril, R. (1993) Induction of HSP70 in cultured rat neonatal cardiomyocytes by hypoxia and metabolic stress. *Circulation* **87,** 2023–2032.
7. Tanaka, M., Ito, H., Adachi, S., et al. (1994) Hypoxia induces apoptosis with enhanced expression of Fas antigen messenger RNA in cultured neonatal rat cardiomyocytes. *Circ Res.* **75,** 426–433.

8. Shimojo, T., Hiroe, M., Ishiyama, S., Ito, H., Nishikawa, T., and Marumo F. (1999) Nitric oxide induces apoptotic death of cardiomyocytes via a cyclic-GMP-dependent pathway. *Exp. Cell Res.* **247,** 38–47.
9. van der Lee, K. A., Vork, M. M., De Vries, J. E., et al. (2000) Long-chain fatty acid-induced changes in gene expression in neonatal cardiac myocytes. *Lipid Res.* **41,** 41–47.
10. Liu, T., Lai, H., Wu, W., Chinn, S., and Wang, P. H. (2001) Developing a strategy to define the effects of insulin-like growth factor-1 on gene expression profile in cardiomyocytes. *Circ. Res.* **88,** 1231–1238.
11. Charron, F., Tsimiklis, G., Arcand, M., et al. (2001) Tissue-specific GATA factors are transcriptional effectors of the small GTPase RhoA. *Genes Dev.* **15,** 2702–2719.
12. Steinhelper, M. E., Lanson, N. A. Jr., Dresdner, K. P., et al. (1990) Proliferation in vivo and in culture of differentiated adult atrial cardiomyocytes from transgenic mice. *Am. J. Physiol.* **259,** H1826–1834.
13. Claycomb, W. C., Lanson, N. A. Jr., Stallworth, B. S., et al. (1998) HL-1 cells: a cardiac muscle cell line that contracts and retains phenotypic characteristics of the adult cardiomyocyte. *Proc. Natl. Acad. Sci. USA* **95,** 2979–2984.
14. Antin, P. B., Mar, J. H., and Ordahl, C. P. (1988) Single cell analysis of transfected gene expression in primary heart cultures containing multiple cell types. *Biotechniques* **6,** 640–649.
15. Kass-Eisler, A., Falck-Pedersen, E., Alvira, M., et al. (1993) Quantitative determination of adenovirus-mediated gene delivery to rat cardiac myocytes in vitro and in vivo. *Proc. Natl. Acad. Sci. USA* **90,** 11498–11502.
16. Kohout, T. A., O'Brian, J. J., Gaa, S. T, Lederer W. J., and Rogers T. B. (1996) Novel adenovirus component system that transfects cultured cardiac cells with high efficiency. *Circ. Res.* **78,** 971–977.
17. Maass, A., Langer, S. L., Oberdorf-Maass, S., Neyses, L., and Leinwand L. A. (2003) Rational promoter selection for gene transfer into cardiac cells. *J. Mol. Cell. Cardiol.* **35,** 823–831.
18. Godbey, W. T., Wu, K. K., and Mikos, A. G. (1999) Tracking the intracellular path of poly(ethylenimine)/DNA complexes for gene delivery. *Proc. Natl. Acad. Sci. USA* **96,** 5177–5181.

19

Adeno-Associated Viral Vector–Delivered Hypoxia-Inducible Gene Expression in Ischemic Hearts

Hua Su and Yuet Wai Kan

Summary

This chapter describes a system using adeno-associated viral (AAV) vector to deliver hypoxia-inducible gene expression to ischemic hearts. The hypoxia induction of gene expression in this system is based on the accumulation of hypoxia-inducible factor-1 (HIF-1) in ischemic hearts and the use of hypoxia-response element (HRE) identified from the enhancers of genes, the expression of which can be induced by hypoxia. The methods of plasmid and AAV vector construction for hypoxia-inducible gene expression, viral vector production and purification, and viral titer determination are described. This chapter also illustrates the methods that can be used to test hypoxia-inducible gene expression in vitro and in vivo, including hypoxia treatment of cultured cells, generation of murine ischemic heart models, and analysis of gene expression.

Key Words: Hypoxia-response element; hypoxia-inducible gene expression; adeno-associated viral vector; virus production; ischemic heart.

1. Introduction

Myocardial infarction is, by nature, an irreversible injury. Despite advances in medical and surgical therapies, coronary heart disease remains a leading cause of morbidity and mortality in the Western world. Administration of exogenous angiogenic factors and potential cardiac protective genes *(1–3)* to induce neoangiogenesis and cardiac protection in ischemic myocardium can reduce postinfarct remodeling and improve cardiac function *(4–7)*. However, unregulated gene expression can cause some unwanted side effects, such as hemangioma formation at injection sites caused by uncontrolled vascular endothelial growth factor (VEGF) expression *(8,9)* and long-term tissue toxicity and cellular damage by constitutive heme-oxygenase-1 (HO-1) expression

(10,11). Examples of inducible promoters that have been used to control gene expression include the tetracycline operons, RU 486, edyasone, and other inducible systems *(12–19)*. However, these systems require the exogenous administration of a ligand to induce expression of the therapeutic gene.

Hypoxic induction of gene expression through activation of hypoxia-inducible factor-1 (HIF-1) is one of the major mechanisms that cells use to regulate gene expression in response to hypoxia stress. HIF-1 is a heterodimeric basic helix-loop-helix protein *(20,21)* and has a labile α-subunit (HIF-1α) and a β-subunit. The β-subunit is constitutively expressed in tissues in normal and hypoxic conditions. The HIF-1α subunit is rapidly degraded in normal conditions and is stabilized in hypoxic tissues. HIF-1 protein activates transcription of several genes, including human erythropoietin, HO-1, and VEGF genes in hypoxic cells *(22)*. Previous experiments have shown that HIF-1α levels were increased in ischemic myocardium from 2 d after the occlusion of the left anterior descending coronary artery (LAD) *(23)*. Thus, hypoxia induction is an ideal system to control gene expression in ischemic hearts. Adeno-associated viral (AAV) vector is an attractive tool for delivering genes to the heart and has been used by many investigators *(23–26)*. AAV vectors carrying nine copies of hypoxia-response elements (HREs) isolated from the erythropoietin gene (Epo) enhancer and a minimum SV40 promoter driving LacZ and VEGF gene expression will be used to illustrate the methods utilized for hypoxia induction of gene expression in ischemic hearts.

2. Materials

1. *Epo* HRE consensus sequence.
2. pBluescript II SK vector (Strategies, La Jolla, CA), *pβgal*-control vector (Clontech, Palo Alto, CA), and SV40 plasmid (New England Biolabs, Beverly, MA).
3. Restriction enzymes and T4 DNA ligase.
4. Agarose gel equipment.
5. *E. coli* strains XL1 blue or DH5α.
6. AAV Helper-Free Gene Delivery and Expression System (Strategies).
7. Dulbecco's modified Eagle's medium (DMEM) with 4.5 g/L glucose and L-glutamine (Cellgro, Herndon, VA).
8. Fetal bovine serum (FBS; Hyclone, Logan, UT).
9. Hot box system (Billups-Rothenberg, Del Mar, CA).
10. Small Animal Volume Controlled Ventilator (Harvard Rodent Ventilator, model 683, Harvard Apparatus, Holliston, MS).
11. DC Protein Assay System (Bio-Red, Hercules, CA).
12. HeLa cells, Hep 3B cells, 293 cells, and NIH3T3 cells (American Type Culture Collection, Gaithersburg, MD).
13. CD1 mice (Charles River, Wilmington, MA).
14. 2X HBS: 280 mM NaCl, 10 mM KCl, 1.5 mM Na$_2$HPO$_4$, 12 mM dextrose, and 50 mM HEPES.
15. Dialysis cassettes (Slide-A-Lyzer 7K, Pierce, Rockford, IL).

A 9 copies of Epo HRE consensus sequence:

CTGCAGGAATTCGATGCACGCGTCCGGGTAGCTGGCG<u>TACGTG</u>CTGCAGCCGGGTAGCT
GGCG<u>TACGTG</u>CTGCAGCCGGGTAGCTGGCG<u>TACGTG</u>CTGAGCTCGAGACTTGACGCGTC
CGGGTAGCTGGCG<u>TACGTG</u>CTGCAGCCGGGTAGCTGGCG<u>TACGTG</u>CTGCAGCCGGGTAG
CTGGCG<u>TACGTG</u>CTGAGCTCGAGACTTGACGCGTCCGGGTAGCTGGCG<u>TACGTG</u>CTGCA
GCCGGGTAGCTGGCG<u>TACGTG</u>CTGCAGCCGGGTAGCTGGCG<u>TACGTG</u>CTGCAGCTCGAG

B pβgal-control vector

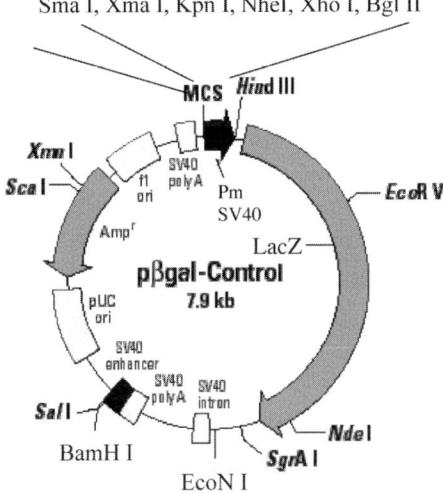

Fig. 1. (**A**) Nine copies of Epo HRE consensus sequence. The HIF-1 binding sites are underlined. (**B**) Schematic drawing of pβgal-control vector. Unique restriction sites are in bold. PmSV40, minimum SV40 promoter. Different HREs can be inserted between the *Sma*I and *Nhe*I sites. The LacZ gene can be cut off by *Hin*dIII and *Nde*I digestion. Genes of interest can be cloned to these sites.

3. Methods
3.1. Plasmid and AAV Construct
3.1.1. Hypoxia-Inducible Gene Expression Plasmid Vector

Hypoxia-inducible gene expression is based on the use of HRE. HRE sequences are found in the enhancers of genes that are inducible by hypoxia. Here, we used Epo HRE as an example. Nine copies of Epo HRE-consensus sequences *(27,28)* (**Fig. 1A**) were cloned between *Sma*I and *Nhe*I sites of the pβgal-control vector (Clontech) (**Fig. 1B**) before the SV40 minimum promoter. Different

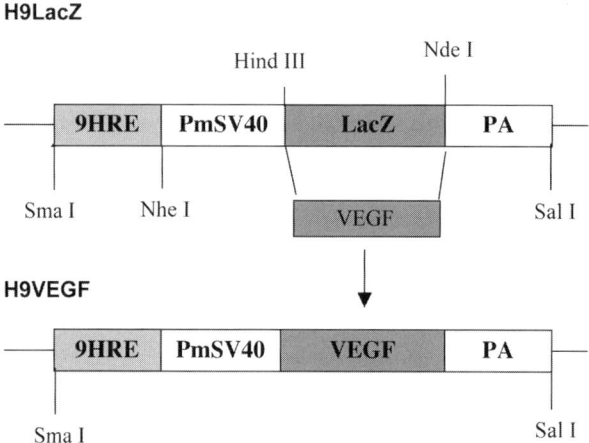

Fig. 2. Schematic drawing of H9LacZ and H9VEGF. Nine copies of Epo HRE (9HRE) were cloned in front of the minimum SV40 promoter (PmSV40). H9VEGF was generated to replace the LacZ sequence between *Hin*dIII and *Nde*I with human VEGF$_{165}$ cDNA.

copies of Epo HRE and HREs isolated from other genes can also be cloned to the same position (*see* **Note 1**). The capacity of the AAV vector is limited to 4.5 kb. The polyadenylation signal in pβgal-control vector is about 1261 bp long and is located between the *Eco*NI and *Bam*HI sites. To give more room for cloning genes of interest, this fragment was deleted from the vector and replaced with a 200-bp polyadenylation signal between the *Bcl*I and *Bam*HI sites of SV40 large T-antigen (SV40 plasmid, New England Biolabs). The resulting plasmid is called H9LacZ (**Fig. 2**). The LacZ gene in the H9LacZ can be replaced by other genes. For example, we have generated a hypoxia-inducible VEGF expression vector by replacing the LacZ gene in the H9LacZ plasmid with human VEGF$_{165}$ cDNA (**Fig. 2**). The expression cassette can be released from the vector by *Sma*I and *Sal*I digestion and cloned to AAV or other vectors.

3.1.2. Recombinant AAV Vector for Hypoxia-Inducible Gene Expression

AAV vectors expressing hypoxia-inducible genes can be created by using pAAV-LacZ (**Fig. 3**; Stratagene). The pCMV, the LacZ gene, and SV40 pA can be cut off from the AAV-LacZ vector by *Not*I enzyme digestion. The gene expression cassettes containing the HREs, minimum SV40 promoter, genes of interest, and an SV40 polyadenylation signal in H9-plasmids can be excised by *Sma*I and *Sal*I digestion and cloned in the pAAV-LacZ vector between the two *Not*I sites.

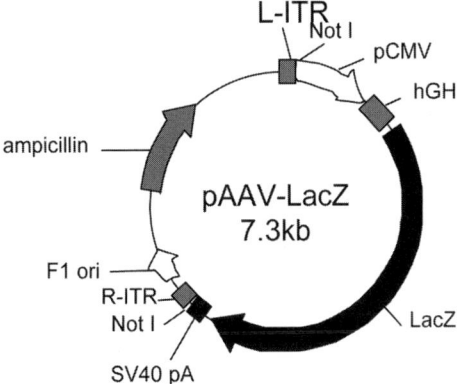

Fig. 3. Schematic drawing of pAAV-LacZ. The expression cassette including the CMV promoter, LacZ gene, and SV40 polyadenylation signal in this vector can be excised out by *Not*I enzyme digestion. The expression cassette excised from the H9-vector can be cloned between two *Not*I sites in this vector to generate an AAV vector for hypoxia-inducible gene expression.

3.2. AAV Virus Production

3.2.1. Transfection

1. First, 4.5×10^6 HEK 293 cells are seeded in 15-cm tissue culture dishes 2 d before transfection. Cells in one 90% confluent dish can be passed to eight dishes. The cells are maintained in 25 mL DMEM plus 10% fetal calf serum (FCS), and 25 mM HEPES.
2. The AAV vector containing the gene of interest is cotransfected with pAAV-RC and pHelper vectors into 293 cells by using the calcium phosphate precipitation method. pAAV-RC contains AAV rep and cap genes (*see* **Note 2**). pHelper has the adenoviral VA, E2A, and E4 regions that mediate AAV vector replication. To transfect one 15-cm dish, a total of 50 µg DNA (17 µg DNA of each plasmid) is mixed with 1 mL of 300 mM CaCl$_2$. A mixture for transfection of four dishes can be prepared in one 50-mL Corning tube.
3. A sterile pipet is placed into the tube. Air is gently bubbled while 1 mL (4 mL for four plates) of 2X HBS is added to the tube drop by drop.
4. The transfection mix is distributed to each plate.
5. The plates are incubated in 37°C for 6 h.
6. The medium is changed to DMEM containing 2% FCS and 25 mM HEPES. Cells are cultured at 37°C for 54 h.

3.2.2. Purification (see **Note 4**)

1. Cells are dislodged from the dishes by gentle pipeting and transferred to 50-mL Corning tubes.
2. Media are removed by centrifugation (1000g for 5 min at 4°C).

3. Cells are resuspended in 100 mM Tris-HCl, 150 mM NaCl, pH 8.0 (1×10^7 cells/mL) and lysed by three freeze and thaw cycles (alternating between dry ice-ethanol and 37°C water baths).
4. The lysates are centrifuged at 10,000g for 15 min to remove the cell debris.
5. The cleared supernatant is precipitated with 25 mM CaCl$_2$ at 0°C for 1 h.
6. Precipitates are removed by centrifugation (10,000g, 15 min at 4°C).
7. NaCl and PEG (8000) are added into the supernatant to make the final concentration of 620 mM NaCl and 8% PEG.
8. The supernatants are incubated on ice for 3 h.
9. AAV vector containing precipitate is collected by centrifugation (300g, for 30 min at 4°C) and resuspended in 5 mL of 50 mM HEPES, 150 mM NaCl, 25 mM EDTA, pH 8.0. (Adjust pH to 8.0 with NaOH.)
10. Insoluble materials are removed by centrifugation (10,000g for 15 min at 4°C).
11. The AAV vector is purified by CsCl$_2$ gradient centrifugation. Solid CsCl$_2$ is added to the supernatant to produce a density of 1.4 g/mL (0.548 g CsCl$_2$/mL). Samples are centrifuged at 15°C for 16 h at 223,000g.
12. The gradient is fractionated (0.5–1 mL/fraction) and assayed by dot blot to detect the viral particles (*see* **Subheading 3.2.3.**).
13. The AAV vector-containing fractions are pooled, put in dialysis cassettes (Slide-A-Lyzer 7K, Pierce) and dialyzed against 1 L (1000 times the volume of viral fraction) buffer containing 10 mM HEPES (pH 7.4), 140 mM NaCl, 0.1% Tween-80, 5% sorbital) three times at 4°C. The dialysis buffer is changed every 2 h.

3.2.3. Titer Determination

Viral titers are determined by dot blot analysis of the DNA content (*see* **Note 3**).

1. First, 2 µL of each fraction collected after CsCl$_2$ centrifugation or dialyzed viral stock are added to 20 µL of 500 mM NaOH and dotted on Hybond H+ Membrane (Strategies).
2. Then 1 ng of gene fragment (about 1 kb long) is serially diluted and dotted on the same membrane as control.
3. The membrane is air-dried and neutralized in 500 mM phosphate buffer, pH 7.4, for 10 min, washed in 2X SSC briefly, and air-dried.
4. The membrane is hybridized with the isotope or digoxin-labeled (Roche Diagnostics, Indianapolis, IN) probes for genes in the AAV vectors using a standard protocol.
5. The density of each fraction or viral stock is compared with that of the control. The relative copy number can be calculated. Copies of a 1-ng gene fragment can be calculated as follows (copies/mL): $(1 \times 10^{-9}/\text{length of the fragment (bp)} \times 660) \times 6 \times 10^{23} \times 2 \times 10^3$.

3.2.4. Toxicity

The viral stocks' toxicity to cells is checked by infection of 293 cells.

1. 293 cells are seeded in 48-well tissue culture dishes, 1×10^5 cells/well. Cells are cultured in 0.5 mL DMEM with 10% FCS for 24 h at 37°C.

2. Then 50 µL of testing viral stock is added to each well and cultured with the cells for 24 h.
3. The medium is changed 24 h later.
4. The status of the infected cells is monitored every day for 3 to 5 d and compared with uninfected control cells. The viral stock will be considered toxic if cell death or slower growth is observed. More dialysis steps can be used to remove the toxic chemicals, if a viral stock is found to be toxic to cells.

3.3. Testing of Hypoxia Induction of Gene Expression In Vitro

3.3.1. Infection

293, HeLa, Hep 3B, and NIH3T3 cells (American Type Culture Collection), which are known to accumulate HIF-1 proteins under hypoxic conditions, are used to test hypoxia induction of gene expression in vitro.

1. Cells are seeded in 48-well tissue culture dishes (10^5 cells/well) and cultured in DMEM supplemented with 10% FBS at 37°C with 95% air, 5% CO_2 for 24 h.
2. AAV vectors with HRE driving gene expression are added to tissue culture dishes (5×10^{10} copies of vectors per well) and incubated with the cells for 24 h at 37°C with 5% CO_2.
3. The medium is changed 24 h later.
4. An AAV vector with CMV driving the corresponding gene expression should also be used to infect the cell and used as control.

3.3.2. Hypoxia Induction

1. Infected cells are split into two or more groups. One group is cultured in normoxic conditions (95% air, 5% CO_2) as a control. The others are placed into a modular incubator chamber (Hot box system, Billups-Rothenberg) and flushed with 95% N_2 and 5% CO_2 (anoxic condition) or 90 to 94% N_2, 5% CO_2, and 1 to 5% O_2 (hypoxic conditions).
2. Cells are incubated in the above conditions for 16 h or longer.

3.3.3. Analysis of Gene Expression In Vitro

Supernatant, cell lysate, or RNA can be collected after the hypoxic or anoxic treatment. Western blot, enzyme-linked immunosorbent assay (ELISA), Northern blot, or reverse transcription polymerase chain reaction (RT-PCR) methods can be used to analyze gene expression. Here we use the cells infected with AAVH9LacZ (an AAV vector with nine copies of Epo HRE and a minimum SV40 promoter driving LacZ expression) as an example. LacZ activities were measured by using a Luminescent β-galactosidase Reporter System (Clontech). Cell lysates were collected 16 h after the anoxic treatment. The protein concentration in each cell lysate was measured by using the Bio-Rad DC Protein Assay System to normalize the LacZ activities for each sample. LacZ activities could be demonstrated to increase 10 times

Table 1
Ratios of LacZ Activity in Anoxic and Normoxic Conditions

	AAVH9LacZ	AAV-LacZ
NIH 3T3	10	1.4
HeLa	28	0.8
Hep 3B	30	1.0

in NIH3T3 cells, 28 times in HeLa cells, and 30 times in Hep 3B cells after AAVH9LacZ transduction in anoxic conditions (**Table 1**). In contrast, LacZ activities remained the same for the AAV-LacZ (an AAV vector with CMV driving LacZ expression) transduced cells under anoxic and normoxic conditions (**Table 1**).

3.4. Hypoxia-Inducible Gene Expression in Ischemic Hearts
3.4.1. Ischemic Heart Model

Hypoxia-inducible gene expression in ischemic hearts can be tested in different animal models. Here we use mouse model as an example.

1. CD1 mice (Charles River) are anesthetized with 15 to 16 µL of 2.5% Avertin/g of body weight by intraperitoneal injection (i.p).
2. A tube is placed in the trachea through the mouth and connected to a Small Animal Volume Controlled Ventilator (Harvard Rodent Ventilator, model 683).
3. After the respiration of the animal has been controlled by the ventilator, a thoracotomy incision is made in the fourth intercostal space.
4. A surgical retractor is put in the incision to expose the heart.
5. The LAD is ligated permanently with a 7-0 nonabsorbable surgical suture.
6. Then 5×10^{10} genomes of viral vectors in 50 µL of HEPES saline (pH 7.4) is injected into the myocardium around the border of the ischemic area.
7. The thoracotomy incision is then closed.
8. The tube in the trachea is gently retracted.
9. The AAV vector with CMV driving corresponding gene expression should also be injected into a separate group of mice in the same way as control.

3.4.2. Analysis of Gene Expression In Vivo

Hearts can be collected from 1 d to 21 d after the LAD ligation and viral vector injection, since previous studies with brains have shown that HIF-1α accumulated in tissues at the onset of hypoxia and decreased gradually to normoxic level by 3 wk, despite the continuous low arterial oxygen tension *(29)*. Hypoxia-induced gene expression may not be observed in the tissues collected 3 wk after the LAD ligation and vector injection. RNA and protein can be isolated from the treated hearts. Gene expression can be analyzed by

Western blot, ELISA, semiquantitative RT-PCR, or real-time RT-PCR. As an example:

1. AAV vectors with nine copies of Epo HRE and a minimum SV40 promoter controlling LacZ (AAVH9LacZ) or VEGF (AAVH9VEGF) gene expression were injected into normal and ischemic mouse hearts.
2. The hearts were collected 2 wk after the vector injection, and LacZ and VEGF gene expression was analyzed by semiquantitative RT-PCR.
3. The TRIzol RNA isolation system (Invitrogen Life Technologies, Philadelphia, PA) was used to isolate RNA. Superscript II reverse transcriptase (Invitrogen) was used to synthesis the cDNA with oligo dT (Roche Diagnostics) as the primer. 1.5 µg total RNA was used for each sample.
4. Semiquantitative RT-PCR was performed by PCR amplification of 5 µL of synthesized cDNA from each sample with limited cycles using human VEGF-specific primers: 5'-GGCAGAAGGAGGAGGGACAGAATC (sense) and 5'-CATTTACACGTCTGCGGATCTTGT (antisense), and LacZ primers: 5'-GGCGTAAGTGAAGCGACCCG (sense) and 5'-GCGTGCAGCAGTGGC GATGG (antisense).
5. The PCR products were separated by electrophoresis in 1.2% agarose gel and transferred onto N+ Hybond nylon membrane.
6. The amplified cDNA fragments were hybridized with isotope-labeled probes of individual genes and quantified by phosphoimager analyses.
7. Mouse hypoxanthine phosphoribosyl transferase (HPRT) was used as internal control using the primers 5'-AAGGACCTCTCGAAGTGTTGGATA-3' (sense) and 5'-CATTTAAAAGGAACTGTTGACAACG-3' (antisense).
8. Both LacZ and VEGF gene expression was induced in AAVH9LacZ- and AAVH9VEGF-transduced ischemic hearts and was 8 times (LacZ) and 15 times (VEGF) higher than in the same vector-injected normal hearts. The gene expression level was about the same in AAV-LacZ and AAV-VEGF (in these vectors, gene expression was controlled by CMV prompter) transduced normal and ischemic hearts.

4. Notes

1. Nine copies of Epo HRE and a minimum SV40 promoter could mediate efficient hypoxia-inducible gene expression in ischemic hearts *(23)* and brain tumor cell lines *(30)*. Other investigators have induced hypoxia-responsive gene expression in hearts and other tissues using four copies of Epo HRE and a minimum SV40 promoter *(24)*. HRE can also be used in combination with CMV minimum promoter to induce hypoxia-responsive gene expression *(31)*. Cardiac-specific promoters can also be used in combination with HRE to mediate cardiac-specific and hypoxia-responsive gene expression.
2. The AAV Helper-Free Gene Delivery and Expression System was generated based on AAV serotype 2 virus. Many new AAV serotypes have been cloned in recent years *(32–37)*. Recombinant cross-packaging of the AAV genome of one serotype into other AAV serotypes to achieve optimal tissue-specific gene trans-

duction is now possible. Studies by us and others *(38)* have shown that AAV serotype 1 results in more efficient transduction of genes in the murine and human adult heart compared with serotypes 2, 3, 4, and 5. The AAV packaged in serotypes other than 2 can be made by replacing the serotype 2 CAP sequence in pAAV-RC with the CAP sequence of the corresponding serotype.
3. AAV titer can also be determined by quantitative PCR assay *(39)*. Infectious particles can be determined by using serial diluted viral stocks to infect 293 or other cells and quantitating the infected cells or transgene expressions. DNase I digestion can be used to eliminate unpackaged DNA in viral stocks *(40)*.
4. AAV can be purified by using other methods, such as nonionic iodixanol gradients, ion exchange, or heparin affinity chromatography by either conventional or high-performance liquid chromatography columns *(40,41)*. Ion exchange or heparin affinity chromatography can also be used in combination with $CsCl_2$ gradients or nonionic iodixanol gradients.

Acknowledgments

We thank Alicia Barcena and Janice Arakawa-Hoyt for preparing the AAV vectors, Yu Huang for performing the mouse surgeries, and Lisa Woldin for critically reviewing the manuscript. This work was supported by a National Institutes of Health Grant, HL067969 to Y.W. Kan.

References

1. Jones, S. P., Greer, J. J., Kakkar, A. K., et al. (2004) Endothelial nitric oxide synthase overexpression attenuates myocardial reperfusion injury. *Am. J. Physiol. Heart Circ. Physiol.* **286,** H276–H282.
2. Jones, S. P., Greer, J. J., van Haperen, R., Duncker, D. J., de Crom, R., and Lefer, D. J. (2003) Endothelial nitric oxide synthase overexpression attenuates congestive heart failure in mice. *Proc. Natl. Acad. Sci. USA* **100,** 4891–4896.
3. Woo, Y. J., Zhang, J. C., Vijayasarathy, C., et al. (1998) Recombinant adenovirus-mediated cardiac gene transfer of superoxide dismutase and catalase attenuates postischemic contractile dysfunction. *Circulation* **98,** II255–II260; discussion II260–II261.
4. Mack, C. A., Patel, S. R., Schwarz, E. A., et al. (1998) Biologic bypass with the use of adenovirus-mediated gene transfer of the complementary deoxyribonucleic acid for vascular endothelial growth factor 121 improves myocardial perfusion and function in the ischemic porcine heart. *J. Thorac. Cardiovasc. Surg.* **115,** 168–176; discussion 176–177.
5. Giordano, F. J., Ping, P., McKirnan, M. D., et al. (1996) Intracoronary gene transfer of fibroblast growth factor-5 increases blood flow and contractile function in an ischemic region of the heart [see comments]. *Nat. Med.* **2,** 534–539.
6. Laham, R. J., Simons, M., and Sellke, F. (2001) Gene transfer for angiogenesis in coronary artery disease. *Annu. Rev. Med.* **52,** 485–502.
7. Banai, S., Jaklitsch, M. T., Shou, M., et al. (1994) Angiogenic-induced enhancement of collateral blood flow to ischemic myocardium by vascular endothelial growth factor in dogs. *Circulation* **89,** 2183–2189.

8. Lee, R. J., Springer, M. L., Blanco-Bose, W. E., Shaw, R., Ursell, P. C., and Blau, H. M. (2000) VEGF gene delivery to myocardium: deleterious effects of unregulated expression. *Circulation* **102,** 898–901.
9. Springer, M. L., Chen, A. S., Kraft, P. E., Bednarski, M., and Blau, H. M. (1998) VEGF gene delivery to muscle: potential role for vasculogenesis in adults. *Mol. Cell* **2,** 549–558.
10. Dennery, P. A., Sridhar, K. J., Lee, C. S., et al. (1997) Heme oxygenase-mediated resistance to oxygen toxicity in hamster fibroblasts. *J. Biol. Chem.* **272,** 14937–14942.
11. Suttner, D. M. and Dennery, P. A. (1999) Reversal of HO-1 related cytoprotection with increased expression is due to reactive iron. *FASEB J.* **13,** 1800–1809.
12. Bohl, D., Salvetti, A., Moullier, P., and Heard, J. M. (1998) Control of erythropoietin delivery by doxycycline in mice after intramuscular injection of adeno-associated vector. *Blood* **92,** 1512–1517.
13. Hofmann, A., Nolan, G. P., and Blau, H. M. (1996) Rapid retroviral delivery of tetracycline-inducible genes in a single autoregulatory cassette [see comments]. *Proc. Natl. Acad. Sci. USA* **93,** 5185–5190.
14. Bohl, D., Naffakh, N., and Heard, J. M. (1997) Long-term control of erythropoietin secretion by doxycycline in mice transplanted with engineered primary myoblasts [see comments]. *Nat. Med.* **3,** 299–305.
15. Gossen, M. and Bujard, H. (1992) Tight control of gene expression in mammalian cells by tetracycline-responsive promoters. *Proc. Natl. Acad. Sci. USA* **89,** 5547–5551.
16. Wang, Y., O'Malley, B. W., Jr., Tsai, S. Y., and O'Malley, B. W. (1994) A regulatory system for use in gene transfer. *Proc. Natl. Acad. Sci. USA* **91,** 8180–8184.
17. Blau, H. M. and Rossi, F. M. (1999) Tet B or not tet B: advances in tetracycline-inducible gene expression [comment]. *Proc. Natl. Acad. Sci. USA* **96,** 797–799.
18. Rossi, F. M., Guicherit, O. M., Spicher, A., et al. (1998) Tetracycline-regulatable factors with distinct dimerization domains allow reversible growth inhibition by p16. *Nat. Genet.* **20,** 389–393.
19. Kringstein, A. M., Rossi, F. M., Hofmann, A., and Blau, H. M. (1998) Graded transcriptional response to different concentrations of a single transactivator. *Proc. Natl. Acad. Sci. USA* **95,** 13670–13675.
20. Wang, G. L., Jiang, B. H., Rue, E. A., and Semenza, G. L. (1995) Hypoxia-inducible factor 1 is a basic-helix-loop-helix-PAS heterodimer regulated by cellular O2 tension. *Proc. Natl. Acad. Sci. USA* **92,** 5510–5514.
21. Wang, G. L. and Semenza, G. L. (1995) Purification and characterization of hypoxia-inducible factor 1. *J. Biol. Chem.* **270,** 1230–1237.
22. Jiang, B. H., Rue, E., Wang, G. L., Roe, R., and Semenza, G. L. (1996) Dimerization, DNA binding, and transactivation properties of hypoxia-inducible factor 1. *J. Biol. Chem.* **271,** 17771–17778.
23. Su, H., Arakawa-Hoyt, J., and Kan, Y. W. (2002) Adeno-associated viral vector-mediated hypoxia response element-regulated gene expression in mouse ischemic heart model. *Proc. Natl. Acad. Sci. USA* **99,** 9480–9485.
24. Pachori, A. S., Melo, L. G., Hart, M. L., et al. (2004) Hypoxia-regulated therapeutic gene as a preemptive treatment strategy against ischemia/reperfusion tissue injury. *Proc. Natl. Acad. Sci. USA* **101,** 12282–12287.

25. Su, H., Lu, R., and Kan, Y. W. (2000) Adeno-associated viral vector-mediated vascular endothelial growth factor gene transfer induces neovascular formation in ischemic heart. *Proc. Natl. Acad. Sci. USA* **97**, 13801–13806.
26. Du, L., Kido, M., Lee, D. V., et al. (2004) Differential myocardial gene delivery by recombinant serotype-specific adeno-associated viral vectors. *Mol. Ther.* **10**, 604–608.
27. Semenza, G. L. and Wang, G. L. (1992) A nuclear factor induced by hypoxia via de novo protein synthesis binds to the human erythropoietin gene enhancer at a site required for transcriptional activation. *Mol. Cell Biol.* **12**, 5447–5454.
28. Wang, G. L. and Semenza, G. L. (1993) General involvement of hypoxia-inducible factor 1 in transcriptional response to hypoxia. *Proc. Natl. Acad. Sci. USA* **90**, 4304–4308.
29. Chavez, J. C., Agani, F., Pichiule, P., and LaManna, J. C. (2000) Expression of hypoxia-inducible factor-1 alpha in the brain of rats during chronic hypoxia. *J. Appl. Physiol.* **89**, 1937–1942.
30. Ruan, H., Su, H., Hu, L., Lamborn, K. R., Kan, Y. W., and Deen, D. F. (2001) A hypoxia-regulated adeno-associated virus vector for cancer-specific gene therapy. *Neoplasia* **3**, 255–263.
31. Shibata, T., Giaccia, A. J., and Brown, J. M. (2000) Development of a hypoxia-responsive vector for tumor-specific gene therapy. *Gene Ther.* **7**, 493–498.
32. Gao, G. P., Alvira, M. R., Wang, L., Calcedo, R., Johnston, J., and Wilson, J. M. (2002) Novel adeno-associated viruses from rhesus monkeys as vectors for human gene therapy. *Proc. Natl. Acad. Sci. USA* **99**, 11854–11859.
33. Rutledge, E. A., Halbert, C. L., and Russell, D. W. (1998) Infectious clones and vectors derived from adeno-assoicated virus (AAV) serotypes other than AAV type 2. *J. Virol.* **72**, 309–319.
34. Muramatsu, S., Mizukami, H., Young, N. S., and Brown, K. E. (1996) Nucleotide sequencing and generation of an infectious clone of adeno-associated virus 3. *Virology* **221**, 208–217.
35. Chiorini, J. A., Yang, L., Liu, Y., Safer, B., and Kotin, R. M. (1997) Cloning of adeno-associated virus type 4 (AAV4) and generation of recombinant AAV4 particles. *J. Virol.* **71**, 6823–6833.
36. Chiorini, J. A., Kim, F., Yang, L., and Kotin, R. M. (1999) Cloning and characterization of adeno-associated virus type 5. *J. Virol.* **73**, 1309–1319.
37. Xiao, W., Chirmule, N., Berta, S. C., McCullough, B., Gao, G., and Wilson, J. M. (1999) Gene therapy vectors based on adeno-associated virus type 1. *J. Virol.* **73**, 3994–4003.
38. Du, L., Sullivan, C. C., Chu, D., et al. (2003) Signaling molecules in nonfamilial pulmonary hypertension. *N. Engl. J. Med.* **348**, 500–509.
39. Clark, K. R., Liu, X., McGrath, J. P., and Johnson, P. R. (1999) Highly purified recombinant adeno-associated virus vectors are biologically active and free of detectable helper and wild-type viruses. *Hum. Gene Ther.* **10**, 1031–1039.
40. Zolotukhin, S., Potter, M., Zolotukhin, I., et al. (2002) Production and purification of serotype 1, 2, and 5 recombinant adeno-associated viral vectors. *Methods* **28**, 158–167.
41. Zolotukhin, S., Byrne, B. J., Mason, E., et al. (1999) Recombinant adeno-associated virus purification using novel methods improves infectious titer and yield. *Gene Ther.* **6**, 973–985.

20

Lentivirus-Mediated Gene Expression

Jing Zhao and Andrew M. L. Lever

Summary

Lentiviruses have the capacity to enter and integrate their genetic material into cells that are not dividing. This property is retained in vectors based on these agents. They can thus effect gene delivery to cells that are difficult to transduce such as cardiac myocytes in vitro and in vivo. They are also relatively efficient at entering dividing cells and can transduce stem cells and vascular endothelium. They have a substantial gene-carrying capacity of up to around 9 kb. They do not trigger an inflammatory response and are thus useful when proinflammatory agents are undesirable, such as in transplantation. Their ease of cloning and well-understood molecular biology have made them highly suitable for gene delivery to the heart.

Key Words: Lentiviral vector; HIV-1 vector; gene therapy; gene transfer; transduction; 293T cells; cardiomyocytes, heart.

1. Introduction

Since the first report that lentiviral vectors could transduce adult and neonatal cardiomyocytes in vitro and in vivo *(1)*, lentiviral vectors have been increasingly used as vehicles for delivery of foreign genes of interest to the heart and vasculature in vitro and in vivo *(2–4)*. Lentiviral vectors possess the desirable, and collectively unique, properties of efficient gene transfer to nondividing cells in vitro and in vivo without evoking a cellular immune response, ease of genetic manipulation, a relatively large capacity to accommodate genes of interest (potentially up to 8 to 9 kb of sequence), stable integration, and long-term gene expression. Animals and (ultimately) patients are also unlikely to have preformed antibody responses to their structural proteins.

Typical of all retroviruses the lentivirus genome (**Fig. 1A**) contains the three genes: *gag*, *pol*, and *env*. In addition, lentiviruses have a number of unique small open reading frames coding for six accessory and regulatory

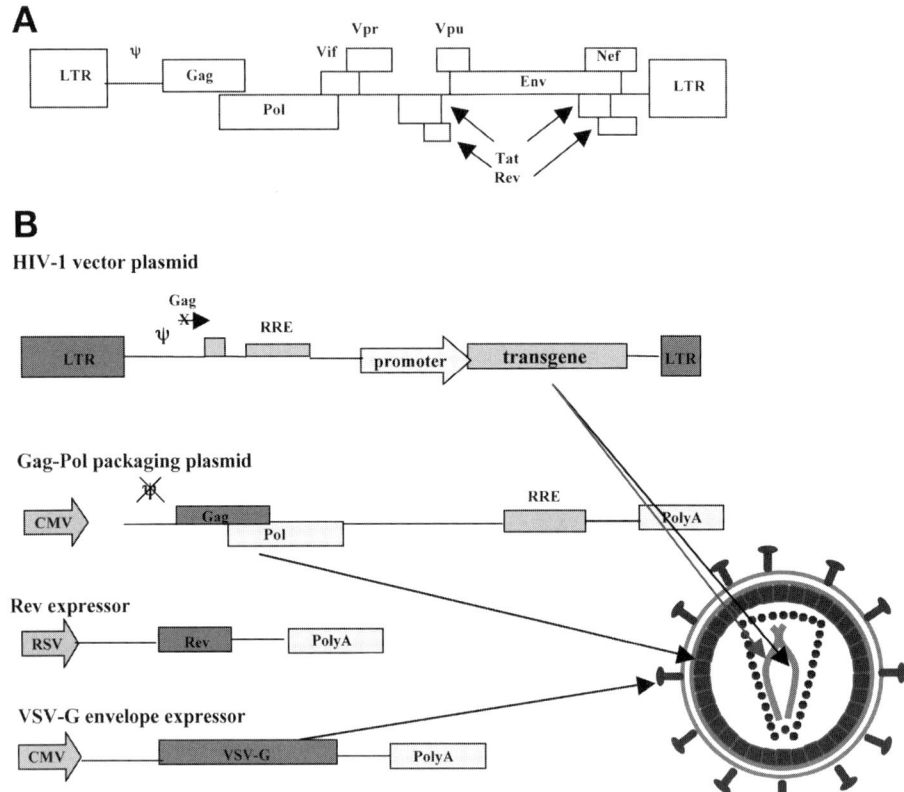

Fig. 1. (**A**) Lentivirus genomic organization. (**B**) Third-generation HIV-1 vector constructs.

proteins: *vif, vpr, vpu, tat, rev,* and *nef* (*5*). These are associated with control of lentivirus gene expression maturation, budding, infectivity, and pathogenesis. Since lentiviruses were first used gene in gene delivery (*6*), many efforts have been made to increase biosafety while at the same time broadening the vector tropism and increasing gene transfer efficiency. Today the most commonly used lentiviral vectors, which are based predominantly on human immunodeficiency virus-type 1 (HIV-1), are the so-called third generation, vectors, whose accessory protein genes are lacking, and in which the structural protein genes and the *rev* gene are encoded *in trans* by separate plasmids (*7,8*). Deletion of nonessential sequences has also been used to minimize sequence identity between constructs to reduce the risk of homologous recombination. In this system, the HIV-1 sequence (**Fig. 1A**) has been deconstructed and segregated into four plasmids: (1) vector plasmid, (2) *gag-pol* packaging plasmid, (3) Rev expressor, and (4) heterologous (usually VSV-G) envelope expressor. Together these four plasmids, when cotransfected, generate a replication-deficient virus containing the RNA of the transgene of interest (**Fig. 1B**), which will be delivered to target cells, reverse-transcribed, and integrated into the genome of the cell. The plasmids are described further as follows:

1. The vector plasmid is the transducing vector, which contains minimal HIV-1 sequence and excludes all viral structural and accessory protein genes. Although it includes a 5' and 3' long terminal repeat (LTR) from the original provirus to permit efficient host cell integration, there is a 400-bp deletion in the U3 region of the 3' LTR. This deletion is copied into the 5' LTR during reverse transcription and prior to integration, resulting in a 5' sequence with a deleted promoter. This eliminates transcription initiation from the LTRs, overcoming the possibility of insertional activation of cellular oncogenes and allowing the transgene to be expressed from a specific heterologous internal promoter. The vector contains the packaging signal (Ψ). This is a *cis*-acting sequence required for the RNA to be captured into the viral particle *(9–11)*. Most of the *gag* sequences and all the *pol* sequences have been deleted. A 5' segment of about 350 bp of the *gag* sequence are usually included in the vector to increase packaging efficiency *(11,12)*. Any potential Gag coding may be disrupted by an introduced frameshift mutation. Most of the *env* gene sequence has been deleted except the Rev response element (RRE), an important element allowing nuclear export of the vector RNA and probably having enhancing effects on translation and encapsidation *(10,13,14)*. The nuclear export function may be substituted or augmented by additional transport elements such as the Mason-Pfizer monkey virus (M-PMV) constitutive transport element (CTE) *(15)* or the woodchuck hepatitis virus posttranscriptional response element (WHPRE) *(16)*. For efficient reverse transcription, the native polypurine tract is retained. Recent papers suggest that the region of nucleic acid that forms a triplex during reverse transcription, the so-called DNA flap *(17)*, enhances vector efficiency possibly (but not conclusively) by facilitating nuclear entry *(18,19)*.
2. The *gag-pol* packaging plasmid is designed to provide the HIV structural proteins Gag, which encodes the core proteins and the nucleocapsid protein, and Pol, which encodes the enzymes needed for replication. Eucaryotic codon optimization of Gag and Pol is claimed to enhance particle production. The genes are commonly expressed from the human cytomegalovirus (hCMV) immediate-early promoter. The 3'LTR sequence is substituted by a heterologous polyadenylation signal (poly A). These two innovations again reduce interplasmid homology. The packaging signal (Ψ) has also been deleted; therefore the transcripts cannot be packaged into the vector particles.
3. The Rev expressor provides the Rev protein, which will bind to the RRE in the vector RNA. It regulates virus RNA transport and splicing.

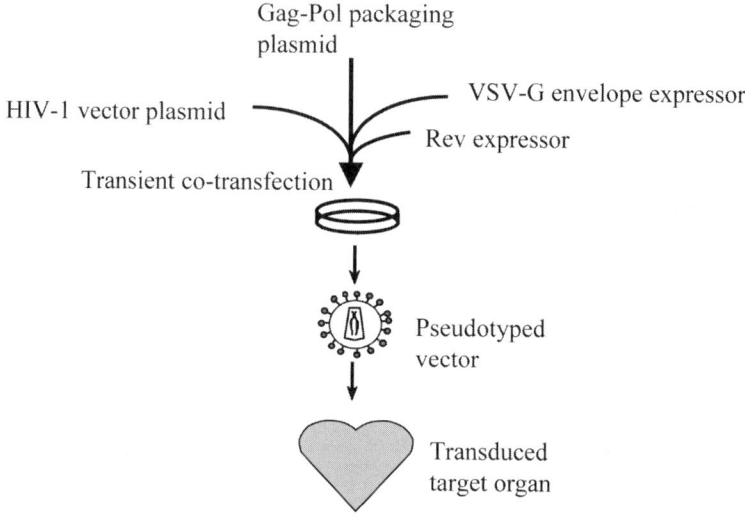

Fig. 2. Schematic of the generation of lentiviral vectors by transient cotransfection of 293T cells.

4. The viral vector particles are pseudotyped by heterologous viral envelope proteins, usually vesicular stomatitis virus G protein (VSV-G). The VSV-G envelope not only broadens the tropism of the vector but also offers an advantage of markedly increasing particle stability, allowing efficient concentration by ultracentrifugation and the ability to store frozen vector stock with relatively little loss of efficacy. It also eliminates potential homologous recombination events with the other constructs or endogenous retroviral sequences. There is evidence that it permits cell entry by an endosomal rather than a fusion pathway [20] and that this reduces the requirement for the HIV Nef protein. Apart from lack of overt sequence homology between these four plasmids, codon usage provides an additional barrier to recombination.

Replication-deficient virus is generated by transient cotransfection of the four plasmids into 293T cells (**Fig. 2**). The use of a four-plasmid combination, inactivation of viral promoters, heterologous envelope pseudotyping, and elimination of *cis*-acting sequences from the packaging vector makes it virtually impossible for replication-competent recombinant virus to emerge without the incorporation of exogenous sequence. When using HIV-1 viral vector-transduced target cells, only vector plasmid RNA will be reverse-transcribed and integrated into the host genomic DNA. All the structural proteins and any transferred Rev protein in the particle will be left in the cell cytoplasm and degraded by cellular endolysosomes (**Fig. 3**).

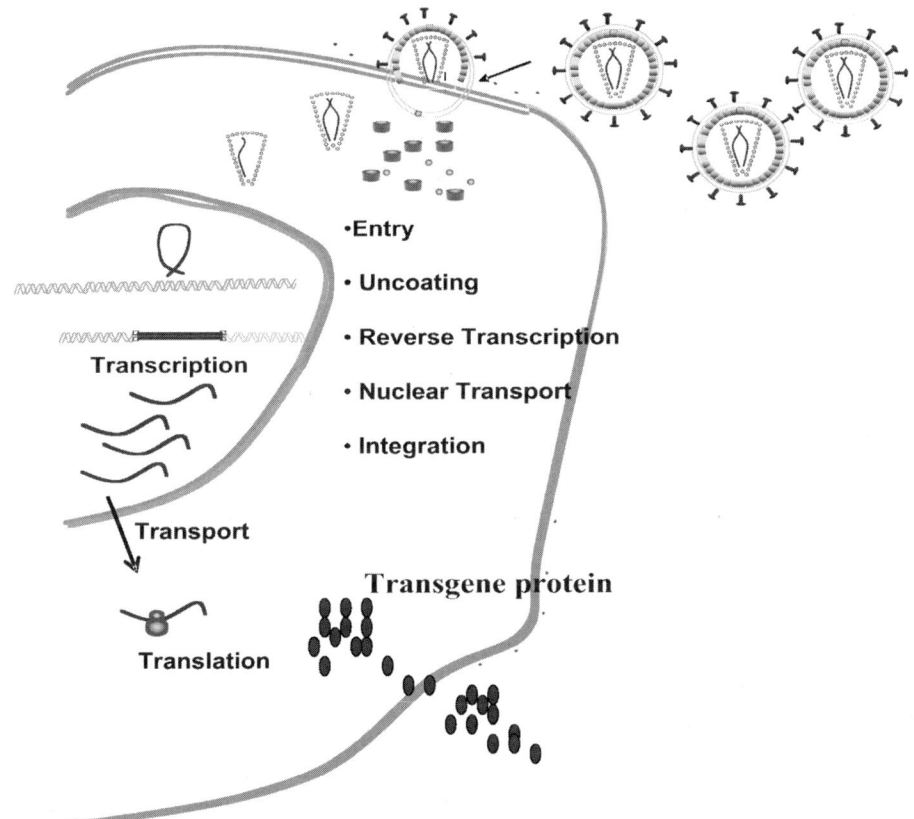

Fig. 3. Schematic of lentiviral vector transduction of target cells.

A number of attempts have been made to produce stable packaging cell lines for lentiviruses, such as those used in other retroviral systems *(21,22)*. A recent development is a line into which the packaging plasmids have been introduced by vectors *(23)*. This appears to be the first line to permit stable expression of packaging proteins for up to 3 mo. Whether long-term production will be achieved except by use of low-passage-number cells remains to be seen. Previous lines have either been low titer and/or expressed only for a few months before protein production has been lost *(24)*.

What follows is a description of our own practice in generating lentiviral vectors, which have so far successfully transduced cardiac myocytes both neonatal and adult, in vitro and in vivo, and with which we have obtained expression of functional genes in cardiac heterotopic transplants in animal studies *(25)*.

2. Materials
2.1. Transfection

1. 293T cells (see **Note 1**) are used as producer cells. They are cultured in Nunc™ (Merck, UK; or equivalent) 10-cm culture dishes in Dulbecco's modified Eagle's medium (DMEM; Gibco-BRL) supplemented with 10% fetal calf serum (FCS), 100 U/mL penicillin, and 100 µg/mL streptomycin.
2. 2.5 M $CaCl_2$ (Sigma, cat. no. C5080). Dissolve 36.7 g $CaCl_2·2H_2O$ in 100 mL Milli Q H_2O. Filter-sterilize through a 0.2-µm filter. Store at –70°C in 20-mL aliquots. Once thawed, $CaCl_2$ solution can be stored at 4°C for several weeks without observing any change in the transfection efficiency.
3. 2X BBS solution, pH 6.95: 50 mM BES (1.06 g/100 mL; Sigma, cat. no. B6266), 280 mM NaCl (1.63 g/100 mL), 1.5 mM Na_2HPO_4 (0.02 g/100 mL).

Dissolve components in 80 mL of Milli Q H_2O and adjust to pH 6.95 with NaOH as required (see **Note 2**). Then make up to 100 mL with Milli Q H_2O. Filter-sterilize through a 0.2-µm filter. Store at –70°C in 15-mL aliquots. One aliquot is enough to transfect 20 10-cm culture dishes. Equilibrate the thawed 2X BBS to room temperature before transfection.

2.2. Concentration of Lentiviral Vector

1. Beckman ultracentrifuge and SW28 rotor.
2. Beckman Ultra-Clear centrifugation tubes (25 × 89 mm, cat. no. 344058).
3. 1% Bovine serum albumin (BSA) in phosphate-buffered saline (PBS). Dissolve 1 g of BSA (Sigma, cat. no. A2153) in 100 mL of PBS. Filter-sterilize through 0.2-µm filter. Store at –70°C in 5-mL aliquots.

2.3. Titration of Lentiviral Vectors

1. SV2 fibroblast cells or other target cells. SV2 cells are cultured in minimum essential medium (MEM; Gibco-BRL) supplemented with 10% FCS, 100 U/mL penicillin, and 100 µg/mL streptomycin.
2. 4% (w/v) Formaldehyde in PBS.
3. 0.1% Toluidine blue in H_2O (Sigma, cat. no. T3260).
4. X-gal Reagents required for β-galactosidase staining.

3. Methods
3.1. Transient Cotransfection of 293T Cells
Using a Modified Calcium Phosphate-Mediated Transfection Procedure

Calcium phosphate is the most widely used method to transfect DNA into mammalian cells. Although the mechanism is not yet clear, it is believed that transfected DNA enters the cells by endocytosis and subsequently transfers to the nucleus. Since Graham and Van de Eb first established this method in 1973 *(26)*, many minor modifications of the procedure have been described. Among them, Chen and Okayama's modified transfection protocol is considered to be highly efficient to achieve stable transformation of mammalian cells with supercoiled plasmid DNA *(27)*. In their modified calcium phosphate transfection protocol, a DNA-calcium phosphate co-precipitate is formed gradually onto the cells in the tissue culture medium during prolonged incubation (16–18 h) under controlled conditions of pH 6.95 and 3% CO_2.

1. Prepare the solutions as described in **Subheading 2.1.** Equilibrate the solution to room temperature if defrosted from –20°C.
2. Day one: split the exponentially growing 293T cells and seed them into 10-cm culture dishes at 40 to 50% confluence with 10 mL of DMEM supplemented with 10% FCS. Make sure the cells are evenly dispersed in the culture dishes. Culture the cells overnight in the humidified 37° C incubator in an atmosphere of 5% CO_2.
3. Day two: change the medium 3 h prior to transfection. In the late afternoon, set up transfection cocktails in bijou tubes (one bijou per 10-cm dish). The following recipe is for one 10-cm dish. The precipitate should be prepared in the following sequential order. The relative concentrations of the different component plasmids are for guidance and may need adjusting depending on the molecular weights of the constructs used.
    a. Prepare, in order: 737 µL 2X BBS, 737 µL Milli Q H_2O, 10 µg vector plasmid *(see* **Note 3**), 6.5 µg Gag-Pol packaging plasmid, 3.5 µg envelope plasmid, 2 µg Rev-expressing plasmid, and 72 µL 2.5 *M* $CaCl_2$.
    b. Mix the reaction by flicking each bijou gently and stand the bijou at room temperature for 20 min. This time is critical for optimal formation of the precipitate. If left too long, they form a coarse precipitate. If left for too short a time, they will form very little precipitate. In both cases the transfection efficiency will be affected.
    c. After 20 min a very fine precipitate will be formed, and the solution should appear lightly opaque. Flick the bijou again prior to adding it to the cells.
    d. Add the mixture to the culture dishes slowly in a dropwise manner while gently rocking the dish a few times to allow even distribution of the precipitate over the cell monolayer. Do not swirl the dish; otherwise the precipitate will collect in the center. Add one bijou mixture to one 10-cm dish.
    e. Then place the dishes in a humidified 37°C incubator with 3% CO_2 tension for 16 to 18 h.
4. Day three: check your cells in the morning; you should see that the DNA-calcium phosphate precipitate on the cells forms a very fine and sandy precipitate all over the cells, indicating successful transfection. Remove the medium by aspiration and add 10 mL of fresh complete medium for each 10-cm dish. Transfer all the dishes to a humidified 37°C incubator in an atmosphere of 5% CO_2.
5. Day five: collect viral particles from the supernatant and filter through a 0.45-µm filter. The viral vector stock can be used immediately to transduce cells or can be concentrated following the protocol in **Subheading 3.2.** It is always advisable to

concentrate the virus before use in vitro and practically essential before in vivo use. By doing so, one obtains not only higher titer of virus but also a cleaner viral vector preparation since some of the cellular debris and vector structural proteins left in the supernatant can be toxic to the cells.

293T cells left on the dish should be examined for green fluorescent protein (GFP; or other reporter) expression. If another transgene was expressed in the vector plasmid, a control transfection with a GFP plasmid should be always used to estimate the transfection efficiency. Only when around 80% of 293T cells turn green does it indicate that cotransfection of the four plasmids was successful.

3.2. Concentration of Lentiviral Vector

For in vivo gene transfer procedure, it is always useful to concentrate the virus to allow for small injection volumes. Lentivirus vector pseudotyped with VSV-G glycoprotein are very stable to mechanical stress, and stocks can be made to very high concentration by ultracentrifugation *(28,29)*.

1. Collect the supernatant from three 10-cm dishes with a 50-mL syringe, place a 0.45-µm filter onto the syringe nozzle, and filter into a 50-mL centrifuge tube. Top up the tube with normal medium and balance the tube in an SW 28 rotor sleeve.
2. Spin the pooled supernatant in Beckman SW 28 swing rotor for $2\,{}^1/_2$ h at 25,000 rpm at 4°C.
3. Aspirate the supernatant carefully since the pellet can be soft and may be difficult to visualize (*see* **Note 4**). Efforts should be made to remove as much supernatant as possible since there will be fragments of structural protein left in the supernatant that are toxic to the cells.
4. Dissolve the pellet in 1% BSA in PBS into 0.2 to 1% of the original volume PBS (or, alternatively, medium with 1% BSA). High concentrations of serum can cause virion aggregation and therefore decrease transduction efficiency. However, some protein is needed to preserve the virus. Try to avoid dissolving the pellet by repeatedly pipeting up and down to minimize bubble formation, rather, stir the pellet gently with a yellow tip. If the pellet is difficult to dissolve, leave on ice for 15 min and reattempt every few minutes using the same yellow tip to get a good suspension. The supernatant at this point contains between 10^5 and 10^6 transducing units per mL (TU/mL).
5. Store the lentiviral vector stock at –70°C. It can be stored at –70°C for over a year without observing any decreasing titer. Do not thaw and refreeze the vector since the virus titer can be reduced up to fivefold after refreezing.

3.3. Titration of Lentiviral Vectors

The titer of the lentiviral vector stocks can be determined by transducing the target cells with serial dilutions of viral vector carrying selectable or detectable

Lentivirus-Mediated Gene Expression

markers and subsequently counting the cells or derived colonies. The commonly used target cells are SV2 cells, Hela cells, or NIH3T3 cells. There are two different way to titer the virus vector, as follows.

3.3.1. Colony-Forming Units

1. Colony-forming-unit per mL (cfu/mL): if the lentiviral vector encodes a selectable marker such as G418 or puromycin resistance, vector titration should be straightforward. After transducing the target cells with serial dilutions of the viral vector stock, selection for the appropriate marker is applied and the number of resistant colonies counted. Virus titer then expresses as cfu/mL).
2. The day before transduction, plate SV2 cells at 20 to 30% confluence in a 10-cm culture dish in 10 mL MEM supplemented with 10% FCS.
3. Transduce separate dishes with serial 10-fold dilutions of viral vector.
4. Apply selection 24 h later after transduction. It is not necessary to change the medium unless desired.
5. Replace the selection medium as soon as it begins to turn yellow (usually 3–4 d post transduction). Gently swirl the plates to bring dead cells into suspension before aspirating the old medium. The dead cells form a heavy sediment, which may take some time to resuspend. Inspect the dishes under a microscope to ensure that most of the dead cells are removed.
6. Repeat this process twice more at 3- to 4-d intervals.
7. Inspect the plates under the microscope to make sure of complete killing. Normally 10 to 12 d after transduction, the colonies are well formed and selection is complete. If there are areas of incomplete killing or dying cells, which have not yet detached from the dishes, add fresh selection medium and incubate the plates for a further 24 h. Patchy killing or background of any kind will make the plates difficult to score.
8. When the selection is complete, aspirate the medium and fix the cells by adding 5 mL of 4% formaldehyde in PBS. Stand at room temperature for 10 min.
9. Aspirate the formaldehyde solution and add 5 mL 0.1% toluidine blue in distilled water. Stand the dishes at room temperature for 5 min.
10. Rinse the dishes twice with 10 mL distilled water and then air-dry and count colonies. Vector titers are calculated by the number of colonies multiplied by the dilution factor, and expressed as cfu/mL.

3.3.2. Transducing Units

Transducing units per mL (TU/mL) or infectious units per mL (IU/mL): this is a commonly used unit for lentiviral vector titer. If a viral vector contains a readily detectable marker gene such as *lacZ* or *GFP*, lentiviral vectors can be titrated directly by serial dilutions of vector and incubated with target cells for a minimum of 72 h (*see* **Note 5**). Then *in situ* histochemical staining of β-gal expression or fluorescence-activated cell sorting (FACS) analysis of GFP expression is performed. The titer is determined to be TU/mL or IU/mL.

1. The day before transduction, plate SV2 cells at 40 to 50% confluence in 6-well plates in 2 mL MEM supplemented with 10% FCS.
2. Next day, aspirate the medium, replace with minimal fresh culture medium, and add viral stock or serial 10-fold dilutions of viral vector.
3. Incubate the plates in the 37°C incubator for a minimum of 6 h. (Overnight achieves better transduction efficiency.)
4. Add 2 mL fresh medium per well and leave for 72 h at 37°C.
5. Aspirate the medium and wash the cells once with 1 mL PBS.
6. Fix the cells and stain with X-gal for β-gal gene expression or FACS analysis of GFP expression.
7. The virus titer can be calculated from the number of β-gal-positive cells or GFP-positive cells times the dilution factor and expressed as TU/mL or IU/mL.

3.4. Transduction

Lentiviral vectors pseudotyped with VSV-G envelope can transduce all non-dividing cells. In line with the interests of the authors, the following protocols describe the transduction of primary aortic endothelial cells, neonatal cardiomyocytes, and adult cardiomyocytes cells in vitro and in vivo.

3.4.1. In Vitro Transduction

1. The primary cells should be plated at 40 to 50% confluence in 6-well plates.
2. Prior to the transduction, replace medium with minimal fresh growth medium, just enough to cover the cells, and then add the viral vector.
3. Incubate the plates at 37°C for a minimum of 6 h. (Overnight achieves better transduction efficiency.)
4. Add 2 mL fresh medium per well and leave for 48 h at 37°C.
5. Aspirate the medium and wash the cells once with 1 mL PBS.
6. Analyze transgene expression.

3.4.2. In Vivo Transduction of Heart

Lentiviral vectors are generally pseudotyped with VSV-G envelope proteins, which mediate cell entry through membrane fusion. It is therefore critical to allow sufficient cell and viron contact time. Efficient delivery of a transgene to transduce the heart can be achieved by direct injection. Intracavity infusion in a hypothermic state has been reported to be successful *(3)*.

1. The volume of fluid that can be injected in animals such as rodents is very limited, and the total should be less than 10% of the volume of the heart. The titer must therefore be sufficient to ensure injection of adequate vector particle number in a volume of 50 to 200 µL.
2. Warm the viral vector to room temperature before injection. To visualize the inoculum better during and after injection, it is helpful to mix the vector preparation with 2% Evans Blue before injection.

3. Viral vectors should be injected between the two muscle layers at the apex of the heart using a 29-gage needle. In our experience, this gives good gene transduction with little toxicity at the site of injection and sustained gene expression for at least 28 d.

If you are using β-gal as the transgene to transduce the heart, steel forceps should not be used to manipulate the heart tissues in the posttransduction staining solution, since the metal reacts to form a blue precipitate, which gives false-positive results. If you are using paraformaldehyde to fix the heart, 2 mM MgCl$_2$ and 1 mM EGTA should be included to reduce endogenous background staining.

4. Notes

1. 293T cells were originally called 293ts A1609 ne[36] which derived from 293 cells, a human embryonic kidney cell line. 293T cells are transformed with sheared type 5 adenovirus DNA and SV40 large T antigen. 293T cells are good DNA recipient cells for transfection; however, they sometimes behave unpredictably. The following suggestions should be considered when you are working with 293T cells:
    a. The medium must be prewarmed to 37°C to avoid thermal shock, which makes 293T cells shrink and detach from the culture dish.
    b. Add medium very gently to the cells along the side of culture dish to avoid stripping cells off the dish surface.
    c. Always split 293T cells before they reach 100% confluence. Overconfluent 293T cells will detach from the culture dish or flask, and the cells left on the dish or flask will not regain normal growth.
    d. Discard the 293T cells if they have been split more than six to eight times, since the transfection efficiency declines at this point.
    e. Freeze down 293T cells at 60% confluence when they are still in exponential growth. Frozen medium should contain 900 μL FCS and 100 μL dimethylsulfoxide (DMSO; Sigma, cat. no. D4540).
    f. 293T cells will not grow well in some commercial culture dishes. In our hands the best culture dishes or flasks are from Nunc.
2. The pH of the solutions in the modified calcium phosphate transfection method is very important. The precipitates will not form below pH 6.90 and will be coarse above 7.05. In both cases this will affect the transfection efficiency dramatically and adversely. (Always calibrate the pH meter before using it.)
3. DNA should be prepared using a commercial kit. In our hands the best DNA is prepared from Qiagen Maxi Kits (cat. no. 12163). At the last step of preparation the DNA should be resuspended in distilled H$_2$O.
4. The vector pellet sometimes is quite soft and easily resuspended again. The supernatant should be aspirated as soon as possible after the ultracentrifugation; otherwise it is easy to lose the pellet.

5. Passive transfer of reporter gene product from the transfection dish on the vector particles can give falsely high impressions of gene transfer. GFP protein may be seen in target cells for up to 3 d. Expression after this time represents true transduction.

References

1. Zhao, J., Pettigrew, G. J., Thomas, J., et al. (2002) Lentiviral vectors for delivery of genes into neonatal and adult ventricular cardiac myocytes *in vitro* and *in vivo*. *Basic Res. Cardiol.* **97,** 348–358.
2. Dishart, K. L., Denby, L., George, S. J., et al. (2003) Third-generation lentivirus vectors efficiently transduce and phenotypically modify vascular cells: implications for gene therapy. *J. Mol. Cell Cardiol.* **35,** 739–748.
3. Bonci, D., Cittadini, A., Latronico, M. V., et al. (2003) 'Advanced' generation lentiviruses as efficient vectors for cardiomyocyte gene transduction in vitro and in vivo. *Gene Ther.* **10,** 630–636.
4. Fleury, S., Simeoni, E., Zuppinger, C., et al. (2003) Multiply attenuated, self-inactivating lentiviral vectors efficiently deliver and express genes for extended periods of time in adult rat cardiomyocytes in vivo. *Circulation* **107,** 2375–2382.
5. Subbramanian, R. A. and Cohen, E. A. (1994) Molecular biology of the human immunodeficiency virus accessory proteins. *J. Virol.* **68,** 6831–6835.
6. Naldini, L., Blomer, U., Gallay, P., et al. (1996) *In vivo* gene delivery and stable transduction of nondividing cells by a lentiviral vector. *Science* **272,** 263–267.
7. Dull, T., Zufferey, R., Kelly, M., et al. (1998) A third generation lentivirus vector with a conditional packaging system. *J. Virol.* **72,** 8463–8471.
8. Zufferey, R., Dull, T., Mandel, R. J., et al. (1998) Self-inactivating lentivirus vector for safe and efficient *in vivo* gene delivery. *J. Virol.* **72,** 9873–9880.
9. Lever, A., Gottlinger, H., Haseltine, W., and Sodroski, J. (1989) Identification of a sequence required for efficient packaging of human immunodeficiency virus type 1 RNA into virions. *J. Virol.* **63,** 4085–4087.
10. Kaye, J. F., Richardson, J. H., and Lever, A. M. (1995) Cis-acting sequences involved in human immunodeficiency virus type 1 RNA packaging. *J. Virol.* **69,** 6588–6592.
11. Berkowitz, R. D., Hammarskjold, M. L., Helga Maria, C., Rekosh, D., and Goff, S. P. (1995) 5' Regions of HIV-1 RNAs are not sufficient for encapsidation: implications for the HIV-1 packaging signal. *Virology* **212,** 718–723.
12. Parolin, C., Dorfman, T., Palu, G., Gottlinger, H., and Sodroski, J. (1994) Analysis in human immunodeficiency virus type 1 vectors of cis-acting sequences that affect gene transfer into human lymphocytes. *J. Virol.* **68,** 3888–3895.
13. D'Agostino, D. M., Felber, B. K., Harrison, J. E., and Pavlakis, G. N. (1992) The Rev protein of human immunodeficiency virus type 1 promotes polysomal association and translation of gag/pol and vpu/env mRNAs. *Mol. Cell. Biol.* **12,** 1375–1386.
14. Mautino, M. R., Ramsey, W. J., Reiser, J., and Morgan, R. A. (2000) Modified human immunodeficiency virus-based lentiviral vectors display decreased sensitivity to trans-dominant Rev. *Hum. Gene Ther.* **11,** 895–908.

15. Srinivasakumar, N. and Schuening, F. G. (1999) A lentivirus packaging system based on alternative RNA transport mechanisms to express helper and gene transfer vector RNAs and its use to study the requirement of accessory proteins for particle formation and gene delivery. *J. Virol.* **73**, 9589–9598.
16. Zufferey, R., Donello, J. E., Trono, D., and Hope, T. J. (1999) Woodchuck hepatitis virus posttranscriptional regulatory element enhances expression of transgenes delivered by retroviral vectors. *J. Virol.* **73**, 2886–2892.
17. Zennou, V., Petit, C., Guetard, D., Nerhbass, U., Montagnier, L., and Charneau, P. (2000) HIV-1 genome nuclear import is mediated by a central DNA flap. *Cell* **101**, 173–185.
18. VandenDriessche, T., Thorrez, L., Naldini, L., et al. (2002) Lentiviral vectors containing the human immunodeficiency virus type-1 central polypurine tract can efficiently transduce nondividing hepatocytes and antigen-presenting cells in vivo. *Blood* **100**, 813–822.
19. Van Maele, B., De Rijck, J., De Clercq, E., and Debyser, Z. (2003) Impact of the central polypurine tract on the kinetics of human immunodeficiency virus type 1 vector transduction. *J. Virol.* **77**, 4685–4694.
20. Aiken, C. (1997) Pseudotyping human immunodeficiency virus type 1 (HIV-1) by the glycoprotein of vesicular stomatitis virus targets HIV-1 entry to an endocytic pathway and suppresses both the requirement for Nef and the sensitivity to cyclosporin A. *J. Virol.* **71**, 5871–5877.
21. Yu, H., Rabson, A. B., Kaul, M., Ron, Y., and Dougherty, J. P. (1996) Inducible human immunodeficiency virus type 1 packaging cell lines. *J. Virol.* **70**, 4530–4537.
22. Carroll, R., Lin, J. T., Dacquel, E. J., Mosca, J. D., Burke, D. S., and St-Louis, D. C. (1994) A human immunodeficiency virus type 1 (HIV-1) based retroviral vector system utilising stable HIV-1 packaging cell lines. *J. Virol.* **68**, 6047–6051.
23. Ikeda, Y., Takeuchi, Y., Martin, F., Cosset, F. L., Mitrophanous, K., and Collins, M. (2003) Continuous high-titer HIV-1 vector production. *Nat. Biotechnol.* **21**, 569–572.
24. Haselhorst, D., Kaye, J. F., and Lever, A. M. L. (1998) Development of cell lines stably expressing human immunodeficiency virus type 1 proteins for studies in encapsidation and gene transfer. *J. Gen. Virol.* **79**, 231–237.
25. Zhao, J., Pettigrew, G. J., Bolton, E. M., et al. (2005) Lentivirus-mediated gene transfer of viral interleukin-10 delays but does not prevent cardiac allograft rejection. *Gene Ther.* **12**, 1509–1516.
26. Graham, F. L. and van der Eb, A. J. (1973) A new technique for the assay of infectivity of human adenovirus 5 DNA. *Virology* **52**, 456–467.
27. Chen, C. and Okayama, H. (1987) High-efficiency transformation of mammalian cells by plasmid DNA. *Mol. Cell. Biol.* **7**, 2745–2752.
28. Yee, J. K., Friedmann, T., and Burns, J. C. (1994) Generation of high-titer pseudotyped retroviral vectors with very broad host range. *Methods Cell Biol.* **43 Pt A**, 99–112.
29. Mochizuki, H., Schwartz, J. P., Tanaka, K., Brady, R. O., and Reiser, J. (1998) High-titer human immunodeficiency virus type 1-based vector systems for gene delivery into nondividing cells. *J. Virol.* **72**, 8873–8883.

Index

A

A/T-rich
   region, 183, 185, 186, 194, 196, 199
   site, 185, 186, 194, 196
absorbance, 21, 27, 66, 111, 115, 117
acetylation, 162, 165, 178
adaptor ligation, 63, 64, 67, 69
adeno-associated viral (AAV) vector, 331–341
adenovirus, 318, 321-323, 327, 329, 353
Affymetrix GeneChip probe arrays, 13, 14, 17, 27, 30, 31, 33
agarose gel, 22, 45, 54, 66, 67, 73, 111, 113, 114, 134, 151, 155, 166, 167, 206, 314, 332, 339
algorithms, 8, 11, 76, 83, 85, 87, 89, 210, 231, 233, 236, 291, 296
alignment, 185, 209, 212, 213, 230, 231, 233-237, 240, 244, 249, 250
amino acid transport, 109, 110, 158
amplicons, 125, 128, 132, 140, 141, 297
amplification cycle, 114, 121
anchoring enzyme (AE), 42, 48, 52, 56–58
anneal, 49, 118, 124, 127, 130, 142, 175, 187, 191, 205, 207–210, 291, 295
annealing temperature, 118
annotation. 9–11, 75, 89, 91, 92, 97, 101, 102, 230, 233, 236, 238, 240, 244, 254, 263, 269
antibody supershift assay, 183, 192, 193
antisense probe, 166, 169, 170, 171, 175
augmentation, 321

autoradiography, 146, 151, 152, 154, 155, 177, 211
Avogadro's number, 115

B

β-actin, 22, 36, 140, 147, 156, 316
β-gal gene expression, 333, 351, 352
β-myosin heavy chain promoter, 218, 221, 223
background, 7, 13, 33–35, 39, 41, 50, 54, 75, 79, 83–85, 98, 125, 127, 132, 151, 154, 165, 166, 168, 172, 176, 246, 318, 351, 353
   correction, 75, 83–85, 98
bioconductor *affy* library, 97–99
bioinformatics, 7, 78, 203, 213
biotinylated cDNA, 42, 48
BLAST, 128, 207, 232, 233–238, 269, 293
blastocyst, 315
BlastZ, 232, 233, 235, 236
B-score, 86

C

calcium phosphate transfection, 349, 353
cardiac
   gene expression, 56, 75, 77, 109, 159, 183, 184, 218, 253–255, 262, 321
   graft infiltrating cells, 3
   myocyte culture, 220, 221, 224, 321
   protection, 331
   -specific
      promoter, 311, 340
      trans-acting factor, 183

troponin T (cTnT) promoter, 183–185
cardiomyocytes, 37, 151, 159, 160, 170, 172, 184, 189, 322, 326, 328, 343, 352
cartridge carrier, 30
CAST (cyclic amplification and selection of targets), 212, 215
cDNA
  library, 57
  subtractive hybridization, 61–63, 65
  synthesis, 5, 15, 24, 25, 42, 45, 48, 49, 57, 73, 123, 124
CEL file, 79, 80, 83, 97–99
cell
  culture, 56, 219, 318, 321–323, 329
  sorting, 5, 351
  strainer, 323, 326
  suspension, 4, 5, 326, 327
centrifugation, 19, 21, 39, 46, 47, 123, 167, 189, 196, 207, 326, 336, 348
chordate genomes, 230
chromatin immunoprecipitation (ChIP), 210, 214
cis-acting elements, 203, 224
  sequences, 225, 346, 354
ClustalW, 236, 237, 244
cluster, 8, 11, 12, 56, 75, 78, 86, 87–95, 100, 101, 255, 260
cluster analysis, 8, 12, 75, 78, 87, 89–93
CMV promoter, 329, 335
coefficient of variation (CV), 7
collagenase, 4, 219
colony-forming unit per mL (cfu/mL), 351
comparative
  delta CT, 121
  genomics, 229–248,
computational
  genomics, 253
  tools, 253, 262
concentration, 11, 20-23, 27–32, 36, 44, 47, 49, 56, 69, 99, 111, 121, 124-130, 134, 139, 141, 147, 153, 156, 160, 168, 172, 176, 189, 198, 210, 267, 270–276, 280, 327, 346, 348, 350
conditional targeting, 309–318
confluence, 349, 351–353
confocal laser scanner, 14
consensus binding sites, 212, 213
constraint-based model, 267, 271, 276
contractile function, 61
conventional RT PCR, 109–112
correlation coefficient, 8, 87, 134
cotransfection, 224, 346, 348, 350
CpG island, 210, 230
Cre recombinase, 309, 310, 312, 319
Cre/LoxP system, 310, 311, 318
crossing point, 115, 116
$CsCl_2$ gradients, 336, 340
cycle threshold (C_T), 6
cyclophilin, 140, 147, 156

D

D module, 183, 186
data acquisition, 268, 303
data analysis, 7, 18, 31, 33, 36, 42, 48, 55, 75, 77, 78, 85, 101–103, 121, 123, 132, 133, 137
database, 229, 230, 234, 237, 248, 253–256, 259–263, 268–272
  Body Maps (http://bodymap.ims.u-tokyo.ac.jp/), 255
  CaGE (Human Cardiac Gene Expression Knowledge Base (CaGE), 253–262
  CardioGenomics (http://www.cardiogenomics.org), 76
  dbSNP (http://www.ncbi.nlm.nih.gov/SNP), 270
  Entrez Gene (www.ncbi.nlm.nih.gov/entrez/query.fcgi?db=gene), 254
  FBA (http://gcrg.ucsd.edu/downloads/index.html), 270, 273, 275, 277, 284
  Kyoto Encyclopedia of Genes and Genomes (KEGG, http://

Index 359

www.genome.jp/kegg), 269
NCBI: http://www.ncbi.nlm.nih.gov/entrez, 269
NCBI RefSeq database (http://www.ncbi.nlm.nih.gov/RefSeq/), 11, 89, 254
Nuclear Protein Database (NPD; http://npd.hgu.mrc.ac.uk), 270
Swiss Prot (http://au.expasy.org/sprot), 270
The SNP Consortium and the International HapMap Project (http://www.hapmap.org/cgi-perl/gbrowse/gbrowse/hapmap), 269
UniGene (http://www.ncbi.nlm.nih.gov/UniGene/), 255
data organization, 259
de novo sequencing, 289
DEDS (Differential Expression via Distance Summary), 8
degradation, 20, 21, 24, 35, 37, 38, 78–80, 124, 140, 149, 151, 156, 12, 163, 172, 178
denaturation, 63, 130, 132, 134, 139, 199
denaturing polyacrylamide gel, 148, 153
densitometric analysis, 147, 154
dephosphorylation, 299,
dewax, 162, 164
differentially expressed genes, 6, 9, 10, 85, 93, 103, 104, 180
dissociation curve, 6, 132, 142
distal enhancer elements, 203
distance summary (DS), 86
distant acting transcriptional enhancers, 229
ditags, 42, 49, 50–52, 55, 58, 59
DNA
  digestion, 48, 64
  footprint, 203, 210
  mutation, 287
  sequence polymorphisms, 287
  –protein complex, 183, 193, 198, 199

–RNA hybrids, 175
DNase, 4, 16, 17, 116, 121, 122, 124, 139, 151, 155, 156, 161, 167, 189, 198, 211, 214, 219, 222, 322, 324, 326, 340
DNase I footprinting, 189, 198
dNTPs, 49, 53, 112, 124, 187, 190, 198, 290, 295, 297, 299
double-stranded cDNA, 14, 24, 26, 39
  synthesis, 62–64, 66

E

ECR Browser, 232, 233, 237
electropherogram, 24, 35, 79
electrophoresis, 23, 44, 45, 50–54, 110, 112, 114, 124, 141, 183, 188, 190, 198, 200, 206, 289, 311, 339
electrophoretic mobility shift assays (EMSAs), 210
electroporation, 45, 46, 53, 314
electrospray ionization (ESI), 288
ELISA, 337, 339,
embedding, 161, 163, 288
embryonic stem (ES) cells, 311, 314
emulsion dipping, 163, 169
endocytosis, 348
endonucleases, 190
enhancer, 186, 200, 203, 211, 215, 219, 221, 224, 229, 234, 241, 244
erythropoietin gene (Epo), 332
ESTs (expressed sequence tags), 3, 204, 230, 255
euclidean distance, 8, 87
Eukaryotic Promoter Database (EPD), 210, 214
evolution, 103, 212, 213, 229, 233, 235, 240, 241, 247
expression index calculation, 80, 83, 85, 98
expression profiling, 13, 75

F

5´-nuclease probes, 121, 139

F module, 183, 186
false discovery rate (FDR), 8, 86, 100
family wise error rate (FWER), 86, 100
FASTA, 237, 244, 269
feature extraction, 79
fibroblast growth factor-8 (FGF8), 237–239, 241
fibroblasts, 196, 328
filtration, 328
flow cytometer, 4
floxed allele, 313, 316
FLP recombinase, 310, 311, 315, 318–320
FLP/CRE, 312, 318
FLPe deleter mouse, 311, 316
fluorescence acquisition, 114, 119
fluorescence-activated cell sorting (FACS), 5, 351
fluorescent dye, 72, 84, 125, 130, 177
reporter, 125, 141
flux balance analysis, 267, 271, 276, 280
freeze/thaw cycles, 26, 27
Frt sites, 310, 312
F-score, 86
functional annotation, 59, 75, 89, 254
functional genomics, 41, 61

G

Gag-Pol packaging plasmid, 345, 349
GC content, 83, 127, 128, 134, 293
gel
 -dye mix, 22, 23
 mobility shift assay, 183–186
 retardation assay, 183
 -shift competition assays, 193
gelatin, 322, 329
GenBank, 43, 58, 89, 233, 234, 238, 244, 255
gene
 delivery, 332, 340, 343, 344, 354, 355
 expression, 3–5, 7, 8, 10, 14, 18, 24, 31, 34, 35, 41, 46, 56, 61, 75, 76, 86, 93, 100, 109, 110, 121, 140, 145–157, 159, 160, 173, 183, 184, 217, 218, 229, 237, 253–255, 257, 269, 262, 343, 344, 352, 353
 Omnibus (GEO), 76
 profiles, 3, 8, 10, 12, 75, 100
 knockout, 309
 Ontology (GO), 9, 91, 254, 263
 regulation, 198, 203, 223, 229, 241, 248
 selection7–9, 12
 therapy. 343, 354
 transcripts, 41, 58, 159, 160
 transfer , 321–323, 327, 329, 343, 344
GeneChip, 4, 6, 13–17, 26–40, 79, 80, 83, 85, 89, 94, 96, 101, 103, 105
genetic variation, 288
genomic DNA, 21, 22, 123, 124, 127, 139, 140, 204, 208, 209, 211, 238, 291, 295, 297, 311, 346
genomics, 41, 56, 61, 76, 229, 230, 231, 233, 234, 235, 238, 246, 248, 253, 269, 288
genotype, 267, 268, 287, 293, 300, 316
Gibbs free energy model, 273
glutaraldehyde, 175
glyceraldehyde-3-phosphate dehydrogenase (GAPDH), 6, 22, 36, 140, 147, 156
GOOD assays, 291
graft rejection, 3, 5, 12
graphical user interface (GUI), 78
green fluorescent protein (GFP), 5, 350

H

hairpins, 127, 293, 304
heart, 4, 13, 15, 18, 19, 37, 41, 55, 56, 61, 63, 76, 109-114, 117–119, 123, 140, 145, 147, 151, 159–165, 169, 173, 174, 184-188, 194, 197, 217, 220, 222, 260, 261, 316, 319, 321, 322, 324, 325, 328, 329, 331, 332, 338–340, 343, 352, 353
exteriorization, 325

Index

transplantation, 3, 12
HEK 293 cell, 335–340
HeLa, 332, 337, 338, 351
heme-oxygenase 1 (HO-1), 332
hemizygotes, 315, 318
hemocytomer, 324
Hep 3B, 332, 337, 338
heparin affinity chromatography, 340
   -agarose column chromatography, 183, 187, 189, 198
herring sperm DNA, 16, 29, 30
heterologous promoters, 217, 220
hierarchical clustering, 8, 76, 87, 88, 92, 101
high throughput, 267–271, 278, 287, 288, 291, 295
   data, 267, 269, 271, 278
HIV-1 vector, 343, 344, 355
HMG2, 183, 186, 196, 199
homogenization, 19, 20, 27, 111, 189, 206
homozygotes, 315, 319
housekeeping gene, 118, 140, 147, 156
HSV-tk gene, 312, 313
Human Genome Project, 253, 255, 289
hybridization, 178, 179, 206, 209, 293
   controls, 16, 28–30, 36
   efficiency, 29, 79
   oven, 6, 29–31, 40
hypoxia
   -inducible
      factor 1 (HIF-1), 331, 332
      gene expression, 331–340
   -response element (HRE), 331, 332
hypoxic, 156, 158, 332, 337

I

Image processing, 79
in silico analysis, 204, 267
   cleavage, 290
in situ hybridization, 63, 72, 158–160, 173, 176–179
in vitro transcription, 6, 13, 14, 24, 149–154, 157, 208, 209, 300

inducible
   cardiac specific Cre construct, 311, 316
   deletion, 309
   promoter, 332
infection, 179, 322, 327, 328, 337
   efficiency, 328
infectious units per mL (IU/mL), 351
infiltrating inflammatory cells, 3
intercalating dye, 121, 125, 126, 129, 132
internal controls, 36, 145–147, 156, 157
intraperitoneal injection (ip), 4, 338
intron/exon, 238, 312
invader assay, 291
ischemia, 3, 4, 61, 109, 110, 113, 117, 118, 140
   reperfusion, 109, 110, 113, 117
ischemic heart, 331, 332, 338–340
isotope labeling, 57
isotope-labeled probes, 339

K

kinetic model, 267, 274, 276, 278, 284
knowledge base, 253, 254, 262

L

lacZ, 193, 332-335, 338, 339, 351
Langendorff mode, 113
lentiviral vector, 343–355
ligation, 42, 46, 47, 49–51, 57, 63, 64, 66–71, 73, 186, 211, 339
linear optimization, 267, 277
long terminal repeat (LTR), 345
lot-to-lot array variation, 14
low-abundance transcripts, 61, 63, 83, 96
low-level analysis, 75, 83, 85
LoxP, 310–312, 318
L-score, 233
luminometer, 222, 223

M

MALDI-TOF MS, 287–305
mass spectrometry (MS), 287, 288

MassCLEAVE, 287, 291, 293, 296, 299, 300, 302
MassEXTEND, 287, 291, 293, 295
mass-to-charge ratio, 288
matrix, 101, 241, 273, 277, 287, 288, 302, 303
MEF2-like motif, 183, 185, 189
melting curve, 114, 117, 118
melting temperature (T_m), 163, 293, 297
messenger RNA (mRNA), 204
metabolic network, 267, 272
metabolite, 267, 268, 271–284
Michaelis binding constants (K_{PEP}), 276
Michaelis–Menten kinetics, 273, 275
microarray, 3, 5–14, 18, 21, 22, 24, 29–31, 35–41, 55–63, 72, 75–79, 83–91, 94–96, 102–105, 160, 173, 178, 212, 213, 255
microarray data analysis, 7, 75, 77, 78, 85, 102
Microarray Gene Expression Data Society (MGED), 7
microplates, 295
microsatellites, 288
microscopy, 170, 171
Minimum Information About a Microarray Experiment (MIAME), 7, 12
minimum SV40 promoter, 332–334, 338–340
mismatch (MM), 7
modeling software, 270
molecular beacons, 121, 126, 127, 130, 139, 143
molecular cardiology, 217
molecular weight, 67, 73, 111, 113, 115, 129, 305, 349
morphology, 164, 172, 175
mRNA, 14, 23, 24, 28, 41–43, 45, 48, 55, 59, 61, 63, 64, 74, 75, 89, 104, 109, 121, 140, 145–149, 153–163, 167, 172, 174, 177, 204–210
multidimensional scaling (MDS), 89

Mus musculus, 230, 231
muscle-specific gene promoters, 212
mutation detection, 287, 305
myocardial infarction, 331

N

Na/Ca exchanger (NCX), 147
negative control, 49, 50, 52, 54, 112, 113, 116, 117, 166, 177, 191, 207, 304
negative selection marker, 312
neomycin (*neo*) resistance gene, 310–312, 315, 316
neonatal cardiomyocytes, 329, 343, 352
neonatal rat ventricular myocytes (NRVMs), 321
NIH3T3 cells, 332, 337, 338, 351
Noise value (Raw Q), 36
nomenclature, 92, 254
nondenaturing polyacrylamide gel, 183, 184, 187, 189, 192
nonionic iodixanol gradients, 340
normalization, 7, 69, 75, 83-85, 98, 137
Northern blotting, 63, 145, 158
novel genes, 61, 63, 73
nuclear extract, 184, 188, 189, 192, 210
nuclease free water, 16, 17, 112, 115, 118, 122, 125, 129, 130, 150, 151

O

oligo-dT primer, 14, 16, 64, 66
oligonucleotide, 13, 30, 34, 40, 57, 103, 121, 122, 125, 129, 175–177, 187, 191, 198, 204, 206–208, 218, 289, 305
oligonucleotide microarrays, 13
oligoprobes, 175
open reading frames (ORFs), 269, 343
optical density (OD), 35
organelle, 270
organism, 270
packaging signal (Ψ), 345
paraformaldehyde, 159, 162, 163, 174

Index   363

PCR, 4–6, 11, 15, 22, 42, 49–57, 62–65, 68, 69, 71, 73, 79, 85, 96, 109, 290, 314–316
PCR amplification, 42, 49, 50, 57, 63, 65, 71, 124, 134, 140, 211, 221, 297, 339
PCR Master Mix, 4, 111–113, 134
penetration, 164, 172, 175
perfect match (PM), 7
Perl script, 253, 256, 257, 262
pGL3-Promoter vector, 223
phenol-chloroform extraction, 18, 19, 47, 142, 208
phenotype, 267, 268, 316
phylogenetic
  footprinting, 212, 216
  shadowing, 235–237, 241, 244, 246, 247
PipMaker, 236, 237
pixel intensity, 96
PLIER (Probe Logarithmic Intensity ERror), 85
poly-A messenger RNA, 62, 63
Poly-A RNA spike-in controls, 16, 25, 28
polyethylenimine (PEI), 321–323
poly-L-lysines, 328
polymorphisms, 287–289, 291, 293, 295, 303
positive
  control, 22, 112, 141, 153, 177, 191
  selection marker, 312
potential binding sites, 212
primary cardiomyocyte, 189, 321, 322
primer, 5, 6, 11, 14, 16, 22, 24, 25, 45, 48, 49, 53, 54, 58, 62, 64–71, 95, 112, 113, 116, 118, 119, 121–135, 140–143, 176, 203–211, 214, 219, 221, 258, 287, 291–304
  design, 121, 122, 125, 127, 206, 297, 299, 304
  dimer, 118, 125, 127, 132, 134, 135, 141, 142, 207, 304
  extension, 176, 203, 204–210, 219, 287, 291, 293, 295

  analysis, 204–206, 209
principal component analysis (PCA), 8
Programs for Genomic Application (PGA), 76
proteinase K, 153, 161, 162, 164, 165, 172, 175
protein-encoding genes, 229
proteomics, 270, 284, 288
proximal promoter, 184, 186, 203, 205, 223
pufferfish, 230, 231, 233, 237, 238
p-value, 8, 9, 35, 86, 99, 100
pyruvate kinase (PK) deficiency, 275

Q

quality assessment, 14, 75, 79, 303
quantification, 6, 27, 55, 109, 110, 121, 142, 145, 149, 152, 155, 209
quantitation, 15, 21, 119, 121, 122, 128, 129, 132, 134, 141, 142, 145, 146, 157, 199
quantitative RT-PCR, 85, 96, 110, 111, 114, 117, 118, 123, 128, 129, 339

R

random primer, 95, 124, 140
randomization test, 9
rat (*Rattus norvegicus*), 230, 231
raw image data, 79
RawQ value, 36, 79
real-time quantitative RT PCR, 110, 111, 114, 118
red blood cell, 267, 275, 277
renature, 194, 197, 200
repetitive DNA, 230
replication-deficient virus, 345, 346
reporter plasmids, 217, 224
reproducibility, 13, 76, 141, 145, 173
restriction enzyme, 45, 57, 161, 166, 167, 187, 211, 218, 311, 312, 314, 332
retroviruses, 343

Rev expressor, 344, 345
reverse transcription, 22, 62-64, 66, 109, 113, 118, 121, 124, 142, 145, 160, 176, 206, 296–298, 338, 345
reverse transcription polymerase chain reaction (RT-PCR), 109, 145, 160, 338
riboprobes, 175, 176, 179, 209
RNA
   amplification, 13–15, 24, 104
   degradation, 20, 37, 80, 149, 151, 156, 163
   –DNA hybrids, 154, 178
   extraction, 5, 11, 15, 18, 63, 64, 206
   isolation, 14, 37, 38, 74, 123, 139, 142, 148, 149, 151, 154, 339
   purity, 5, 21, 35
   quality, 13, 15, 21, 35, 38, 78, 151, 155
   –RNA hybrids, 154, 168
RNAse
   A, 290, 296, 301, 303
   digestion, 146, 148, 153, 156, 168
   inhibitor, 151, 161,167, 207
   Protection Assay, 110, 145–148, 151, 157, 203
   T1, 290

S

SAM statistic score, 86
scaling factors (SF), 36
scintillation counting, 152
SDS-PAGE, 183, 186, 192, 194, 196, 197
self-organizing maps (SOMs), 8
semiquantitative RT-PCR, 338, 339
sense probe, 166, 169–171, 175, 177
sequencing, 41, 42, 48, 52, 54, 56, 58, 61, 63, 71, 72, 118, 150, 158, 205, 208, 210–213, 229, 233–237, 246, 287, 289, 304, 305, 312, 314
SEQUENOM, 293, 295, 296, 303
serial analysis of gene expression (SAGE), 41

single-nucleotide polymorphisms (SNPs), 121–144, 267, 288, 291, 293, 303
single-stranded DNA probes, 176
single-stranded RNA, 146, 178, 208, 209, 290, 296
skeletal muscle, 37, 56, 184, 187, 188, 196, 200
SNAT3, 109–118
SNP-OMICS, 269
SOC media, 44, 53
Southern blotting, 311, 314–316
spectra, 187, 291, 300, 302, 303, 304
SpectroCHIP array, 293
spectrophotometer, 27, 110, 117, 122, 124
standard
   curve, 116, 117, 121, 123, 128, 129, 133, 134, 136, 137, 141, 142
   deviations, 8
statistical
   comparison, 75, 83
   filtering, 85, 86, 91
   validation, 9, 10
statistics, 8, 78, 85, 86
steady-state mRNA, 145
stoichiometric matrix, 271–273, 277, 282
streptomycin, 319, 322, 327, 348
subcloning, 63, 71
SV40 promoter, 223, 332-334, 338–340
SYBR green I, 6, 45, 47, 50, 51, 111, 114, 121, 125-127, 129, 131–134, 139, 141, 142
systematic evolution of ligands by exponential amplification (SELEX), 212

T

260:280 nm ratio, 27, 35
293T cells, 343–353
T7 promoter, 24, 290, 292, 293, 295, 297, 299
T7 RNA polymerase, 24, 290, 296
T_m, 118, 127, 128, 134, 135
tag

Index 365

frequency, 55
per million (TPM), 56
tamoxifen-inducible Cre (MerCreMer), 310
Taq DNA Polymerase, 45, 53, 112, 118, 119,122, 125, 130, 139, 140, 142, 218, 295
TaqMan, 121, 125, 126, 143
targeting vector, 311, 312, 314
TATA binding site, 220
TCTG(G/C) direct repeat, 183, 185, 186, 194, 196, 199
temporal and spatial control of gene deletion, 310
thoracotomy incision, 338
threshold, 6, 86,132, 133, 237
  cycle, 6, 132
tissue
  fixation, 159, 162, 173, 174
  -specific promoters, 310
titer determination, 336
titration , 210, 348, 350, 351
total RNA, 5, 13, 18, 21-23, 25, 28, 37, 39, 48, 62, 63, 94, 111, 113, 117, 123, 124, 139, 147, 149–152, 154–156, 206, 207, 339
toxicity, 12, 315, 331, 337, 357
trans-acting factors, 183, 184
transcription, 6, 12–14, 22, 24, 26, 36, 62–66, 109, 113, 118, 119, 124, 142, 145, 149–160, 166, 183, 184, 189, 193, 203–225, 229, 236, 239–244, 248, 290, 293, 296, 300, 301, 304
transcription factor binding sites (TFBSs), 198, 205, 210, 236, 240, 243
transcriptional
  factors, 184
  repressors, 184
  start site, 184, 203, 204, 223, 241
transcriptome, 41, 55, 56, 160, 255

transducing
  units per mL (TU/mL), 350, 351
  vector, 345
transduction, 338, 340, 343, 347, 350–354
transfection, 321, 323, 327, 329, 335, 345, 348, 349
  efficiency, 328, 348–350, 353
transgene, 5, 310, 316, 322, 329, 340, 345, 350, 352, 353
transgenic mouse, 311, 315, 316, 318
transient cotransfection, 346, 348
TRIZOL reagent, 5, 15, 18, 19, 38
trypsin digestion, 324, 328
t-score, 86, 100
T-stretches, 297, 304
two-way clustering, 87, 90

U

ultracentrifugation, 350, 353
UV transilluminator, 110, 113

V

validation, 9, 10, 61, 72, 96
vasculature, 343
VEGF gene, 331, 332, 334, 339
ventilator, 338
viral vector, 322, 331, 339, 346, 349, 351, 353
virus production, 331, 335
VISTA Genome Browse, 233, 237, 240
VSV-G envelope expressor, 344, 345

W

Western blotting, 311, 316

X

X-gal, 58, 71, 348, 352
X-ray film, 168, 169, 188, 190, 192

Z

zebrafish (*Danio rerio*), 230, 231